THIRD EDITION

ॐ ॐ ॐ

THE
WORKING
WRITER

Toby Fulwiler

University of Vermont

Prentice
Hall

Upper Saddle River, New Jersey 07458

Library of Congress Cataloging-in-Publication Data

Fulwiler, Toby (date)
  The working writer / Toby Fulwiler.—3rd ed.
    p.   cm.
  Includes index.
  ISBN 0-13-028912-4
  1. English language—Rhetoric.   2. Report writing.   I. Title.

PE1408.F82   2001
808'.042—dc21

Editor in Chief: Leah Jewell
Acquisitions Editor: Corey Good
Editorial Assistant: Joan Polk
Managing Editor: Mary Rottino
Production Editor: Joan E. Foley
Prepress and Manufacturing Buyer: Mary Ann Gloriande
Senior Marketing Manager: Brandy Dawson
Marketing Assistant: Christine Moodie
Cover Designer: Bruce Kenselaar
Cover Art: Connie Hayes/Stock Illustration Source, Inc.

Acknowledgment
Page 146. Brooks, Gwendolyn. "We Real Cool" from *Blacks* by Gwendolyn Brooks.
The Third World Press, 1987. Reprinted by permission.

This book was set in 10/12 Berkeley Book by Publications Development Company
and was printed and bound by Hamilton Printing Company.
Covers were printed by Phoenix Color Corp.

© 2002, 1999, 1995 by Pearson Education, Inc.
Upper Saddle River, New Jersey 07458

Printed in the United States of America
10 9 8 7 6 5 4 3 2 1

ISBN 0-13-049290-6

Prentice-Hall International (UK) Limited, *London*
Prentice-Hall of Australia Pty. Limited, *Sydney*
Prentice-Hall Canada Inc., *Toronto*
Prentice-Hall Hispanoamericana, S.A., *Mexico*
Prentice-Hall of India Private Limited, *New Delhi*
Prentice-Hall of Japan, Inc., *Tokyo*
Pearson Education Asia Pte. Ltd., *Singapore*
Editora Prentice-Hall do Brasil, Ltda., *Rio de Janeiro*

FOR **LAURA**
My Partner in Life, Love, and Work

❧ Contents ❧

ॐ ॐ ॐ

PART THREE
CONDUCTING RESEARCH, 169

ॐ ॐ ॐ

PART FOUR
REVISING AND EDITING, 253

WRITER REFERENCES, 329

ᔔ Preface ᔔ

The basic approach of *The Working Writer* remains the same from edition to edition: to address the needs of working undergraduate writers in a friendly, no-nonsense writer-to-writer manner, and to present the process of writing as both rigorous and delightful. *The Working Writer* presents a process approach to the teaching of writing, examining the different but overlapping stages of writing we call planning, drafting, researching, revising, and editing. The book addresses rhetorical issues of purpose, audience, and voice as well as the details of field, library, and Internet research—with particular attention to evaluating sources. All chapters emphasize that writing well is a matter of making wise choices rather than following formulaic rules.

The third edition of *The Working Writer* includes up-to-date strategies for using the Internet for research and for documenting Internet sources correctly according to newly revised MLA and APA conventions. The research chapters guide writers through the many stages of the whole process, which is viewed here as yet another matter of making choices: from keeping a project log and learning how to find sources (including the proliferation of electronic choices), to conducting field research, to using and documenting sources. These chapters offer strategies for planning, organizing, and writing major research papers.

New to this edition is Chapter 18, "Conducting Internet Research," since online research is now the most commonly used (and misused) form for college papers. Because of the increased reliance on often unreliable electronic sources, I've also added Chapter 20, "Evaluating Research Sources" covering electronic, library, and field sources. It is, I believe, one of the most thorough treatments available in assessing the value of research sources.

Also new is Chapter 24, "Creative Nonfiction," which suggests that all undergraduates might profit from learning to write in lively experimental forms and styles. I've found that first-year students welcome the chance to write snapshot-style as well as to experiment with multiple voices and with the effective uses of both labyrinthine and fragment sentences. This chapter is heavily indebted to the pioneering work of Winston Weathers and his concept of alternate style.

To make room for the three new chapters, this edition of *The Working Writer* has condensed and combined three separate chapters on purpose, audience, and voice into a briefer and more pointed treatment of these interrelated matters in Chapter 4, "The Elements of Composition." Combining matters of purpose, audience, and voice into a single chapter makes it easier to explain how these concepts are mutually dependent on each other.

SUPPLEMENTS

Instructor's Manual to accompany *The Working Writer* (0-13-041635-5)

The New American Webster Handy College Dictionary (0-13-032870-7)
FREE when packaged with the text.

English on the Internet 2002: Evaluating Online Resources
FREE when packaged with the text.

NEW! The Writer's Guide Series FREE when packaged with the text:

The Writer's Guide to Document and Web Design (0-13-018929-4)

The Writer's Guide to Writing Across the Curriculum and Oral Presentations
(0-13-018931-6)

The Writer's Guide to Writing About Literature (0-13-018932-4)

turnitin www.turnitin.com   This online service makes it easy for teachers to find out
if students are copying their assignments from the Internet, and it is now free to professors
using the third edition of *The Working Writer*. In addition to helping educators easily
identify instances of Web-based student plagiarism, Turnitin.com also offers a digital
archiving system and an online peer review service. Professors set up a "drop box" at
the Turnitin.com Web site where their students submit papers. Turnitin.com then cross-
references each submission with millions of possible online sources. Within 24 hours,
teachers receive a customized, color-coded "Originality Report," complete with live links
to suspect Internet locations, for each submitted paper. To access the site for free, professors must visit the site via the faculty resources section of Prentice Hall English Central at
www.prenhall.com/english.

ACKNOWLEDGMENTS

I'd like to acknowledge the continued stimulation and ideas I receive from my students at
the University of Vermont. I'd also like to thank my thoughtful editors at Prentice Hall,
Corey Good and Joan Foley, for their help and support. Thanks also to the following re-
viewers who helped guide my revisions: William Condon, Washington State University;
T.A.E. Fishman, Purdue University; Brad Gambil, Oklahoma State University; Britton
Gildersleve, Oklahoma State University; Dan Holt, Lansing Community College; Kip
Knott, Oklahoma State University; John Marlin, The College of St. Elizabeth; and Scott
Rodger, Oklahoma State University.

Toby Fulwiler

PART ONE

✺ ✺ ✺

THE ELEMENTS OF COMPOSITION

Chapter

❧ 1 ❧

College Writing

I am the absolute worst writer. I will never forget when I had to write my college essay. I thought it was good and then I brought it downstairs to have my parents read it and they tore it apart. By the time they were finished with it, I had to rewrite it five times. I was so mad, but the weird part was, the final copy was exactly what I wanted to say. They had to tear it out of me.

JESSICA

I like to write. Writers are nothing more than observant, perceptive, descriptive people.

PAT

Y ou can count on one thing, attending college means writing papers—personal and critical essays in English, book reviews and reports in history, research reports in psychology and sociology, position papers in political science, lab reports in biology, and so on. Some of these assignments will be similar to those you've done in high school, some will be new, and all will be demanding.

When I asked a recent class of first-year college students to talk about themselves as writers, several began by describing their habits and attitudes: John, for example, said he wrote best "under pressure," while Kevin wanted to write at his own pace, on his own time, and "hated deadlines." Becky preferred writing when she "felt strongly or was angry about something," Doug when the assignment "asked for something personal," but Lisa enjoyed writing any paper so long as the "assignment was clear and fair."

Other students talked about where and when they wrote. Amy, for example, said she did most of her writing "listening to classical music and, if it is a nice day, under trees." Jennifer felt "most comfortable writing on her bed and being alone." Dan said he could write anywhere, so long as he had "a good computer," and José, "as long as it was after midnight." In fact, there proved to be as many different perspectives on being a writer as there were students in the class.

✍ WHY IS WRITING HARD?

Even professional authors admit that writing is not easy. In writing this book, I encountered numerous problems, from organizing material to writing clearly to finding time and meeting publication deadlines. What, I wondered, did first-year college writers find difficult about writing? Were their problems similar to or different from mine? Here is what they told me:

Jennifer: "I don't like being told what to write about."

Amy: "I never could fulfill the page requirements. My essays were always several pages shorter than they were supposed to be."

Jill: "I always have trouble starting off a paper . . . and I hate it when I think I've written a great paper and I get a bad grade. It's so discouraging and I don't understand what I wrote wrong."

Omar: "Teachers are always nitpicking about little things, but I think writing is for communication, not nitpicking. I mean, if you can read it and it makes sense, what else do you want?"

Cara: "I hate revising. I had this teacher in high school who insisted we rewrite every paper over and over again, and that got really boring. Once I've said what I have to say, I don't have anything else to say."

Mike: "If I'm in a bad mood or don't have the right beginning, I find myself stumbling and not giving a hoot about whether it's right or not."

Kennon: "Putting thoughts down on paper as they are in your mind is the hardest thing to do. It is like in music—to make the guitar make the sound you imagine in your head, to make the words on the page paint the picture in your head."

I wasn't surprised by these answers, since I, too, remember wondering: What did teachers want? How long was enough? How do you get thoughts into words? Why all the nitpicking?

✍ WRITING 1

What do you find difficult about writing? Do you have a problem finding subjects to write about? Or do you have trouble getting motivated? Or does something about the act of writing itself cause problems for you? Explain in your own words by writing quickly for five minutes without stopping.

✍ WHAT DO YOU LIKE ABOUT WRITING?

Though any writer will tell you writing isn't easy, most writers will also describe it as interesting and exciting. So I asked our first-year students what it was about writing that gave them pleasure.

Jolene: "If I have a strong opinion on a topic, it makes it so much easier to write a paper."

Rebecca: "On occasion I'm inspired by a wonderful idea. Once I get going, I actually enjoy writing a lot."

Casey: "I enjoy most to write about my experiences, both good and bad. I like to write about things when I'm upset—it makes me feel better."

Darren: "I guess my favorite kind of writing is letters. I get to be myself and just talk in them."

Like my students, I prefer to write about topics that inspire or interest me, and I find personal writing such as letters especially easy, interesting, and enjoyable.

↬ WRITING 2: EXPLORATION

What kind of writing do you most enjoy doing? What do you like about it: Communicating? Exploring a subject? Playing with words? Something else?

↬ WHAT SURPRISES ARE IN STORE?

At the same time I was teaching these first-year students, I was also teaching an advanced writing seminar to seniors. Curious about their attitudes toward writing, I asked them: "What has surprised you the most about writing in college?"

Scott: "Papers aren't as hellish as I was told they'd be. In fact, I've actually enjoyed writing a lot of them—especially after they were done."

Aaron: "My style has changed a lot. Rather than becoming more complex, it's become simpler."

Kerry: "The most surprising and frustrating thing has been the different reactions I've received from different professors."

Rob: "I'm always being told that my writing is superficial. That I come up with good ideas but don't develop them."

John: "The tutor at our writing lab took out a pair of scissors and said I would have to work on organization. Then she cut up my paper and taped it back together a different way. This really made a difference, and I've been using this method ever since."

Chrissie: "Sharing papers with other students is very awkward for me. But it's extremely beneficial when I trust and like my group, when we all relax enough to talk honestly about one another's papers."

As you can see, most advanced students found ways to cope with and enjoy college writing. Several reported satisfying experiences sharing writing with each other. I'm sorry that some students, even in their last year, could not figure out what their instructors wanted—there *are* ways to do that.

〜 **WRITING 3: EXPLORATION**

Think about your experience with writing in the last school you attended. What surprised you—pleasantly or not—about the experience? What did you learn or not learn?

〜 WHY IS WRITING SO IMPORTANT?

I also asked these advanced students why, in their last year, they had enrolled in an elective writing class: "What made the subject so important to you?"

Kim: "I have an easier time expressing myself through writing. When I'm speaking, my words get jumbled—writing gives me more time, and my voice doesn't quiver and I don't blush."

Rick: "Writing allows me to hold up a mirror to my life and see what clear or distorted images stare back at me."

Glenn: "The more I write, the better I become. In terms of finding a job after I graduate, strong writing skills will give me an edge over those who are just mediocre writers."

Amy: "I'm still searching for meaning. When I write I feel I can do anything, go anywhere, search and explore."

Angel: "I feel I have something to say."

Carmen: "It's simple, I love words."

I agree, easily, with virtually all of these reasons. At times writing is therapeutic, at other times it helps us clarify our ideas, and at still other times it helps us get and keep jobs.

〜 **WRITING 4: EXPLORATION**

Look over the various answers given by the college seniors and select one. Do you agree or disagree with the student? Explain.

〜 WHAT THE SENIORS ADVISE

Since my advanced students had a lot to say about writing, I asked them to be consultants: "What is your advice to first-year college writers?" Here are their suggestions:

Aaron: "Get something down! The hardest part of writing is starting. Forget the introduction, skip the outline, don't worry about a thesis—just blast your ideas down, see what you've got, then go back and work on them."

Christa: "Plan ahead. It sounds dry, but planning makes writing easier than doing laundry."

Victor: "Follow the requirements of the assignment to the T. Hand in a draft for the professor to mark up, then rewrite it."

Allyson: "Don't think every piece you write has to be a masterpiece. And sometimes the worst assignment turns into the best writing. Don't worry about what the professor wants—write what you believe."

Carmen: "Imagine and create, never be content with just retelling a story."

Rick: "When someone trashes your writing, thank them and listen to their criticism. It stings, but it helps you become a better writer."

Jason: "Say what you are going to say as clearly and as straightforwardly as possible. Don't try to pad it with big words and fancy phrasing."

Angel: "Read for pleasure from time to time. The more you read, the better you write—it just happens."

Kim: "When choosing topics, choose something that has a place in your heart."

These are good suggestions to any writers: Start fast, think ahead, plan to revise and edit, listen to critical advice, consider your audience, be clear, read a lot. I hope, however, that instructors respond to your writing in critically helpful ways and don't "trash" it or put it down. Whether or not you take some of the advice will depend on what you want from your writing: good grades? self-knowledge? personal satisfaction? clear communication? a response by your audience? When I shared these suggestions with first-year students, they nodded their heads, took some notes, and laughed—often with relief.

☞ WRITING 5: EXPLORATION

What else would you like to ask advanced college students about writing? Find a student and ask; report back.

☞ WHAT ELSE DO YOU WANT TO KNOW?

I knew that my first-year students had already received twelve years' worth of "good advice" about learning to write, so I asked them one more question: "What do you want to learn about writing that you don't already know?" In parentheses, I've provided references to chapters of this text that answer these questions.

Emma: "Should I write to please the professor or to please myself?" *(See Chapter 4, "The Elements of Composition.")*

José: "I'm always being told to state my thesis clearly. What exactly is a thesis and why is it so important?" *(See Chapters 11, "Explaining Things"; 12, "Arguing For and Against"; and 13, "Interpreting Texts.")*

Jolene: "How do I develop a faster way of writing?" *(See Chapter 7, "Strategies for Starting.")*

Amy: "Is there a trick to making a paper longer without adding useless information?" *(See Chapters 22–24, which address revising.)*

Sam: "How do I learn to express my ideas so they make sense to common intelligent readers and not just to myself?" *(See Chapters 25–27, which address editing.)*

Scott: "How can I make my writing flow better and make smooth transitions from one idea to the next?" *(See Chapter 26, "Working Paragraphs.")*

Terry: "I want to learn to like to write. Then I won't put off assignments until the last minute." *(See Chapter 5, "The Working Writer.")*

Jennifer: "I have problems making sentences sound good. How can I learn to do that?" *(See Chapter 27, "Working Sentences.")*

John P.: "I would like to develop some sort of personal style so when I write people know it's me." *(See Chapter 4, "The Elements of Composition.")*

Jeff: "I want to become more confident about writing research papers. I don't want to have to worry about whether my documentation is correct or if I've plagiarized or not." *(See Part Three, "Conducting Research.")*

Woody: "Now that I'm in college, I would like to be challenged when I read and write, to think, and ask good questions, and find good answers." *(See Chapter 2, "College Reading.")*

Pat: "I don't want to learn nose-to-the-grindstone, straight-from-the-textbook rules. I want to learn to get my mind into motion and pencil in gear." *(See Chapters 5, "The Working Writer," and 7, "Strategies for Starting.")*

Heidi: "I would love to increase my vocabulary. If I had a wider range of vocabulary, I would be able to express my thoughts more clearly." *(I'm not sure this book addresses that directly, but the more reading and writing you do, the more words you'll learn!)*

Jess: "I'm always afraid that people will laugh at my writing. Can I ever learn to get over that and get more confident about my writing?" *(See Chapter 6, "Sharing and Responding.")*

I can't, of course, guarantee that if you read *The Working Writer* your writing will get easier, faster, longer, clearer, or more correct. Or, for that matter, that your style will become more personal and varied, or that you will become a more confident and comfortable writer—no handbook can do that for you. Becoming a better writer depends on your own interest and hard work. It will also depend on your college experience, the classes you take, and the teachers with whom you study. However, whether in class or on your own, if you read this text carefully and practice its suggestions, you should find possible answers to all these questions and many more.

I admit that there was at least one student's concern for which I really had no good response. Jessica wrote, "My biggest fear is that I'll end up one semester with four or five courses that all involve writing and I'll die." Or maybe I do have a response: If you become comfortable and competent as a writer, you'll be able to handle all the writing assignments thrown your way. Even if you can't, Jessica, you won't die. It's just college.

ꕔ SUGGESTIONS FOR WRITING AND RESEARCH

Individual

1. Interview a classmate about his or her writing experiences, habits, beliefs, and practices. Include questions such as those asked in this chapter as well as others you think may be important. Write a brief essay profiling your classmate as a writer. Share your profile with a classmate.
2. Over a two-week period, keep a record of every use you make of written language. Record your entries daily in a journal or class notebook. At the end of two weeks, enumerate all the specific uses as well as how often you did each. What activities dominate your list? Write an essay based on this personal research in which you argue for or against the centrality of writing in everyday life.

Collaborative

As a class or in small groups, design a questionnaire to elicit information about people's writing habits and attitudes. Distribute the questionnaire to both students and instructors in introductory and advanced writing classes. Compile the results. Compare and contrast the ideas of students at different levels and disciplines and write a report to share with the class. Consider writing a feature article for your student newspaper or faculty newsletter reporting what you found.

Chapter

✤ 2 ✤

College Reading

Now that I'm in college, I would like to be challenged when I read and write,
to think, and ask good questions, and find good answers.

WOODY

To read and write well in college means to read and write critically. In fact, a major goal of most college curricula is to train students to be critical readers, writers, and thinkers so they can carry those habits of mind into the larger culture beyond college. What, you may ask, does it mean to be critical? How does being a critical reader, writer, and thinker differ from being a plain, ordinary, everyday reader, writer, and thinker?

Being critical in writing means making distinctions, developing interpretations, and drawing conclusions that stand up to thoughtful scrutiny by others. Being critical in reading means knowing how to analyze these distinctions, interpretations, and conclusions. Becoming a critical thinker, then, means learning to exercise reason and judgment whenever you encounter the language of others or generate language yourself.

Most of *The Working Writer* explores strategies for helping you become an accomplished critical writer. This chapter, however, explores strategies for helping you become a more accomplished critical reader and emphasizes as well the close relationship between critical reading and critical writing.

✤ WRITING 1

Describe yourself as a reader, answering some of these questions along the way: How often do you read on your own? What kinds of reading do you do when the choice of reading material is up to you? Where and when do you most commonly do your reading? What is the last book you read on your own? Who is your favorite author? Why?

☞ UNDERSTANDING WRITTEN TEXTS

To understand a text, you need some context for the new ideas you encounter, some knowledge of the text's terms and ideas, and knowledge of the rules that govern the kind of writing you're reading. It would be difficult to read Mark Twain's novel *The Adventures of Huckleberry Finn* with no knowledge of American geography, the Mississippi River, or the institution of slavery. It would also be difficult to read a biology textbook chapter about photosynthesis but know nothing of plants, cell structure, or chemical reactions. The more you know, the more you learn; the more you learn, the more careful and critical your reading, writing, and thinking will be.

Many college instructors will ask you to read about subjects that are new to you; you won't be able to spend much time reading about what you already know. To graduate, you've got to keep studying new subjects that require, first, that you understand what you read and, second, that you can critically assess and write about this new understanding. As you move through the college curriculum, you will find yourself an expert reader in some disciplines, a novice reader in others, and somewhere in between in the rest—often during the same semester.

If getting a college degree requires that you read one unfamiliar text after another, how can you ever learn to read successfully? How do you create a context, learn a background, and find the rules to help you read unfamiliar texts in unfamiliar subject areas? What strategies or shortcuts can speed up the learning process? Let's consider some strategies for doing this.

As an experiment, read the following short opening paragraph from an eight-paragraph *New York Times* story titled "Nagasaki, August 9, 1945." When you have finished, pause for a few moments, and think about (1) what you learned from it, (2) how you learned what you learned, and (3) what the rest of the story will be about.

> In August 1945, I was a freshman at Nagasaki Medical College. The ninth of August was a clear, hot, beautiful summer day. I left my lodging house, which was one-and-one-half miles from the hypocenter, at eight in the morning, as usual, to catch a tram car. When I got to the tram stop, I found that it had been derailed in an accident. I decided to return home. I was lucky. I never made it to school that day.
>
> MICHAITO ICHIMARU

How did you do? It is possible that your reasoning went something like mine, which I reconstruct here. Note, however, that although the following sequence presents ideas one after the other, that's not how it seemed to happen when I read the passage for the first time. Instead, meaning seemed to occur in flashes, simultaneously and unmeasurably. Even as I read a sentence for the first time, I found myself reading backward as much as forward to check my understanding. Here are the experiences that seemed to be happening.

1.   I read the first sentence carefully, noticing the year 1945 and the name of the medical college, "Nagasaki." My prior historical knowledge kicked in as I *identified* Nagasaki, Japan, as one of the cities on which the United States dropped an atomic bomb at the end of World War II—though I did not remember the precise date.

2. I noticed the city and the date, August 9, and wondered if that was when the bomb was dropped. I *asked* (silently), "Is this a story about the bomb?"

3. Still looking at the first sentence, a reference to the writer's younger self ("I was a freshman"), I guessed that the author was present at the dropping of this bomb. I *predicted* that this would be a survivor's account of the bombing of Nagasaki.

4. The word *hypocenter* in the third sentence made me pause again; the language seemed oddly out of place next to the "beautiful summer day" described in the second sentence. I *questioned* what the word meant. Though I didn't know exactly, it sounded like a technical term for the place where the bomb went off. Evidence was mounting that the narrator may have lived a mile and a half from the exact place where the atomic bomb detonated.

5. In the next-to-last sentence of the paragraph, the author says that he was "lucky" to miss the tram. Why, unless something unfortunate happened to the tram, would he consider missing it "lucky"? I *predicted* that had the author gone to school "as usual" he would have been closer to the hypocenter, which I now surmise was at Nagasaki Medical College.

6. I then *tested* my several predictions by reading the rest of the story, which you, of course, could not do. My predictions proved correct: Michaito Ichimaru's story is a firsthand account of witnessing and surviving the dropping of the bomb, which in fact killed all who attended the medical college, a quarter of a mile from the hypocenter.

7. Finally, out of curiosity, I looked up Nagasaki in the *Columbia Desk Encyclopedia* and *confirmed* that 75,000 people were killed by this second dropping of an atomic bomb, on August 9, 1945; the first bomb had been exploded just three days earlier, on August 6, at Hiroshima.

You'll notice that in my seven-step example some parts of the pattern of identifying/questioning/predicting/testing/confirming occur more than once, perhaps simultaneously, and not in a predictable order. This is a slow-motion description—not a prescription or formula—of the activities that occur in split seconds in the minds of active, curious readers. No two readers would—or could—read this passage in exactly the same way, because no two readers are ever situated identically in time and space, with identical training, knowledge, and experience to enable them to do so. However, my reading process may be similar enough to yours that the comparison will hold up: reading is a messy, trial-and-error process that depends as much on prior knowledge as on new information to lead to understanding.

Whether you read new stories or watch unfamiliar events, you commonly make meaning by following a procedure something like mine, trying to identify what you see, question what you don't understand, make and test predictions about meaning, and consult authorities for confirmation or information. Once you know how to read successfully for basic comprehension, you are ready to read critically. Learn the following reading strategies to improve your reading comprehension:

1. *Identify.* Read first for what you recognize, know, and understand. Identify what you are reading about. Read carefully—slowly at first—and let meaning take hold where it can.

2.   *Question*. Pause, and look hard at words and phrases you don't know or understand. See if they make sense when you reread them, compare them to what you do know, or place them in a context you understand.

3.   *Predict*. Make predictions about what you will learn next: How will the essay, story, or report advance? What will happen? What theme or thesis will emerge? What might be the point of it all?

4.   *Test*. Follow up on your predictions by reading further to see if they are correct or nearly correct. If they are, read on with more confidence; if they are not, read further, make more predictions, and test them. Trial and error are good teachers.

5.   *Confirm*. Check your reading of the text with others who have also read it and see if your interpretations are similar or different. If you have questions, ask them. Share answers.

↲ WRITING 2

Select a book you have been assigned to read for one of your courses and find a chapter that has not yet been covered in class. Read the first page of the chapter and then stop. Write out any predictions you have about where the rest of the chapter is going. (Ask yourself, for example, "What is its main theme or argument? How will it conclude?") Finish reading the chapter and check its conclusion against your predictions. If your predictions were close, you are reading for understanding.

↲ READING CRITICALLY

How people read depends on what they're reading; people read different materials in different ways. When they read popular stories and magazine articles for pleasure, they usually read not to be critical but to understand and enjoy. In fact, while pleasure readers commonly go through a process similar to the one described in the last section—identifying, questioning, predicting, and testing—they usually do so rapidly and unconsciously. Since such reading is seldom assigned in college courses, whether they go further to confirm and expand their knowledge depends solely on their time, energy, and interest.

When people read college textbooks, professional articles, technical reports, and serious literature, they read more slowly and carefully to assess the worth or validity of an author's ideas, information, argument, or evidence. The rest of this chapter describes the strategies that lead readers from *understanding* texts to *interpreting* and *evaluating* them *critically,* paying special attention to the strategies of *previewing, responding,* and *reviewing.*

Although critical reading is described here as a three-stage process, it should be clear that these activities seldom happen in a simple one-two-three order. For example, one of the best ways to preview a text is to respond to it briefly as you read it the first time; as you respond, you may find yourself previewing and reviewing, and so on. But if you're not engaging in all three activities at some time, you're not getting as much from your reading as you could.

Previewing Texts

To be a critical reader, you need to be more than a good predictor. In addition to following the thread of an argument, you need to evaluate its logic, weigh its evidence, and accept or reject its conclusion. You read actively, searching for information and ideas that you both understand and can make use of—to further your own thinking, speaking, or writing. To move from understanding to critical awareness, you plan to read a text more than once and more than one way, which is why critical readers *preview* texts before reading them from start to finish.

To understand a text critically, plan to preview before you read, and make previewing the first of several steps needed to appraise the value of the text fully.

First Questions

Ask questions of a text (a book, an article, a Web page) from the moment you look at it. Ask first questions to find general, quickly gleaned information, such as that provided by skimming the title, subtitle, subheads, table of contents, or preface.

- What does the title suggest?
- What is the subject?
- What does the table of contents promise?
- What is emphasized in chapter titles or subheads?
- Who is the author? (Have I heard of him or her?)
- What makes the author an expert or authority?
- How current is the information in this text?
- How might this information help me?

You may not ask these first questions methodically, in this order, or write down all your answers, but if you're a critical reader you'll ask these types of questions before you commit too much time to reading the whole text. If your answers to these first questions suggest that the text is worth further study, you can continue with the preview process.

Second Questions

Once you've determined that a book or article warrants further critical attention, it's very helpful to skim selected parts of it to see what they promise. Skim reading leads to still more questions, the answers to which you will want to capture on note cards or in a journal.

- Read the prefatory material: What can I learn from the book jacket, foreword, or preface?
- Read the introduction, abstract, or first page: What theme or thesis is promised?
- Read a sample chapter or subsection: Is the material about what I expect?
- Scan the index or chapter notes: What sources have informed this text? What names do I recognize?
- Note unfamiliar words or ideas: Do I have the background to understand this text?
- Consider: Will I have to consult other sources to obtain a critical understanding of this one?

Previewing *Iron John*

One of my students gave me a book called *Iron John*. To find out more about the book, I previewed it by asking first questions and second questions, the answers to which I've reproduced here for illustration.

ANSWERS TO FIRST QUESTIONS

- The title *Iron John* is intriguing and suggests something strong and unbreakable.
- I already know and admire the author, Robert Bly, for his insightful poetry, but I've never read his prose.
- The table of contents raises interesting questions but doesn't tell me much about where the book is going:
  1. The Pillow and the Key
  2. When One Hair Turns Gold
  3. The Road of Ashes, Descent, and Grief
  4. The Hunger for the King in Time with No Father
  5. The Meeting with the God-Woman in the Garden

ANSWERS TO SECOND QUESTIONS

- The jacket says, "*Iron John* is Robert Bly's long-awaited book on male initiation and the role of the mentor, the result of ten years' work with men to discover truths about masculinity that get beyond the stereotypes of our popular culture."
- There is no introduction or index, but the chapter notes in the back of the book (260–267) contain the names of people Bly used as sources in writing the book. I recognize novelist D. H. Lawrence, anthropologist Mircea Eliade, poet William Blake, historian/critic Joseph Campbell, and a whole bunch of psychologists, but many others I've never heard of. An intriguing mix.

This preview, which took maybe ten minutes, confirmed that *Iron John* is a book about men and male myths in modern American culture by a well-known poet writing a serious prose book in friendly style. Apparently, Bly not only will examine current male mythology but will make some recommendations about which myths are destructive and which are constructive.

Previewing is only a first step in a process that now slows down and becomes more time-consuming and critical. As readers begin to preview a text seriously, they often make notes in the text's margin or in a journal or notebook to mark places for later review. In other words, before the preview stage of critical reading has ended, the *responding* stage has probably begun.

✑ **WRITING 3**

Select any unfamiliar book about which you are curious and preview it, using the strategy of first and second questions discussed in this section. Stop after ten minutes, and write what you know about the text.

Responding to Texts

Once you understand, through a quick critical preview, what a text promises, you need to examine it more slowly, evaluating its assumptions, arguments, evidence, logic, and conclusion. The best way to do this is to *respond,* or "talk back," to the text in writing.

Talking back can take many forms, from making margin notes to composing extensive notebook entries. Respond to passages that cause you to pause for a moment to reflect, to question, and to read again, or to say "Ah!" or "Aha!" At points of high interest, take notes.

If the text is informational, try to capture the statements that pull together or summarize ideas or are repeated. If the text is argumentative (and many of the texts you'll be reading in college will be), examine the claims the text makes about the topic and each piece of supporting evidence. If the text is literary (a novel, play, or poem), pay extra attention to language features such as images, metaphors, and crisp dialogue. In any text, notice words the author puts in **boldface** or *italic* type: They have been marked for special attention.

Note what's happening to you as you read. Ask about the effect of the text on you: How am I reacting? What am I thinking and feeling? What do I like? What do I distrust? Do I know why yet? But don't worry too much now about answering all your questions. (That's where reviewing comes in.)

The more you write about something, the more you will understand it. Using a reading journal is a good way to keep your responses together in one place that you can return to when writing an essay or research paper. Write each response on a fresh page and include the day's date, the title, and author. Write any and all reactions you have to the text, including summaries, notes on key passages, speculations, questions, answers, ideas for further research, and connections to other books or events in your life. Note especially ideas with which you agree or disagree. Explore ideas that are personally appealing. Record memorable quotations (with page numbers) as well as the reasons they strike you as memorable.

The following brief passage from Bly's *Iron John* is an example of a text to respond to.

> **The dark side of men is clear. Their mad exploitation of earth resources, devaluation and humiliation of women, and obsession with tribal warfare are undeniable. Genetic inheritance contributes to their obsessions, but also culture and environment. We have defective mythologies that ignore masculine depth of feeling, assign men a place in the sky instead of earth, teach obedience to the wrong powers, work to keep men boys, and entangle both men and women in systems of industrial domination that exclude both matriarchy and patriarchy. . . .**
>
> **I speak of the Wild Man in this book, and the distinction between the savage man and the Wild Man is crucial throughout. The savage soul does great damage to soul, earth, and humankind; we can say that though the savage man is wounded he prefers not to examine it. The Wild Man, who has examined his wound, resembles a Zen priest, a shaman, or a woodsman more than a savage.**

When you want to read a text such as this critically, do so with pen or pencil in hand. Mark places to examine further, but be aware that mere marking (underlining, checking, highlighting) does not yet engage you in a conversation with the text. To converse with the text, you need to engage in one or more of the following activities actively: probing, annotating, cross-referencing, and outlining. The following sections illustrate full responses for each activity; in reality, however, a reader would use no more than one or two of these techniques to examine a single text critically.

Probing

You probe a text when you raise critical questions about the text and see if you can answer them. *Probing* is, in essence, asking deeper questions than those asked in previewing. What you ask will depend, of course, on your reason for reading in the first place. Here, for example, are the questions I raised about the *Iron John* passage:

- Bly refers to the dark side of men; does he ever talk about the dark side of women? How would women's darkness differ from men's? What evidence for either does he provide?
- Bly suggests that part of men's dark behavior is genetic, part cultural; where does he get this information? Does he think it's a 50/50 split?
- What "defective mythologies" is Bly talking about? Does he mean things like religion and politics, or is he referring to nursery rhymes and folktales?
- Bly generalizes in his opening sentence, "The dark side of men is clear"—in most sentences actually. Will subsequent chapters support these statements or are we asked to accept them on faith?
- I like the dimension Bly makes between "Wild" and "savage" men. Did he coin the terms or are they used pervasively in mythology in the same way? I wonder how sharp the line really is between the two.

Those are five good questions to ask about the passage; however, any other reader could easily think of five or more. These questions are "critical" in the sense that they not only request further information from the book—which all readers need to request—but also challenge the text's terms, statements, and sources to see if they will stand up under sharp scrutiny.

The questions are written in my own language. Using your own words helps in at least three ways: it forces you to articulate precisely; it makes the question *your* question; and it helps you remember the question for future use.

Annotating and Cross-Referencing

Annotating, or talking back to the author in the margins of the text, is an excellent way to make that text your own, a necessary step in understanding it fully. Annotating is easier if you have your own copy of the text; otherwise you can make your annotations on Post-it Notes or in a notebook with page numbers marked. As a critical reader, you can annotate the following:

- Points of agreement and disagreement
- Exceptions and counterexamples
- Extensions and further possibilities
- Implications and consequences
- Personal associations and memories
- Connections to other texts, ideas, and courses
- Recurring images and symbols

To move beyond annotating (commenting on single passages) to *cross-referencing* (finding relationships among your annotations), devise a coding system to note when one annotation is related to another and thus identify and locate different patterns in the text. Some students write comments in different-colored ink—red for questions, green for nature images, blue for speculations, and so on. Other students use numbers—1 for questions, 2 for images, and so on.

In *Iron John*, for example, the term *Wild Man* occurs on pages 6, 8–12, 14, and 26–27, in other chapters, and in the title of the book's epilogue. A critical reader would mark all of these. In addition, the related term *Hairy Man* occurs on pages 5, 6, and 11, and so on. In cross-referencing, I noted in the margins when the two terms occurred together.

Outlining

Another way of talking back to a text is *outlining*. This involves simply writing out a condensed version of the opening sentence or topic sentence of each paragraph, capturing its essence, as I did for the two paragraphs from *Iron John:*

1.   The dark side of men
2.   The savage man versus the Wild Man

Of course, two paragraphs are simply a start; outlining ten or more paragraphs provides a real clue to the author's organizational pattern. Once you have outlined an article or chapter, you will remember that text better, be able to find key passages more quickly, and see larger patterns more easily.

ॐ **WRITING 4**

Keep a reading journal for one article, chapter, or book that you are assigned to read this semester. Be sure to write something in the journal after every reading session. In addition, annotate and cross-reference the text as you go along to see what patterns you can discover. Finally, make a paragraph outline of the text. Write about the result of these response methods in your journal. Did they help? Which ones worked best?

Reviewing Texts

To *review* you need both to reread and to "re-see" a text, reconsidering its meaning and the ideas you have about it. You need to be sure that you grasp the important points within the text, but you also need to move beyond that to a critical understanding of the text as a whole. In responding, you started a conversation with the text so you could put yourself into its framework and context; in reviewing, you should consider how the book can fit into your own framework and context. Review any text you have previewed and responded to as well as anything you've written in response: journal entries, freewriting, annotations, outlines. Keep responding, talking back to the text even as you review, writing new journal entries to capture your latest insights.

Reviewing can take different forms depending on how you intend to use the text—whether or not you are using it to write a paper, for example. In general, when reviewing a text you have to understand what it means, to interpret its meaning, to evaluate its soundness or significance, and to determine how to use it in your own writing.

Reviewing to Understand

Reviewing to understand means identifying and explaining in your own words the text's main ideas. This task can be simplified if you have outlined the text while responding or have cross-referenced your annotations to highlight relationships among ideas. In reviewing to understand, you can reread portions of articles that you previewed, considering especially abstracts, if there are any; first and last paragraphs; and sections titled "Summary," "Observations," or "Conclusions." In a book, you can reconsider the table of contents, the introductory and concluding chapters, and central chapters that you recognize as important to the author's argument or theme.

Reviewing to Interpret

Reviewing to interpret means moving beyond an appreciation of what the text *says* and building your own theory of what the text *means*. An interpretation is an assertion of what you as a reader think the text is about.

In reviewing to interpret, look over any of your journal entries that articulate overall reactions to the text's main ideas. What did you see in the text? Do you still have the same interpretation? Also reread key passages in the text, making sure that your interpretation is reasonable and is based on the text and is not a product of your imagination.

If you plan to write a critical paper about a text, it's a good idea to confirm your interpretation by consulting what others have said about that text. The interpretations of other critics will help put your own view in perspective as well as raise questions that may not have occurred to you. Try to read more than one perspective on a text. It is better to consult such sources in this reviewing stage after you have established some views of your own, so that you do not simply adopt the view of the first expert you read.

Reviewing to Evaluate

Reviewing to evaluate means deciding whether you think the text accomplishes its own goals. In other words, is the text any good? Different types of texts should be judged on different grounds.

ARGUMENTS.   Many texts you read in college make arguments about ideas, advancing certain *claims* and supporting those claims with *evidence*. A claim is a statement that something is true or should be done. Every claim in an argument should be supported by reliable and sufficient evidence.

At the responding stage, you probably started to identify and comment on the text's claims and evidence. In reviewing, you can ask the following questions to examine and evaluate each part of the argument to see whether it is sound:

- Is the claim based on facts? A *fact* is something that can be verified and that most readers will accept without question. (Fact: The title of the book is *Iron*

John; the author is Robert Bly; it was published in 1990; the myth of Iron John is found in several ancient folktales that have been written down and can be found in libraries; and so on.)

- Is the claim based on a credible inference? An *inference* is a conclusion drawn from an accumulation of facts. (Bly's inferences in *Iron John* about the warrior in modern man are based on his extensive study of ancient mythology. His inferences have a basis in the facts, but other readers might draw other inferences.)

- Is the claim based on opinion? An *opinion* reflects an author's personal beliefs and may be based on faith, emotion, or myth. Claims based on opinion are considered weak in academic writing. (Bly's "dark side of men" is metaphorical and not factual. Some readers would consider it a fair inference based on the savage history of humankind; others would dismiss it as Bly's opinion, based on emotion rather than on facts and careful reasoning.)

All three types of evidence—facts, inference, and opinions—have their place in argumentative writing, but the strongest arguments are those that are based on accurate facts and reasonably drawn inferences. Look out for opinions that are masquerading as facts and for inferences that are based on insufficient facts.

INFORMATIONAL TEXTS. In reviewing informational texts, like reviewing argumentative texts, you need to make sure that the facts are true, that inferences rely on facts, and that opinions presented as evidence are based on expertise, not emotion. Informational texts don't make arguments, but they do draw conclusions from the facts they present. You must decide whether there are enough reliable facts to justify these conclusions. Consider also whether you think the author is reliable and reasonable: Is the tone objective? Has all the relevant information been presented? Is this person an expert?

LITERARY TEXTS. Short stories, poems, and plays don't generally make arguments, but they do strive to be believable, to be enjoyable, and to be effective in conveying their themes. One way to evaluate literature is to reread journal entries in which you responded to the author's images, themes, or overall approach. Then look through the text again, guided by any annotations you've made, and ask whether you think the author's choices were good ones. Look in particular for repeated terms, ideas, or images that will help you see the pattern of the text as a whole. Evaluating literature is often very personal, relying on individual associations and responses, but the strongest critical evaluations are based on textual evidence.

Reviewing to Write

Reviewing a text to use in writing your own paper means locating specific passages to quote, paraphrase, or summarize in support of your own assertions about the text. When you quote, you use the exact language of the text; when you paraphrase, you restate the text in your own words; when you summarize, you reduce the text to a brief statement in your own words. When you identify a note card that contains a passage to quote, paraphrase, or summarize, make sure that you have recorded the page on which the passage occurs in the text so you can find it again and so you can prepare correct documentation.

✍ READING AND WRITING

Reading and writing, like producing and consuming, are two sides of the same coin. When you study one, you inevitably learn more about the other at the same time. The more you attend to the language of published writers, the more you will learn about your own language. The more you attend to your own written language, the more you will learn about the texts you read.

In fact, many of the reading strategies you use to understand and evaluate published texts work equally well when reading your own writing. You can preview, respond to, and review your own or your classmates' writing to gain a critical understanding of your writing and to discover strategies for effective revision.

✍ SUGGESTIONS FOR WRITING AND RESEARCH

Individual

Select a short text. First, read it quickly for understanding. Second, read it critically as described in this chapter. Finally, write a short (two-page) critical review of the text, recommending or not recommending it to other readers. (For more detailed information about writing critically about texts, see Chapter 13, "Interpreting Texts.")

Collaborative

As a class or in small groups, agree on a short text to read and write about according to the preceding directions. Share your reviews in small groups, paying particular attention to the claims and evidence each writer uses in his or her review. Rewrite the reviews based on the responses in the groups. (For more information about responding to others' texts, see Chapter 6, "Sharing and Responding.")

College Journals

Journal writing forces me to think about the problems I'm having with a paper. It's almost a relief, like talking to a friend. I have a conversation with myself and end up answering my own questions.

PETER

J ournals allow people to talk to themselves without feeling silly. Writing in a journal helps college students think about what is happening in their personal and academic lives—an especially important activity for first-year students coping with a new, often bewildering and exciting environment. Sometimes students focus their journal writing narrowly, on the subject matter of a single discipline; at other times they speculate broadly, on the whole range of academic experience; at still other times they write personally, exploring their private thoughts and feelings.

College instructors often require or recommend that students keep journals to monitor what and how they are learning. Just as often, however, students require journals of themselves, realizing that journals are useful and easy to keep whether they're handed in or not.

❧ CHARACTERISTICS OF JOURNALS

In simplest terms, journals are daily records of people's lives (*jour* is French for "day"). Journals are sometimes called *diaries, daybooks, logs, learning logs,* or *commonplace books.* No matter what you call them, the entries written in them are likely to include whatever thoughts, feelings, activities, or plans are on your mind when you sit down to write. In this sense, a journal can be whatever you want it to be, recording whatever snippets of life you find interesting and potentially useful. Certain characteristics, however, remain true for most journals.

Sequence

You use a journal to capture your thoughts sequentially, from one day to the next, though you may not write in it every day. Over time the entries form a cumulative record of what's on your mind. Dating each entry allows you to compare ideas to later and earlier ones and provides an ongoing record of your constancy, change, or growth. You thus end up documenting your learning over the course of a semester or a project.

Audience

Journals are written to help writers rather than readers. A journal is a place for you to explore what's important to you, not to communicate information or ideas to someone else. While you may choose to share entries with readers whom you trust, that is not the reason you keep a journal. A journal assigned by an instructor who intends to read it may initiate an informal conversation between you and the instructor. As such, it has much in common with notes, letters, and other informal means of communication. Some instructors ask to see sample entries rather than read the whole journal. In most cases, required journals receive credit but not a specific grade.

Language

Journal writing is whatever writers want it to be. There usually are no rules; you choose your own language and your own subjects. (The exception may be an assigned journal.) Your focus should be on *ideas* rather than on style, spelling, or punctuation. In journal writing, simply concentrate on what you want to say and use the word, spelling, or punctuation that comes most readily to mind.

Freedom

Students usually are free to practice, discover, rehearse, and even get things wrong in journals without being penalized. Used in this way, journals are practice and discovery books: You can put new concepts into your own language, try out new lines of reasoning or logic, and not worry about completing every thought. If something doesn't work the first time, you can try it again in subsequent entries—or abandon it entirely. In a journal, you always have the freedom to try again.

✍ WRITING 1

Describe your experiences or associations with journals. Have you ever kept one for school before? In which class? With what result? Have you ever kept one on your own? With what result? Do you still keep one? What is it like?

৵ USING JOURNALS IN COLLEGE

Academic journals differ from diaries, daybooks, and private journals in important ways. Whereas diaries and the like may record any and all events of the writer's day, academic journals focus more consistently on ideas under study in college.

Academic journals also differ in important ways from class notebooks, which record the instructor's words rather than the writer's. Academic journals might be called "learning logs" because they record the writer's own perceptions about the business of learning, including reactions to readings, impressions of class, and ideas for writing papers.

Academic journals might be described as a cross between private diaries and class notebooks. Like diaries, journals are written in the first person about ideas important to the writer; like class notebooks, they focus on a subject under study in a college course. I might diagram academic journals like this:

Diary → Academic journal ← Class notebook

Academic journals are most often associated with writing classes, but they can be worthwhile in other classes, too, because they help students become better thinkers and writers.

Journals in the Writing Class

Journals are often assigned in writing classes both to help students discover, explore, advance, and critique their specific writing projects and to help instructors monitor and informally assess students' development as writers.

You can use your journal to find topics to write about, to try out introductions and arguments to use in a paper, to record relevant research and observations, to assess how a paper is turning out, and to make plans for what to do next. In the following journal entry, John tells himself what to do in the next draft of a paper describing his coaching of an eighth-grade girls' soccer team:

> 9/16 I'm going to try to use more dialogue in my paper. That is what I really think I was missing. The second draft is very dull. As I read it, it has no life. I should have used more detail.
>
> I'll try more dialogue, lots more, in draft #3. I'll have it take place at one of my practices, giving a vivid description of what kids were like.
>
> I have SO MUCH MATERIAL. But I have a hard time deciding what seems more interesting.

John's entry is an excellent example of a writer critically evaluating himself and, on the basis of that evaluation, making plans to change something.

Use your journal to record regularly what you are learning in class as you read the textbook, participate in class discussion, read other student papers and models of good professional writing, and review your own writing. Near the end of John's writing course, he reflected in his journal about what he'd learned so far.

> 11/29 I've learned to be very critical of my own work, to look at it again and again, looking for big and little problems. I've also learned from my

writing group that other people's comments can be extremely helpful—so now I make sure I show my early drafts to Kelly or Karen before I write the final draft. I guess I've always known this, but now I actually do it.

⌘ WRITING 2

Keep a journal for the duration of a writing project, recording in it all of your starts, stops, insights, and ideas related to the project. At the end, consider whether the journal presents a fair portrait of your own writing process.

Journals Across the Curriculum

Journals are good tools for learning any subject better. They are especially useful in helping you clarify the purposes of a course, pose and solve problems, keep track of readings, raise questions to ask in class, practice for exams, and find topics for paper assignments.

In science or mathematics, when you switch from numbers to words, you often see the problem differently. In addition, putting someone else's problem into your own words makes it your problem and so leads you one step further toward a solution. Ross made the following entry in a journal for a first-year biology course. He was trying to connect what he was learning in the class to what he knew from fishing.

> 10/7 I noticed that saltwater barracudas resemble freshwater pike, pickerel, and muskies. As a matter of curiosity, are these different species analogous—that is, equally successful forms but of different evolution, which converged toward similitude? Or are they of common heritage, homologous?

One of the best uses of a journal is to make connections between college knowledge and personal knowledge—each reinforces the others, and the connections often lead to greater total understanding. Once Ross finds the answer to his questions, he will be more likely to remember this information than information about which he cannot make personal connections.

When you record personal reflections in a literature or history journal, you may begin to identify with and perhaps make sense of the otherwise distant and confusing past. When you write out trial hypotheses in a social science journal, you may discover good ideas for research topics, designs, or experiments. Whether or not an instructor assigns a

---

⌘ WRITING 3

Think of a course you are taking that does not require a journal. Could you find a use for a journal in that class? What topics would you explore? Write about something in the course that you have not fully figured out. Or keep a journal for a week or two and see if it helps your understanding of the course. After doing so, consider how it worked. Did you find out something interesting? Explain.

journal, keeping one will help you raise, reflect on, and answer your own questions in almost any course.

Double-entry Journals

A double-entry journal can help you separate initial observations from later, more reflective observations. To make such a journal, divide each page in a notebook with a vertical line down the middle. On the left side of the page, record initial impressions or data observations; on the right side, return as often as necessary to reflect on the meaning of what you first recorded.

While the idea of a double-entry journal originated in the sciences as a way for lab scientists to collect data at one time and to speculate about them later, these notebooks also serve well in other courses. In a literature class, for example, you can make initial observations about the plot of a story on the left, while raising questions and noting personal reflections on the right.

The example on page 28 is a sample journal entry by Susan, a first-year college student, who read Alice Walker's novel *The Color Purple* for the first time. In the left column, she recorded the plot; in the right column, she noted her personal reaction to what she was reading.

The reason Susan took such careful notes in a double-entry journal is that she intended to write an interpretive paper about the novel. You can see the value of a reader monitoring his or her reactions with such care, even noting which pages raise which questions. When Susan began to write her paper, these journal entries helped her to find a thesis and to locate particular passages in the novel to support her thesis.

☞ WRITING 4

Keep a double-entry journal for two weeks. On one side of the page, include notes from books you are reading or lectures you are attending. On the other side, write your own thoughts or reactions to those notes. At the end of two weeks, assess the value of this technique for your own understanding of the course material.

Personal Journals

Personal journal writing also has many powerful benefits for students and other writers. In personal journals writers can explore their feelings about any aspect of their lives—being in college, prospective majors, getting along with a roommate, the frustration of receiving a low grade on a paper, the weekend party, or a new date. Anne, a student in a first-year writing class, put it this way:

> Writing is a release, a way of expressing myself, and a way for me to be introspective. It helps me find meaning in my thoughts and gets me through hard times.

Summary	What I think
pp. 3–12. Celie's mother is dying so her father starts having sex with her. She got pregnant by him twice, and he sold both of her babies.	Why did Celie's father sell her kids?
Celie's mother died and he got married again to a very young girl.	How could Mr. take Celie if he wanted Nettie so much?
Mr. is a man whose wife died and he has a lot of children. He wants to marry Celie's sister Nettie. Their father won't let him.	I think Celie's father is lowdown and selfish. A very cruel man.
He says Nettie has too much going for her so he lets him have Celie.	
pp. 13–23. Celie got married to Mr., and his kids don't like her. While Celie was in town she met the lady who has her kids. She was a preacher's wife. Nettie ran away and came to stay with Celie. Mr. still likes her and puts her out because she shows no interest in him. Celie tells her to go to the preacher's wife's house and stay with them because she was the only woman she saw with money.	I think it's wrong to marry someone to take care of your children and to keep your home clean.
	I think Celie was at least glad to know one of her children was in good hands.
	I am glad Nettie was able to get away from her Dad and Mr., hopefully the preacher & wife will take her in.
pp. 24–32. Shug Avery, Mr.'s old friend and also an entertainer, came to town. Mr. got all dressed up so he could go see her, he stayed gone all weekend. Celie was very excited about her.	How could he go and stay out with another woman all weekend? Why didn't he marry Shug?
	Why was Celie so fascinated with Shug?

When you keep a journal in a writing class, it's a good idea to mark off a section for personal entries. Whether you share these with your instructor should be your choice. In the following example, Amy was writing more about herself than her writing class; however, she chose to share the entry with her instructor anyway.

> 11/12 I think I should quit complaining about being misunderstood . . . since I don't try very hard to be understandable, it's no wonder people don't. I just get ticked because more people don't even seem to try to understand others. So many people talk instead of listening. (I think I'm scared of the ones who listen.)

♺ WRITING 5

Keep a personal journal for two weeks, writing faithfully for at least ten minutes each day. Write about your friends, family, future, work, money, frustrations, successes, failures, plans, dates, movies—whatever is on your mind. After two weeks, reread all of your entries and assess the value of such a journal to you.

♺ EXPERIMENTING WITH JOURNALS

If you are keeping a journal for the first time, write often and regularly on a wide variety of topics, and take risks with form, style, and voice. Notice how writing in the early morning differs from writing late at night. Notice the results of writing at the same time every day, regardless of inclination or mood. Try to develop the habit of using your journal even when you are not in an academic environment. Good ideas, questions, and answers don't always wait for convenient times. Above all, write in your journal in your most comfortable voice, freely, and don't worry about someone evaluating you. The following selection of journal entries illustrates some of the ways journals can be especially helpful.

Inventing

Journals can help you plan and start any project by providing a place to talk it over with yourself. Whether it's a research paper, a personal essay, or a take-home exam, you can make journal notes about how to approach a project, where to start, or whom to consult before beginning a draft. Here are two entries from Peter's journal in which he tried to discover a research paper topic for his first-year writing class:

> 10/8 The first draft of this research paper is really difficult: how can you write about something you aren't even interested in? It was not a good idea to pick "legalization of marijuana" just because the issue came up in class discussion. I'm afraid my paper will be all opinion and no facts, because I really don't feel like digging for these facts—if there are any.

> 10/12 Well, I switched my research topic to something I'm actually inter-
> ested in, a handicapped children's rehabilitation program right here on
> campus. My younger brother was born deaf and our whole family has
> pitched in to help him—but I've never really studied what a college pro-
> gram could do to help. The basis of my research will be interviews with
> people who run the program—I have my first appointment tomorrow with
> Professor Stanford.

Sometimes planning means venting frustration about what's going wrong; at other
times it means trying a new direction or topic. Peter does both. Journal writing is ultimately unpredictable: your writing doesn't come out neat and orderly, and sometimes it
doesn't solve your problem, but your journal provides a place where you can keep trying.

Learning to Write

Part of the content of a writing course is the business of learning to write. In other
courses, part of the content is learning to write papers about specific topics. You can use a
journal to document how your writing is going and what you need to do to improve it. In
the following example, Bruce reflects on his experience of writing a report:

> 10/3 I'm making this report a lot harder than it should be. I think my prob-
> lem is I try to edit as I write. I think what I need to do is just write whatever
> I want. After I'm through, then edit and organize. It's hard for me though.

Bruce chastises himself for making his writing harder than need be but at the same time
reminds himself about the process he learned in class that would help his report writing.
Journals are good places to monitor your own writing process and document what helps
you the most.

Writing to Learn

Journal writing can help you discover what you think. The act of regular writing cer-
tifies thoughts and even causes new ones to develop. In that sense, journal writing is strategy for starting in new directions (see Chapter 9). In the following example, Julie, who
kept a journal about all the authors she studied in her American literature course, noticed
a disturbing pattern and wrote in her journal to make some sense of it:

> 5/4 So far, the first two authors we have to read have led tragic, unhappy
> lives. I wonder if this is just a coincidence or if it has something to do with
> the personality of successful writers. Actually, of all people, writers need a
> lot of time alone, by themselves, thinking and writing, away from other
> people, including, probably, close family members. The more I think about
> it, writers would be very difficult people to live with, that's it—writers
> spend so much time alone and become hard to live with. . . .

Julie used the act of regular journal writing to discover and develop ideas, make interpretations, and test hypotheses. Writing to learn requires you to trust that as you write, ideas
will come—some right, some wrong; some good, some bad.

Questioning and Answering

A journal is a place to raise questions about ideas or issues that don't make sense. Raising questions is a fundamental part of all learning: the more you ask, the more you learn as you seek answers. In the following example, Jim wrote in his journal to figure out a quotation written on the blackboard in his technical writing class:

> 9/23 "All Writing Is Persuasive"—It's hard to write on my understanding of this quotation because I don't think that all writing is persuasive. What about assemblies for models and cookbook recipes? I realize that for stories, newspaper articles, novels, and so forth that they are persuasive. But is all writing persuasive? I imagine that for assemblies and so forth that they are persuading a person to do something a particular way. But is this really persuasive writing?

While Jim began by writing "I don't think that all writing is persuasive," he concluded that even assembly instructions "are persuading a person to do something a particular way." The writing sharpened the focus of Jim's questioning and made him critically examine his own ideas, leading him to reconsider his first response to the quotation.

Catching Insights

College is a good place to develop a wider awareness of the world, and a journal can help you examine the social and political climate you grew up in and perhaps took for granted. Jennifer used her journal to reflect on sexist language, recording both her awareness of sexist language in society as well as her own difficulty in avoiding it:

> 3/8 Sexist language is everywhere. So much so that people don't even realize what they are saying is sexist. My teacher last year told all the "mothers-to-be" to be sure to read to their children. What about the fathers? Sexist language is dangerous because it so easily undermines women's morale and self-image. I try my hardest not to use sexist language, but even I find myself falling into old stereotypes.

Evaluating Classes

Journals can be used to capture and record feelings about how a class is going, about what you are learning and not learning. In the following entry, Brian seemed surprised that writing can be fun:

> 9/28 English now is more fun. When I write, the words come out more easily and it's not like homework. All my drafts help me put together my thoughts and retrieve memories that were hidden somewhere in the dungeons of my mind. Usually I wouldn't like English, like in high school, but I pretty much enjoy it here. I like how you get to hear people's reactions to your papers and discuss them with each other.

Entries like this can help you monitor your own learning process. Instructors also learn from candid and freely given comments about the effects of their teaching. Your journal is one place where you can let your instructor know what is happening in class from your point of view.

Clarifying Values

Your journal can be a record of evolving insight as well as the tool to gain that insight. You might ask yourself questions that force you to examine life closely: "If my house were on fire and I could save only one object, what would it be?" or "If I had only two more days to live, how would I spend them?" I used my journal to wrestle with the next direction my life would take:

> 3/12 Do I really want to switch jobs and move to North Carolina? The climate is warmer—a lot longer motorcycle season—and maybe this time we'd look for a farm. But Laura would have to start all over with her job, finding new contacts in the public school system, and we'd both have to find new friends, new doctors, dentists, auto mechanics, get new driver's licenses.
>
> In truth, we really like Vermont, the size, the scale, the beauty, our house, and Annie is just starting college. Money and sunshine aren't everything. . . .

What you read here is only one entry from nearly a month's worth of writing as I tried to figure out what to do with an attractive job offer. In the end, and with the clarifying help of my journal, I stayed put.

Letting Off Steam

Journals are good places to vent frustration over personal or academic difficulties. College instructors don't assign journals to improve students' mental health, but they know that journals can help. Kenyon wrote about the value of the journal experience:

> 10/14 This journal has saved my sanity. It got me started at writing. . . . I can't keep all my problems locked up inside me, but I hate telling others, burdening them with my problems—like what I'm going to do with my major or with the rest of my life.

In many ways, writing in a journal is like talking to a sympathetic audience; the difference, as Kenyon noted, is that the journal is always there, no matter what's on your mind, and it never gives you grief.

Finding Patterns

The very nature of the journal—sequential, chronological, personal—lends it to synthesizing activities, such as finding patterns or larger structures in your learning over time. Rereading journal entries after a few weeks or months can provide specific material from which you can make generalizations and hypotheses. Each individual act of summary becomes a potential thread for weaving new patterns of meaning. Near the end of an American literature course, Maureen summarized the journal's cumulative power this way:

> 5/2 I feel that through the use of this journal over the weeks I have been able to understand certain aspects of each story by actually writing down what I like, and what I don't. . . . Many times I didn't even realize that something bothered me about a story until I put down my feelings in words. I wasn't even sure how I even felt about *The Sun Also Rises* until I kicked a few ideas around on paper. Now I plan to write my take-home exam about it. In short, this journal has really helped me understand this class.

Recording Change

Sometimes it's hard to see how a journal functions overall until you reread it at the end of a term and notice where you began and where you ended. All along your writing may have been casual and fast, your thoughts tentative, your assessments or conclusions uncertain. But the journal gives you a record of who you were, what you thought, and how you changed. Rereading a term's entries may be a pleasant surprise, as Jeff found out:

> 11/21 The journal to me has been like a one-man debate, where I could write thoughts down and then later read them. This seemed to help clarify many of my ideas. To be honest there is probably fifty percent of the journal that is nothing but B.S. and ramblings to fulfill assignments, but that still leaves fifty percent that I think is of importance. The journal is also a time capsule. I want to put it away and not look at it for ten or twenty years and let it recall for me this period of my life.

✑ LETTERS FROM JOURNALS

Sometimes, instead of turning in your journal for an instructor to check, you will be asked to write a class letter to your instructor, addressing him or her personally, about issues related to the course—perhaps issues captured in your journal. What I like about this use of journals is that all your thoughts and insights remain private unless you choose to share them with a specific and known audience. An audience is the only real difference between a class letter and a journal entry—the former written to someone else, the latter to yourself.

As an instructor, I appreciate getting letters that are informal, personal, honest, and that contain some references to and insights about course materials: readings, class discussion, subject matter. In other words, I enjoy hearing from my own students about all the ideas for journal writing already explored in this chapter, the only difference being that these are intentionally shared with me. If you are asked to share letters with an instructor, consider the following:

1. Address your instructor personally, including references to common experience or shared ideas, single-space, and sign your name.
2. Focus on course ideas rather than your private life, but include incidents from your private life when they are relevant to course material.

3.   Avoid being overly general: Include references to specific passages in the readings or specific incidents in the class.
4.   Ask real questions or pose real problems that you hope to have answered.
5.   Write informally, in your natural voice, but revise and edit enough so that your language is clear and coherent.

When I receive letters, I either write back a single letter to the whole class, quoting passages from different students so the whole class can be brought into the conversation, or I write brief responses on each letter and return every letter to the sender. I count the letters as I do journals, quantitatively: Students get credit for simply doing them, not grades on specific content.

Increasingly, my students and I share e-mail messages, which is a wonderfully quick and efficient way of conversing about specific ideas, questions, or problems related to the course. E-mail writing may resemble journal writing in that it is usually informal and unrevised; however, like talking on the telephone and writing letters, the addition of a specific audience means that even this most informal language needs to be clear, correct, and respectful. A loose version of the guidelines above would cover e-mail writing as well.

↪ SUGGESTIONS FOR WRITING AND RESEARCH

Individual

1.   Select a writer in your intended major who is known for having written a journal (for example, Mary Shelley, Ralph Waldo Emerson, Virginia Woolf, or Anaïs Nin in literature; Leonardo da Vinci, Georgia O'Keeffe, or Edward Weston in the arts; B. F. Skinner or Margaret Mead in the social sciences; Charles Darwin or Marie Curie in the sciences). Study the writer's journals to identify the features that characterize them and the purpose they served. Write a report on what you find and share it with your class.
2.   At the end of the semester, review your journal and do the following: (a) put in page numbers, (b) write a title for each entry, (c) make a table of contents, and (d) write an introduction to the journal explaining how it might be read by a stranger (or your instructor).
3.   Review your journal entries for the past five weeks, select one entry that seems especially interesting, and write a reflective essay of several pages on it. How are they different? Which is better? Is that a fair question?

Collaborative

Have each student agree to bring a typed copy of one journal entry written during the term. Exchange entries in writing groups or in the whole class and discuss interesting features of the entries.

Chapter

❧ 4 ❧

The Elements of Composition

I find it very confusing moving from one professor to another.
They all expect different things. I still haven't learned yet what
makes a "good" paper as opposed to a "bad" paper.

JENNIFER

Why do teachers always make you write about what they want you
to and never what you want to? What is the writing for anyway?

ERIC

I would like to develop some sort of personal style so
when I write people know it's me.

JOHN

The focus, structure, and style of every paper you write are determined by your situation—why you are writing (purpose) and to whom (audience). Taken together, purpose and audience largely determine the voice in which you write. While this chapter asks you to consider each of these fundamental elements of composition in isolation, in truth, experienced writers usually think about them subconsciously and simultaneously. In any case, I believe the following discussion may be useful when you are assigned to write an academic paper.

❧ WRITING FOR A PURPOSE

People write to discover what's on their minds, figure things out, vent frustrations, keep records, remember things, communicate information, shape ideas, express feelings, recount experiences, raise questions, imagine the future, create new forms—and simply for pleasure. They also write when they're required to in school, to demonstrate knowledge and solve problems. But no matter what the task, writers write better when they do so purposefully—when they know what they want to accomplish. This section examines three broad and overlapping stages of writing: discovering, communicating, and creating. It discusses strategies to accomplish each one effectively.

Writing to Discover

Writing helps people discover ideas, relationships, connections, and patterns in their lives and in the world. In college, students write to discover paper topics, develop those topics, expand and explain ideas, and connect seemingly unrelated material in coherent patterns. In this sense, writing is one of the most powerful learning tools available.

Writing is especially powerful because it makes language and therefore thoughts stand still, allowing thoughts to be examined slowly and deliberately, allowing ideas to be elaborated, critiqued, rearranged, and corrected. Playwright Christopher Fry once said, "My trouble is that I'm the sort of writer who only finds out what he is getting at by the time he's got to the end of it." In other words, his purpose and plan become clear only after he has written a whole draft; he knows that the act of writing will help him find his way. But rather than considering this inventive power of writing "troublesome," to use Fry's words, you can consider it a solution to many other problems. Once you know that writing can generate ideas, advance concepts, and forge connections, then you can use it deliberately and strategically to help you write college papers.

Discovery can happen in all writing. Any time you write, you may find new or lost ideas, implications, and directions. However, sometimes it pays to write with the specific intention of discovering. Discovery writing is often used before actual drafting to explore the subject and purpose of a paper or to solve writing problems once drafting and revising have begun.

↜ WRITING 1

Describe a time when you used writing for discovery purposes. Did you set out to use writing this way, or did it happen accidentally? Have you used it deliberately since then? With what results?

Writing to Communicate

The most common reason for writing in college is to transmit ideas or information to an audience. College students write essays, exams, and reports to instructors, as well as letters, applications, and résumés to potential employers. The general guidelines for such writing are well known: Communicative writing needs to be obviously purposeful so both writer and reader know where it's going. It needs to be clear in order to be understood. It needs to include assertions supported by evidence in order to be believable. And it needs to be conventionally correct in terms of spelling, mechanics, and grammar in order to be taken seriously. While there are interesting exceptions to these guidelines (see Chapter 24), they are the rule in most academic writing situations.

Thesis-Based Writing

Many academic assignments ask you to write *thesis-based* papers, that is, papers that assert, explain, support, or defend a position or idea. Your assertion about the idea or position is called the paper's *thesis*. Since most academic papers are assigned so that instructors can witness and assess student knowledge, stating a thesis makes clear what, exactly, the paper's claim or position is. The thesis, broadly speaking, is the place to summarize the main idea to make it explicit to readers. For example, the thesis of this chapter is stated in the first paragraph: that *the focus, structure, and style of papers are determined by purpose and audience.* This claim is then supported throughout the rest of the chapter.

Some thesis-based papers present the thesis first, usually in the opening paragraph to tell readers what's coming. Such *thesis-first* papers are common in academic as well as technical and scientific writing because they emphasize the transmission of an idea or information clearly, directly, and economically, thus helping readers get rapidly to the point. In contrast, *delayed-thesis* papers do not state conclusions up front but examine a variety of conditions or circumstances to be considered before a decision is made. Such papers emphasize the process by which the writer discovered knowledge as much as the knowledge itself. Whether first or delayed, a thesis-driven paper explicitly answers the critical reader's question "So what?" Why does this paper exist? What's it about? Common thesis-based assignments in college include the following, which are examined in more detail in subsequent chapters:

- *Explaining ideas.* The purpose of explaining something is to make it clear to somebody who knows less about the subject than you do. You explain best by following a logical order, using simple language, and providing illustration and examples of what you mean. (See Chapter 11.)
- *Arguing positions.* The purpose of arguing is to persuade readers to agree with your position. College assignments frequently ask you to explore opposing sides of an issue or several different interpretations of a text and then to take a stand advocating one point of view. (See Chapter 12.)
- *Interpreting texts.* The purpose of interpreting a text is to explain to others what the text means, to tell why you believe it means this, and to support your reading with reasons based on evidence from the text. (See Chapter 13.)

Question-Based Writing

Still other papers assigned in college may have, as their larger purpose, more personal or reflective dimensions, and never directly state a thesis at all. Such exploratory papers might pose questions to which specific answers are illusive or examine dimensions of the writer's self in relation to the larger world. These *question-based* papers might be said to emphasize the play of the writer's mind more than the direct transmission of knowledge. However, these papers, too, answer though less directly—the critical reader's questions "So what? Why does this paper exist? What's it about?" Common question-based assignments include the following, which are examined in more detail in subsequent chapters:

- *Recounting experience.* People narrate personal stories to share common experiences with others and to learn, in the telling, the meaning of those experiences for themselves. (See Chapter 8.)

- *Exploring identity.* People examine their own lives to find out important things about who they are, how they got that way, or where they're going next. (See Chapter 9.)
- *Profiling people.* Writers profile other people to find out how other lives are lived, valued, and expressed, and, in the process, to learn about our shared culture. (See Chapter 10.)
- *Reflecting on the world.* Writers speculate, muse, and ponder about an infinite number of ideas and issues in our world, and share them with others, to find out what these ideas mean and what, along the way, the world means. (See Chapter 11.)

☞ WRITING 2

When is the last time you wrote to communicate something? Describe your purpose and audience. How successful were these acts of communication? How do you know?

Writing to Create

When you write to imagine or create, you often pay special attention to the way your language looks and sounds, its form, shape, rhythm, images, and texture. Though the term *creative writing* is most often associated with poetry, fiction, and drama, it's important to see any act of writing, from personal narratives to research essays, as creative.

When you write to create, you pay less immediate attention to your audience and subject and more to the act of expression itself. Your goal is not so much to change the world or to transmit information about it as to transform an experience or idea into something that will make your readers pause, see the world from a different angle and perhaps reflect upon what it means. You want your writing itself, not just the information it contains, to affect your readers emotionally or esthetically as well as intellectually.

In most college papers, your primary purpose will be to communicate, not to create. However, nearly every writing assignment has room for a creative dimension. When writing for emotional or esthetic effect in an otherwise communicative paper, be especially careful that your creativity serves a purpose and that the communicative part is strong on its own. You want your creative use of language to enhance, not camouflage, your ideas. (See Chapter 24.)

INTENSIFYING EXPERIENCE.   When Amanda recounted her experience picking potatoes on board a mechanical potato harvester on her father's farm, she made her readers feel the experience as she did by crafting her language to duplicate the sense of hard, monotonous work:

> Potatoes, mud, potatoes, mud, potatoes, that was all I saw in front of me. They moved from my right side to my left, at hip level. A conveyor belt never stopping. On and on and on. The potatoes passed fast, a constant stream. My hands worked

deftly, pulling out clods of dirt, rotten potatoes, old shaws, and anything else I found that wasn't a potato. It was October, the ground was nearly frozen, the mud was hard and solid. Cold. Dirt had gotten into my yellow and yet brown rubber gloves, had wedged under my nails, increasing my discomfort.

This is a creative approach to essay writing because the writer uses a graphic, descriptive style to put readers at the scene of her experience rather than summarizing it or explaining explicitly what it meant to her.

EXPERIMENTING WITH FORM.   Keith created a special language effect in an otherwise traditional and straightforward academic assignment by writing a poetic prologue for a research essay about homeless people in New York City. The full essay contains factual information derived from social workers, agency documents, and library research.

The cold cement
no pillow
The steel grate
no mattress
But the hot air
of the midnight subway
Lets me sleep.

Using the poetic form creates a brief emotional involvement with the research subject, allowing readers to fill in missing information with their imaginations. Note, however, that the details of the poem (*cold cement, steel grate, subway*) spring not from the writer's imagination but from his research notes and observations.

ᠵ WRITING 3

Describe a time when your primary purpose in writing was to create rather than to discover or communicate. Were you pleased with the result? Why or why not?

ᠵ ADDRESSING AUDIENCES

The better you know your audience, the better you're likely to write. Whether your writing is judged "good" or not depends largely on how well it's received by the readers for whom it's intended. Just as you change the way you speak depending on the audience you're addressing—your boss, mother, instructor, friend, younger brother—so you change the way you write depending on the audience to whom you're writing. You don't want to overexplain and perhaps bore the audience or underexplain and leave it wanting.

Speakers have an advantage over writers in that they see the effect of their words on their listeners and can adjust accordingly. A puzzled look tells the speaker to slow down, a smile and nod says keep going full speed ahead, and so on. However, writers can only imagine the reactions of the people to whom they're trying to communicate.

I believe all college papers need to be written to at least two audiences, maybe more: first, to yourself, so you understand it; second, to your instructor who has asked you to write it in the first place. In addition, you may also be writing to other students or for publication to more public audiences. This chapter examines how expectations differ from one audience to the next.

Understanding College Audiences

It might help to think of the different audiences you will address in college as existing along a continuum, with those closest and best known to you (yourself, friends) at one end and those farthest from and least known to you (the general public) at the other end:

Self—Family—Friends—Instructor—Public

While the items on your continuum will always differ in particulars from somebody else's, the principle that you know some audiences better than others will always be the same and will influence how you write. The audience of most concern to most college students is the instructor who will evaluate their learning on the basis of their writing.

> ✌ **WRITING 4**
>
> Think back over the past several weeks and list all the different audiences to whom you have written. To whom did you write most often? Which audiences were easy for you to address? Which were difficult? Why?

Shaping Writing for Different Audiences

To shape your writing for a particular audience, you first need to understand the qualities of your writing that can change according to audience. The context you need to provide; the structure, tone, and style you use; and your purpose for writing can all be affected by your audience. (Structure, tone, and style are important elements of voice.)

CONTEXT.   Different audiences need different contexts—different amounts or kinds of background information—in order to understand your ideas. Find out whether your readers already know about the topic or whether it's completely new to them. Consider whether any terms or ideas need explaining. For example, other students in your writing group might know exactly whom you mean if you refer to a favorite singer, but your instructor might not. Also consider what sort of explanation would work best with your audience.

STRUCTURE.   Every piece of writing is put together in a certain way: some ideas are discussed early, others late; transitions between ideas are marked in a certain way; similar ideas are either grouped together or treated separately. How you structure a paper depends in large part on what you think will work best with your particular audience. For example, if you were writing an argument for someone who disagrees with your position, you might begin with the evidence with which you both agree and then later introduce more controversial evidence.

TONE. The tone of a piece of writing conveys the writer's attitude toward the subject matter and audience. How do you want to sound to your readers? You may, of course, have a different attitude toward each audience you address. In addition, you may want different audiences to hear in different ways. For example, when writing to yourself, you won't mind sounding confused. When writing to instructors, though, you will want to sound confident and authoritative.

STYLE. Style is largely determined by the formality and complexity of your language. You need to determine what style your readers expect and what style will be most effective in a given paper. Fellow students might be offended if you write in anything other than a friendly style, but some instructors might interpret the same style as disrespectful.

PURPOSE. The explicit purpose of your writing depends more on you and your assignment than on your audience. However, certain purposes are more likely to apply to particular audiences than others. Also, there are unstated purposes embedded in any piece of writing, and these will vary depending on whom you're addressing. For example, is it important that your readers like you? Or that they respect you? Or that they give you good grades? Always ask yourself what you want a piece of writing to do for or to your audience and what you want your audience to do in response to your writing.

Let's follow the way writing generally needs to change as you move along the scale away from the audience you know best, yourself.

WRITING TO YOURSELF. Every paper you write is addressed in part to yourself, and some writing, such as journals, is addressed primarily to yourself. However, most reports, essays, papers, and exams are also addressed to other people: instructors, peers, parents, or employers. Journal writing is your opportunity to write to yourself and yourself alone. When you write to yourself alone, you don't need to worry about context, structure, tone, or style; only purpose matters if you are the sole reader. However, if you make a journal entry that you might want to refer to later, it's a good idea to provide sufficient background and explanation to help you remember the event or the idea described if you do return to it. When you are the reader of your own writing, choose words, sentences, rhythms, images, and punctuation that come easiest and most naturally to you.

WRITING TO PEERS. Your peers are your equals, your friends and classmates, people of similar age, background, or situation. Some of your assignments will ask you to consider the other students in the class to be your audience, for example, when you read papers to each other in writing groups or exchange papers to edit each other's work.

The primary difference between writing to yourself and writing to peers is the amount of context and structure you need to provide to make sure your readers understand you. If your paper is about a personal experience, you need to provide the explanations and details that will allow readers who did not have your experience to fully understand the events and ideas you describe.

If your paper is about a subject that requires research, be sure to provide background information to make the topic comprehensible and interesting in a structure (for example, chronological order, logical order, cause-effect sequence) that makes sense. Be direct, honest, and friendly; peers will see right through any pretentious or stuffy language.

You usually write to peers to share a response to their writing, to recount an experience, to explain an idea, or to argue a position. In a writing class, the most important implicit purpose is probably to establish a good working rapport with your classmates by being honest, straightforward, and supportive.

WRITING TO INSTRUCTORS.   Instructors are among the most difficult audiences for whom to write. First, they usually make the assignments, so they know what they want, and it's your job to figure out what that is. Second, they often know more about your subject than you do. Third, different instructors may have quite different criteria for what constitutes good writing. And fourth, each instructor may simultaneously play several different roles: a helpful resource, a critic, an editor, a coach, and, finally, a judge.

It is often difficult to know how much context to provide in a paper written for an instructor unless the assignment specifically tells you. For example, in writing about a Shakespearean play to an English professor, should you provide a summary of the play when you know that he or she already knows it? Or should you skip the summary information and write only about ideas original with you? The safest approach is to provide full background, explain all ideas, support all assertions, and cite authorities in the field. Write as if your instructor needed all this information and it were your job to educate him or her.

When writing papers to instructors, be sure to use a structure that suits the type of paper you are writing. For example, personal experience papers are often chronological, reports may be more thematic, and so on. (The chapters in Part Two describe conventional structures for each type of paper discussed there.)

One of your instructor's roles is to help you learn to write effective papers. But another role is to evaluate whether you have done so and, from a broader perspective, whether you are becoming a literate member of the college community. Therefore, your implicit purpose when you write to instructors is to demonstrate your understanding of conventions, knowledge, reasoning ability, and originality.

WRITING TO PUBLIC AUDIENCES.   Writing to a public audience is difficult for all writers because the audience is usually both diverse and unknown. The public audience can include both people who know more and those who know less than you; it can contain experts who will correct the slightest mistake and novices who need even simple terms explained; it can contain opponents looking for reasons to argue and supporters looking for reasons to continue support. And you are unlikely to know many of these people personally.

You usually have some idea of who these anonymous readers are or you wouldn't be writing to them in the first place. Still, it is important to learn as much as you can about any of their beliefs and characteristics that may be relevant to the point you intend to make. What is their educational level? What are their political, philosophical, or religious beliefs? What are their interests?

When you don't know who your audience is, provide context for everything you say. If you are referring to even well-known groups such as the NCAA or ACLU, write out the full names the first time you refer to them (National Collegiate Athletic Association or

American Civil Liberties Union). If you refer to an idea as postmodern, define or illustrate what the term means. (Good luck!) Your writing should be able to stand by itself and make complete sense to people you do not know.

Your purpose and structure should be as clear as possible, with your opening paragraph letting this audience know what's to come. Your tone will depend on your purpose, but generally it should be fair and reasonable. Your style will depend on the publication for which you are writing.

↬ WRITING 5

How accurate do you find the preceding discussion about different college audiences? Describe circumstances that confirm or contradict the description here. If instructors are not your most difficult audience, explain who is.

↬ FINDING A VOICE

Each individual speaks with distinctive voice. Some people speak loudly, some softly, others with quiet authority. Some sound assertive or aggressive while others sound cautious, tentative, or insecure. Some voices are clear and easy to follow while others are garbled, convoluted, and meandering. Some create belief and inspire trust while others do not.

An individual's voice can also be recognized in the writing he or she produces. A writer's voice, like a person's personality, is determined by factors such as ethnic identity, social class, family, or religion. In addition, some elements of voice evolve as a writer matures, such as how one thinks (logically or intuitively) and what one thinks (a political or philosophical stance). Writers also can exert a great deal of control over their language. They create the style (simple or complex), the tone (serious or sarcastic), and many other elements. Writers try to be in control of as many elements of their writing voice as they can.

Defining Voice

The word *voice* means at least two distinctly different things. First, it is the audible sound of a person speaking (*He has a high-pitched voice*). Speaking voices distinguish themselves by auditory qualities such as pitch (high, low, nasal), pace (fast, slow), tone (angry, assertive, tentative), rhythm (regular, smooth, erratic), volume (soft, loud), and accent (Southern, British, Boston). Applied to writing, this meaning cannot be taken literally; unless writers read their work aloud, readers don't actually hear writers' voices. However, the language on the page can re-create the sound of the writer talking. Careful writers control, as much as they can, the sound of their words in their readers' heads.

Second, *voice*, especially when applied to writing, suggests who a person is and what he or she stands for. Written voices convey something of the writers behind the words, including their personal, political, philosophical, and social beliefs. In addition, writers' beliefs and values may be revealed in the way they reason about things, whether they do so in an orderly, scientific manner or more intuitively and emotionally.

> ☙ **WRITING 6**
>
> In your own words, describe the concept of voice. Do you think writers have one voice or many? Explain what you mean.

☙ ANALYZING VOICE

Readers experience a writer's voice as a whole expression, not a set of component parts. However, to help you understand and gain control of your own voice, let's examine the individual elements that combine to make the whole.

TONE.   Tone is your attitude toward the subject and audience: angry, anxious, joyous, sarcastic, puzzled, contemptuous, respectful, friendly, and so on. Writers control their tone, just as speakers do, by adopting a particular perspective or point of view, selecting words carefully, emphasizing some words and ideas over others, choosing certain patterns of inflection, and controlling the pace with pauses and other punctuation. For example, note how your tone might change as you speak or write the following sentences:

- The English Department was unable to offer enough writing courses to satisfy the demand this semester.
- Why doesn't English offer more writing courses?
- It's outrageous that so many students were closed out of first-year writing courses!

To gain control of the tone of your writing, read drafts of your paper aloud and listen carefully to the attitudes you express. Try to hear your own words as if you were the audience: decide whether the overall tone is the one you intended, and reread carefully to make sure every sentence contributes to this tone.

STYLE.   Style is the way writers express themselves according to whom they are addressing and why. Style is found in the formality or informality, simplicity or complexity, clarity or muddiness of a writer's language. For example, in writing to a friend, you may adopt an informal, conversational style, characterized by contractions and simple, colloquial language:

John. Can't make it tonight. Spent the day cutting wood and am totally bushed.

In writing to instructors, however, you may be more formal, careful, and precise:

Dear Professor James,

I am sorry, but my critical essay will be late. Over the weekend, I overextended myself cutting, splitting, and stacking two cords of wood for my father and ended up with a sprained back. Would you be willing to extend the paper deadline for one more day?

In other words, the style you adopt depends on your audience, purpose, and situation. To gain control of your style, think about how you wish to present yourself, and shape your words, sentences, and paragraphs to suit the occasion.

STRUCTURE.   The structure of a text is how it's put together: where it starts, where it goes next, where the thesis occurs, what evidence fits where, how it concludes. Structure is the pattern or logic that holds together thoughtful writing, revealing something of the thought process that created it. For example, a linear, logical structure may characterize the writer as a linear, logical thinker, while a circular, digressive structure may suggest more intuitive, less orderly habits of mind. Skillful writers, of course, can present themselves one way or the other depending on whom they're addressing and why.

The easiest way to gain control of an essay's structure is to make an outline that reveals visually and briefly the organization and direction you intend. Some writers outline before they start writing and stick to the outline all the way through the writing. Others outline only after writing a draft or two to help control their final draft. And still others start with a rough outline they continue to modify as the writing modifies thought and direction.

VALUES AND BELIEFS.   Your values include your political, social, religious, and philosophical beliefs. Your background, opinions, and beliefs will be part of everything you write, but you must learn when to express them directly and when not to. For example, including your values would enhance a personal essay or other autobiographical writing, but it may detract attention from the subject of a research essay.

To gain control of the values in your writing, consider whether the purpose of the assignment calls for an implicit or explicit statement of your values. Examine your drafts for words that reveal your personal biases, beliefs, and values; keep them or take them out as appropriate for the assignment.

AUTHORITY.   Your authority comes from confidence in your knowledge and is projected through the way you handle the material about which you are writing. An authoritative voice is often clear, direct, factual, and specific, leaving the impression that the writer is confident about what he or she is saying. You can exert and project real authority only over material you know well, whether it's the facts of your personal life or carefully researched information. The more you know about your subject, the more clearly you will explain it, and the more confident you will sound.

To gain control over the authority in your writing, do your homework, conduct thorough research, and read your sources of information carefully and critically.

↭ WRITING 7

Describe your own writing voice in terms of each of the elements outlined in this section (tone, style, structure, values, authority). Then compare your description with a recent paper you have written. In what ways does the paper substantiate your description? In what ways does it differ from your description? How do you account for any differences?

◌ SUGGESTIONS FOR WRITING AND RESEARCH

Individual

1. Select a topic that interests you and write about it in each of the three modes described in this chapter. First, begin with discovery writing to yourself, perhaps in a journal. Second, write a letter to communicate with somebody about this interest. Third, write creatively about it in a short poem, story, or play. Finally, describe your experience writing in these different modes.

2. Select a paper you have written recently for an instructor audience and rewrite it for a publication, choosing either a student newspaper or a local magazine. Before you start, make notes about what elements need to be changed: context, structure, tone, style, or purpose. When you finish recasting the paper to this larger, more public audience, complete collaborative assignment 3. Make final revisions, taking into account your partner's observations, and send your paper to the publication.

3. Collect and examine as many samples of your past writing as you have saved. Also look closely at the writing you have done during this term. Write a paper in which you describe and explain this history and evolution of your voice and the features that most characterize your current writing voice.

Collaborative

1. Select a topic that your whole writing group is interested in writing about. Divide your labors so that some of you do discovery writing, some do communicative writing, and some write creatively. With scissors and tape, combine your efforts into a single coherent, creative piece of college writing, making sure that some of every member's writing is included in the finished product. Perform a reading of this collage for the other groups; listen to theirs in return.

2. In a group of five students, select a topic of common interest. Write about the topic (either as homework or for fifteen minutes in class) to one of the following audiences: yourself, a friend who is not attending your school, your instructor, an appropriate magazine or newspaper. Share your writing with one another, and together list the choices you needed to make for each audience.

3. Exchange recently written papers with a partner. Examine your partner's paper for the elements of voice. In a letter, each of you describe what you find. How does your partner's perception of your voice match or differ from your own? Now do individual assignment 2, including your partner's assessment as part of your analysis.

The Working Writer

Is writing a matter of learning and practice, or are good writers born?

Ross

While there is no one best way to write, some ways do seem to work for more people in more situations than do others. Learning what these ways are may save you some time, grief, or energy—perhaps all three. This chapter describes the sometimes exciting, sometimes tedious, and always messy process of working writers, taking a close look at the various phases most writers go through, from planning and drafting through researching, revising, and editing. If you're interested in improving your writing, examine closely your own writing process: Describe how you do it, identify what works and what doesn't, then study the ideas and strategies that work for others—some of these are bound to help you.

What are your writing habits? What do you do, for example, when you are assigned to write a paper due in one week? Do you sit down that day and start writing the introduction? Or do you sit down but do something else instead? If you don't work on the assignment right away, do you begin two days before deadline, or is your favorite time the night before the paper is due? Do you write a few pages a day, every day, and let your paper emerge gradually? Or do you prefer to draft it one day, revise the next, and proofread it just before handing it in?

What writing conditions do you seek? Do you prefer your own room? Do you like to listen to certain kinds of music? Do you deliberately go somewhere quiet, such as the library? Or do you prefer a coffee shop, a cafe, or a booth at McDonald's?

With what do you write? Your own computer or the school's? Your old Smith Corona portable typewriter or a pencil on tablets of lined paper? Or do you first write with a favorite pen and then copy the result onto a computer?

Which of the habits or methods described here is the right one? Which technique yields the best results? These are trick questions, since different ones work best for different individuals. There is no single best way to write. People manage to write well under wildly different conditions.

The rest of this chapter identifies five discrete but overlapping and often nonsequential phases of the way writers work—planning, drafting, researching, revising, and editing—and explains how computers have become critical tools in this process.

> ↷ **WRITING 1**
>
> Answer the questions posed on the opening pages of this chapter and describe your own work as a writer: Where, when, and how do you usually write? What are the usual results? With what do you need some extra help?

↷ PLANNING

Planning consists of creating, discovering, locating, developing, organizing, and trying out ideas. Writers are doing deliberate planning when they make notes, turn casual lists into organized outlines, write journal entries, compose rough drafts, and consult with others. They also are doing less deliberate planning while they walk, jog, eat, read, browse in libraries, converse with friends, or wake up in the middle of the night thinking. Planning involves both expanding and limiting options, locating the best strategy for the occasion at hand, and focusing energy productively.

Planning comes first. It also comes second and third. No matter how careful your first plans, the act of writing usually necessitates that you keep planning all the way through the writing process, that you continue to think about why you are writing, what you are writing, and for whom. When writers are not sure how their ideas will be received by someone else, they often write to themselves first, testing their ideas on a friendly audience, and find good voices for communicating with others in later drafts.

During the planning process for the first edition of *The Working Writer,* for instance, I was trying out ideas and exploring broadly and also narrowing my thinking to focus both on my purpose as a writer and the purposes that handbooks serve. I had to consider the audience: Who uses textbooks? I had to find my voice not only as a classroom teacher but as a writer: Would I be friendly and casual or authoritative and serious? I spent some time inventing and discovering ideas, figuring out what kind of information readers require, how much of this information I already knew, and where to find what I didn't know.

Strategies for Planning

1.   Make planning a first, separate stage in your writing process. You will feel freer to explore ideas and directions for your writing if you compose tentative outlines, take notes, and write journal entries before attending to matters of neatness, correctness, and final form.
2.   Play with ideas while you draft, research, revise, and edit. Just as it's important for invention and discovery to be a first, discrete stage, it's also important to continue inventing and discovering as your draft moves toward its final shape.
3.   When you begin writing, write out crazy as well as sane ideas. While the wild ones may not in themselves prove useful, they may suggest others that are.
4.   When stuck for ideas, try to articulate, in writing or speech, how you are stuck, where you are stuck, and why you think you are stuck. Doing so may help you get unstuck.

5. When searching for direction, *read* to find new information, *talk* to find out how your ideas sound to others, *listen* to the responses you receive, and *keep writing* to test the directions you find.

✍ WRITING 2

Describe the strategies you commonly use when you plan papers. How much does your planning vary from time to time or assignment to assignment? Now use your favorite planning strategy for twenty minutes to plan one currently assigned paper.

✍ DRAFTING

At some point all writers need to move beyond thinking, talking, and planning and actually start writing. Many writers like to schedule a block of time, an hour or more, to draft their ideas, give them shape, see what they look like. One of the real secrets to good writing is simply learning to sit down and write. *Drafting* is the intentional production of language to convey information or ideas to an audience. First drafts are concerned with ideas, with getting the direction and concept of the piece of writing clear. Subsequent drafting, which includes revising and editing, is concerned with making the initial ideas ever sharper, more precise and clearer.

While most writers hope their first draft will be their final draft, it seldom is. Still, try to make your early drafts as complete as possible at the time—that is, give each draft your best shot: Compose in complete sentences, break into paragraphs where necessary, and aim at a satisfying form. At the same time, allow time for second and third drafts and maybe more.

Sometimes it's hard to separate drafting from planning, researching, revising, and editing. Many times in writing *The Working Writer,* I sat down to explore a possible idea in a notebook and found myself drafting part of a chapter instead. Other times, when trying to advance an idea in a clear and linear way, I kept returning instead to revise a section just completed. While it's useful to separate these phases of writing, don't worry too much if they refuse to stay separate. In most serious writing, every phase of the process is as much circular and haphazard as linear and direct as you move back and forth almost simultaneously from planning to revising to editing to drafting, back to planning, and so on. While Part Three will discuss the details of drafting specific papers, the following ideas may prove useful now.

Strategies for Drafting

1. Sit down and turn on your computer, or place paper in your typewriter, or open your writing notebook and pick up a pen. Once you have done any of these initial acts, your chances of starting your paper increase dramatically. Plan to sit still and write for one hour.

2.   Start writing by writing. Do not sit and stare at the blinking cursor or the blank page. Instead, put words and sentences in front of you and see where they lead. Do this for at least fifteen minutes and then pause and see what you've got.

3.   Plan to throw away your first page. This simple resolution will take off a lot of pressure, help you relax, and let the momentum of the writing take over. Later, you may even decide to keep some of this page—a clear bonus.

4.   Compose in chunks. It's hard to write a whole term paper; it's fairly easy to write a section of it; it's easier still to write a paragraph; it's a breeze to write a sentence or two. In other words, even large projects start with single words, sentences, and paragraphs.

5.   Allow time to revise and edit. Start drafting any writing assignment as soon as you can, not the night before it's due.

ᴕ **WRITING 3**

Describe the process you most commonly use to draft a paper. Is your way of starting consistent from paper to paper? Now write the first draft for the paper you planned in Writing 2. Sit down, and for half an hour compose as much of the paper as you can, noting in brackets as you go along where you need to return with more information or ideas.

ᴕ **RESEARCHING**

Unless writers are drafting completely from memory, they need ideas and information to make their writing interesting. However, even personal essays, experiential papers, and self-profiles benefit from additional factual information that substantiates and intensifies what the writer remembers. In other words, good working writers are researching writers.

Don't confuse using real and useful research in your writing with writing what are sometimes called *research papers*—school exercises meant to demonstrate how to find information in libraries, and how to include it in your papers, but that often focus more on the mechanics of collecting information than on writing with information about something important to you. Actually, as a college student you're researching every time you write an analysis or an interpretation of a text, comparing one text to another, tracking down the dates of historical events or conducting laboratory experiments, visiting museums, interviewing people in the college community, or surfing the World Wide Web. Research is an integral part of learning and learning to write well.

Consider this. Whenever you write about unfamiliar subjects, you have two choices: to research and find things out, or to bluff with unsupported generalizations. Which kind of paper would you prefer to read? Which kind of writing will you profit by doing? Part Four of *The Working Writer* explains what you need to know to write effectively with research in a college community. Meanwhile, the brief ideas listed below may prove useful.

Strategies for Writing with Research

1. Consider incorporating research information into every paper you write. For interpretive papers, revisit your texts; for argument and position papers, visit the library. For experiential papers, revisit places and people.
2. Research in the library. Visit the library, look around, hang around, ask questions, and take a tour. The library is the informational center of the university; using it well will make all of your writing and learning more substantial.
3. Research people. Interview experts to add a lively and local dimension to your papers. Consider who in your college community can provide current information, ideas, or insider stories to enhance your paper.
4. Research places. Visit settings in which you can find real, concrete, current information. Where appropriate, visit local sites—stores, institutions, streets, neighborhoods, farms, factories, and lakes—to connect local people with your topic.
5. Learn to document sources. Whenever you do research, write down who (the author) said what (the title), where (publication), and when (date). Then, as needed, look up the specific forms required in specific disciplines.

↻ WRITING 4

Describe the kind of research assignments you have done in the past. See if you can add one piece of research information to the paper you began drafting in Writing 3, using any research method with which you are familiar.

↻ REVISING

Somewhere in the midst of their writing, most writers revise the drafts they have planned, started, and researched. Revising involves rewriting to make the purpose clearer, the argument stronger, the details sharper, the evidence more convincing, the organization more logical, the opening more inviting, or the conclusion more satisfying.

I consider revising to be separate from editing, yet the two tasks may not always be separable. Essentially, revising occurs at the level of ideas, whereas editing occurs at the level of the expression. Revising means re-seeing the drafted paper and thinking again about its direction, focus, arguments, and evidence. In writing this third edition of *The Working Writer,* I revised, to some extent, every chapter to make it sharper, simpler, and more direct than it was in the first and second editions.

While it is tempting to edit individual words and sentences as you revise major ideas, try to ignore the small stuff and keep your focus on the big stuff—in the long run, revising before you edit saves time and energy. Revising to refocus or redirect often requires that you delete paragraphs, pages, and whole sections of your draft, actions that can be painful if you have already carefully edited them. Revising is discussed in greater depth in Part Five; meanwhile, consider the ideas listed below.

Strategies for Revising

1. Plan to revise from the beginning. Allow time to examine early drafts for main points, supporting evidence, and logical direction from first to last. (In addition, allow time later to edit and proofread.)
2. Revise by limiting your focus. Many first drafts are too ambitious. When you revise, make sure your topic is narrow enough for you to do it justice, given the time and space you have available. Often, this means omitting points that aren't really relevant to your main focus.
3. Revise by adding new material. An excellent time to do additional research, regardless of the kind of paper you are writing, is *after* you've written one draft and now see exactly where you need more information.
4. Revise by reconsidering how you tell your story. Consider the effect other points of view may have on your subject. Consider the effect of past tense versus present tense.
5. Revise for order, sequence, and form. Have you told your story in the only form you can, or are there alternative structures that would improve how you tell it?

☙ WRITING 5

Does your usual process for revising a paper include any of the ideas discussed in this section? Describe how your process is similar or different. Now revise the paper to which you added research information in Writing 4, using any revision techniques with which you are comfortable.

☙ EDITING

Whether writers have written three or thirteen drafts, they want the last one to be perfect—or as near perfect as time and skill allow. When editing, writers pay careful attention to the language they have used, striving for the most clarity and punch possible. Many writers edit partly to please themselves, so their writing sounds right to their own ear. At the same time, they edit hoping to please, satisfy, or convince their intended readers.

You edit to communicate as clearly as possible. After you've spent time drafting and revising your ideas, it would be a shame for readers to dismiss those ideas because they were poorly expressed. Check the clarity of your ideas, the logic and flow of paragraphs, the precision and power of your words, and the correctness and accuracy of everything, from facts and references to spelling and punctuation.

In finishing *The Working Writer*, I went over every word and phrase to make sure each one expressed exactly what I wanted to say. Then my editors did the same. They even sent the manuscript to other experts on writing, and they, too, went over the whole manuscript. Then I revised and edited once again. While editing is discussed in greater depth in Part Five, the ideas listed below may also be helpful:

Strategies for Editing

1. Read your draft out loud. Does it sound right? Your ear is often a trustworthy guide, alerting you to sentences that are clear or confused, formal or informal, grammatically correct or incorrect.
2. "Simplify. Simplify. Simplify." Henry David Thoreau offers this advice about both life and writing in his book *Walden*. I agree. When you edit, simplify—words, sentences, paragraphs, the whole paper—so that you make your point clearly and directly.
3. Delete unnecessary words. The easiest of all editing actions is to omit words that do not carry their own weight. Cut to improve clarity, simplicity, and directness as well as sentence rhythm.
4. Proofread by reading line by line with a ruler to mask out the following sentences. This forces your eyes to read word by word and allows you to find mistakes you might otherwise miss. This is important advice even if you use a spell checker on a computer, for a spell checker will not catch all mistakes (such as the use of *their* for *there*).

ᨠ **WRITING 6**

Do you edit using any of the ideas mentioned in this section? Describe your usual editing techniques. Now edit the paper you revised in Writing 5, referring to Part Five.

ᨠ **WRITING WITH COMPUTERS**

Computers are the greatest writing tools ever invented. Unlike typewriters, pens, and pencils, computers allow writers to change their writing infinitely and easily before the words are ever printed on paper. Writers are not committed to final copy until they print it out, and even then they can work on it again and again without retyping the whole thing over.

All word-processing programs work in pretty much the same way, so it doesn't matter if you're using a Macintosh, DOS, or a Windows-based system. You type the words on the computer screen, then store the file on your hard drive or on a portable disk, giving this file a brief descriptive name (paper.1). When you want to work on the document again, you call the file back to the screen and start all over, not needing to print it until you are ready. But the advantages of computers go far beyond paper-saving efficiency, giving you the freedom to play with ideas and sentences as well as add solid research without leaving your writing desk. The following suggestions may be useful:

1. **Planning.** Use the computer to invent and discover ideas by writing out lists of topics or tentative outlines. The computer's ability to add, delete, and rearrange makes these planning and organizing strategies easy.
2. **Freewriting.** Write rapidly on your computer whether you're exploring ideas or simply stuck. Don't worry about spelling, punctuation, grammar, or style.

Let your words trigger new thoughts and suggest new plans and directions. Save these entries on disk to create a computer journal.

3. **Drafting.** Use the computer to compose initial ideas, taking advantage of the ease with which words, sentences, and paragraphs can be modified and moved around as your ideas and direction become clearer. It's also easy to lift whole passages from computer-written freewrites or journal writing and incorporate them directly into your draft.

4. **Researching.** Networked computers allow you to gain access to either your campus library or the Internet at the stroke of a key. Instead of traveling physically to locate books, periodicals, and special collections in the library, you search for material and receive printouts right in your room. And for the most up-to-date, but sometimes questionable, information, the World Wide Web is loaded with information about every possible research topic.

5. **Revising.** Typing on a computer creates instant distance from your words and ideas, allowing you to view them more objectively. Whole paragraphs can be deleted, added, changed, and moved around—all useful activities when revising drafts. You can add new information or evidence in appropriate sections. Your computer will repaginate and reformat instantly. But remember that your paper won't be good simply because it looks good in an attractive font or after laser printing.

6. **Editing.** Computers allow you to try out numerous possibilities when editing sentences and paragraphs. The search-and-replace function can make each sentence start a new line so it can be judged on its own merits. Periodically, print out a hard copy to review your text; changes you make on paper can then be incorporated easily on screen for a new printout.

7. **Referencing.** With a keystroke you can consult CD or online dictionaries, encyclopedias, thesauruses, grammar books, and style manuals to check and change your text.

8. **Proofreading.** The built-in spell checker becomes your first line of spelling defense, as it will rapidly identify which words you've mistyped or misspelled. You must still proofread with your own eyes, since the computer will not identify misspelled words that spell other words. The grammar checker, likewise, identifies missing or misused grammatical components, though it may also mark long sentences that are otherwise grammatically correct.

9. **Formatting.** Word-processing programs provide type styles, fonts, graphic images, and page layouts that can produce professional-looking and visually exciting papers with improved readability and aesthetic appeal. Keep in mind, however, that an attractive presentation is no substitute for clear language, logic, and organization. You'll also find that the ways in which you use technology can benefit from their combination with more "traditional" methods— for example, printing hard copy for editing and for preliminary revisions. Also, visiting the library and hunting for information there bolsters Web-generated research sources.

10. **Saving.** When you write with a computer, save all versions of your papers on the hard drive or external disks so that you can always return to them. And be

sure to make duplicate or backup copies of all work on separate disks in case an accident destroys one disk or wipes out your hard drive. If you make radical revisions on one draft and don't like them, simply go back to the saved original and try again.

↬ SUGGESTIONS FOR WRITING AND RESEARCH

Individual

Study your own writing process as you work on one whole paper from beginning to end, taking notes in your journal to document your habits and practices. Write an analytic sketch describing the way you write and speculating about the origins of your current habits.

Collaborative

With your classmates, form interview pairs and identify local professional writers or instructors who publish. Make an appointment with one of these practicing writers, interview him or her about the writing process he or she practices, and report back to the class. Write a collaborative report about writers in your community; make it available to other writing classes or interested instructors.

Chapter

ᔥ **6** ᔥ

Sharing and Responding

Listening to other people's criticism is really helpful, especially when they stop being too nice and really tell you what they think about your paper.

KELLY

All writing profits from help. Most published writing has been shared, explored, talked over, revised, and edited somewhere along the way to make it as readable, precise, and interesting as it is. The published writing you read in books, magazines, and newspapers that is signed by individual authors is seldom the result of either one-draft writing or one-person work. This is not to say that authors do not write their own work, for of course they do. But even the best of writers begin, draft, and revise better when they receive suggestions from friends, editors, reviewers, teachers, and critics. This chapter explores specific ways to ask for help with your own writing as well as to provide it to others.

ᔥ **SHARING YOUR WRITING**

Writers can ask for help at virtually all stages of the writing process. Sometimes they try out ideas on friends and colleagues while they are still planning or drafting. More often, however, writers ask for help while they are revising, as they try to make their ideas coherent and convincing, and while they are editing, as they try to make their language clear and precise. In addition, when a draft is nearly final, many writers ask for proofreading help since most writers are their own worst proofreaders. Here are some suggestions for getting good help as you revise your writing.

Specify what you want. When you share a draft with a reader, be sure to specify what kind of help you are looking for. If you consider your draft nearly finished and want only a quick proofreading, it can be very frustrating if your reader suggests major changes that you do not have time for or interest in. Likewise, if you share an early draft and want help organizing and clarifying ideas because you intend to do major rewriting, it is annoying to have your paper returned with every sentence edited for wordiness, misspellings, and typographical errors. You can usually head off such undesirable responses by being very clear about exactly what kind of help you want. If you do want a general reaction, say so, but be prepared to hear about everything and anything.

Ask specific questions. Tell your reader exactly what kinds of comments will help you most. If you wonder whether you've provided enough examples, ask about that. If you want to know whether your argument is airtight, ask about that. If you are concerned about style or tone, ask about that. Also mark in the margins your specific concerns about an idea, a phrase, a sentence, a conclusion, or even a title.

Ask global questions. If you are concerned about larger matters, make sure you identify what these are. Ask whether the reader understands and can identify your thesis. Ask whether the larger purpose is clear. Ask whether the paper seems right for its intended audience. Ask for general reactions about readability, style, evidence, and completeness. Ask whether your reader can anticipate any objections or problems other readers may have.

Don't be defensive. Whether you receive responses about your paper orally or in writing, pay close attention to them. You have asked somebody to spend time with your writing, so you should trust that that person is trying to be helpful, that he or she is commenting on your paper, not on you personally. While receiving oral comments, stay quiet, listen, and take notes about what you hear, interrupting only to ask for clarity, not to defend what your reader is commenting on. Remember, a first draft may contain information and ideas that are clear to you; what you want to hear is where they are less clear to someone else.

Maintain ownership. If you receive responses that you do not agree with or that you consider unhelpful, do not feel obliged to heed them. It is your paper; you are the ultimate judge of whether the ideas in it represent you. You will have to live with the results; you may, in fact, be judged by the results. Never include someone else's idea in your paper if you do not understand it or believe it.

✥ WRITING 1

Describe the best response to a piece of your writing that you remember. How old were you? What were the circumstances? Who was the respondent? Explain whether you think the response was deserved or not.

✥ GIVING HELPFUL RESPONSES

When you are asked to respond to other writers' work, keep these basic ideas in mind.

Follow the golden rule. The very best advice is to give the kind of response to others' writing that you would like to receive on your own. Remember how you feel being praised, criticized, or questioned. Remember what comments help advance your own papers as well as comments that only make you defensive. Keeping those in mind will help you help others.

Attend to the text, not the person. Word your comments so that the writer knows you are commenting on his or her writing and not his or her person. The writer is vulnerable, since he or she is sharing with you a product of individual thinking and reasoning. Writers, like all people, have egos that can be bruised easily with careless or cruel comments. Attending to the text itself helps you avoid these problems. Point out language constructions that create pleasure as well as those that create confusion, but avoid commenting on the personality or intelligence of the writer.

Praise what deserves praise. Tell the writer what is good about the paper as well as what is not good. *All* writers will more easily accept critical help with weaknesses if you also acknowledge strengths. But avoid praising language or ideas that do not, in your opinion, deserve it. Writers can usually sense praise that is not genuine.

Ask questions rather than give advice. Ask questions more often than you give answers. You need to respect that the writing is the writer's. If you ask questions, you give the writer room to solve problems on his or her own. Of course, sometimes it is very helpful to give advice and answers or to suggest alternatives when they occur to you. Use your judgment about when to ask questions and when to give advice.

Focus on major problems first. If you find a lot of problems with a draft, try to focus first on the major problems (which are usually conceptual), and let the minor ones (which are usually linguistic and stylistic) go until sometime later. Drafts that are too marked up with suggestions can overwhelm writers, making them reluctant to return to the job of rewriting.

↭ WRITING 2

What kind of response do you usually give to a writer when you read his or her paper? How do you know what to comment on? How have your comments been received?

↭ RESPONDING IN WRITING

The most common written responses that college students receive to their writing are those that instructors make in the margins or at the end of a paper, usually explaining how they graded the paper. Many of these comments—except the grade—are similar to those made by professional editors on manuscripts. In writing classes you will commonly be asked by classmates to read and write comments on their papers. Here are some suggestions to help you do that:

Use a pencil. Many writers have developed negative associations from teachers covering their writing in red ink, primarily to correct what's wrong rather than to praise what's right. If you comment with pencil, the message to the writer is more gentle— in fact, erasable—and suggests the comments of a colleague rather than the judgments of a grader.

Use clear symbols. If you like, you can use professional editing symbols to comment on a classmate's paper. Or you can use other symbols that any writer can figure out. For example, underline words or sentences that puzzle you and put a question mark next to them. Put brackets where a missing word or phrase belongs or around a word or phrase that could be deleted. Circle misspellings.

There are many advantages to written responses. First, writing your comments takes less time and is therefore more efficient than discussing your ideas orally. Second, written comments are usually very specific, identifying particular sentences, paragraphs, or examples that need further thought. Third, written comments leave a record to which the writer can refer later—after days, weeks, and even months—when he or she gets around to revising.

There are also disadvantages to writing comments directly on papers. First, written comments may invite misunderstandings that the respondent is not present to help clarify. Second, written comments that are too blunt may damage a writer's ego—and it's easier to unintentionally make such comments in writing than face to face. Third, written comments do not allow the writer and reader to clear up simple questions quickly and so risk allowing misinterpretations to persist.

᧵ WRITING 3

Describe your most recent experience in receiving written comments from a reader. Were the comments helpful? Did the respondent use an approach similar to that detailed in this section or some other approach? In either case, were the reader's comments helpful?

᧵ RESPONDING THROUGH CONFERENCES

Most writers and writing teachers believe that one-to-one conferences provide the best and most immediate help that writers can get. Sitting together, a reader and a writer can look at a paper together, read passages aloud, and ask both general and specific questions about the writing: "What do you want to leave me with at the end?" or "Read that again, there's something about that rhythm that's especially strong" or "Stop. Right there I could really use an example to see what you mean." Often an oral conference helps as a follow-up to written comments.

The suggestions for making effective written responses in the previous section also apply to oral conferences. In addition, here are a few other things to keep in mind:

Relax. Having your conference in a comfortable place can go a long way toward creating a friendly, satisfying discussion. Don't be afraid of digressions. Very often a discussion about a piece of writing branches into a discussion about the subject of the paper instead of the paper itself. When that happens, both writer and reader learn new things about the subject and about each other, some of which will certainly help the writer. Of course, if the paper is not discussed specifically, the writer may not be helped at all.

Ask questions. If you are the reader, ask follow-up questions to help the writer move farther faster in his or her revising. If you have written your responses first, the conference can be a series of follow-up questions, and together you can search for solutions.

Listen. If you are the writer, remember that the more you listen and the less you talk, the more you will learn about your writing. Listen attentively. When puzzled, ask questions; when uncertain, clarify misunderstanding. But keep in mind that your reader is not your enemy and that you and your work are not under attack, so you do not need to be defensive. If you prefer the battle metaphor, look at it this way: Good work will defend itself.

One advantage of one-to-one conferences is that they promote community, friendship, and understanding between writer and reader. Also, conferences can address both global and specific writing concerns at the same time. In addition, conferences allow both writer and reader to ask questions as they occur and to pursue any line of thought until both parties are satisfied with it. And finally, writer and reader can use their facial expressions, body language, and oral intonation to clarify misunderstandings as soon as they arise.

There are, however, a few disadvantages to one-on-one conferences. First, it is harder to make tough, critical comments face to face, so readers are often less candid than when they write comments. Second, conferring together in any depth about a piece of writing takes more time than communicating through written responses.

꒐ **WRITING 4**

Confer with a writer about his or her paper, using some of the techniques suggested in this section. Describe in a journal entry how each technique worked.

꒐ **RESPONDING IN WRITING GROUPS**

Writing groups provide a way for writers to both give and receive help. When the group considers a particular writer's work, that writer receives multiple responses; and the writer is also one respondent among several when another writer's work is considered.

All of the suggestions in previous sections about responding to writing apply, with appropriate modifications, to responding in writing groups. But writing groups involve more people, require more coordination, take more time, and, for many people, are less familiar. Here are some guidelines for organizing writing groups:

Form a group along common interests. Most commonly, writing groups are formed among classmates, often with the instructor's help, and everyone is working on the same or similar class assignments. Membership in a group may remain fixed over a semester, and members may meet every week or two. Or membership may change with every new assignment. Writing groups can also be created outside of class by interested people who get together regularly to share their writing.

Focus on the writing. The general idea for all writing groups is much the same: to improve one another's writing and encourage one another to do more of it. Writers pass out copies of their writing in advance or read it aloud during the group meeting. After members have read or heard the paper, they share, usually orally but sometimes in writing, their reactions to it.

Make your group the right size for your purpose. Writing groups can be as small as three or as large as a dozen. If all members are to participate, smaller groups need less time than larger groups and provide more attention to each member. Groups that meet outside of classroom constraints have more freedom to set size and time limits, but more than a dozen members will make it hard for each member to receive individual attention and will require several hours, which may be too long to sustain constructive group efforts.

Appoint a timekeeper. Sometimes group meetings are organized so that each member reads a paper or a portion of a paper. At other times a group meeting focuses on the work of only one member, and members thus take turns receiving responses. If papers are to be read aloud, keep in mind that it generally takes two minutes to read a typed, double-spaced page out loud. Discussion time should at least match the oral reading time for each paper. If group members are able to read the papers before the meeting, length is not as critical an issue because group time can be devoted strictly to discussion. Independently formed groups can experiment to determine how much they can read and discuss at each session, perhaps varying the schedule from meeting to meeting.

There are many advantages to discussing writing in groups. First, writing groups allow a single writer to hear multiple perspectives on his or her writing. Second, writing groups allow an interpretation or consensus to develop through the interplay of those perspectives; the result can be a cumulative response that existed in no single reader's mind before the session. Third, writing groups can give both writers and readers more confidence by providing each with a varied and supportive audience. Fourth, writing groups can develop friendships and a sense of community among writers that act as healthy stimuli for continuing to write.

One disadvantage is that groups meeting outside of class can be difficult to coordinate, set up, and operate, as they involve people with varied schedules. Also, at the outset, the multiple audiences provided by groups may be more intimidating and threatening to a writer than a single person responding.

↶ WRITING 5

Imagine a writing group you would like to belong to. What subjects would you write about? Whom would you invite to join your group? How often and where would you like to meet? Explain in a journal entry why you would or would not join a writing group voluntarily.

↶ RESPONDING ELECTRONICALLY

In the event that you respond to classmates' papers via e-mail or by using the Web, you usually have two choices for response: (1) to mark the text itself with internal comments (using boldface type, italics, capital letters, or colors to distinguish your comments from their language) and send it back, or (2) to send back a written response only, in which you generalize about the language in the text and your response. As an instructor, I have done both, depending on the time and circumstances. Here is what I might suggest:

- When you read classmates' discovery drafts, freewrites, and early drafts, send back summary responses with your reactions to the paper as a whole—what aspects are strong, where you have questions, and so on. Don't mark up the text itself to send back, since at the early stages, writers need to attend to the larger idea level of their texts and not so much to the specific language of sentences and paragraphs.
- When you read later drafts, it may be more helpful—so long as it is convenient and you have the time—to mark up the text itself with your questions and suggestions about specific paragraphs or sentences.

It may go without saying—but I'll say it anyway—that most of the preceding guidelines still apply: Try to ask questions rather than give advice, be specific rather than general, and so on. However, with electronic responses, since people see only your disembodied words and not accompanying facial expressions, hand gestures, or other cues that might moderate written language, the Golden Rule is more important than ever.

↬ SUGGESTIONS FOR WRITING AND RESEARCH

Individual

Recently a large body of literature has developed concerning the nature, types, and benefits of peer responses in writing. Go to the library and see what you can find about writing groups or peer response groups. Check, in particular, for work by Patricia Belanoff, Kenneth Bruffee, Peter Elbow, Anne Ruggles Gere, Thom Hawkins, and Tori Haring-Smith. Write a report to inform your classmates about your discoveries.

Collaborative

Form interview pairs and interview local published writers about the ways in which responses by friends, family, editors, or critics affect their writing. Share results orally or by publishing a short pamphlet.

✌ *7* ✌

Strategies for Starting

Get something down! The hardest part of writing is starting. Forget the
introduction, skip the outline, don't worry about a thesis—just blast your
ideas down, see what you've got, then go back more slowly and work on them.

AARON

Good writing depends on good ideas. When ideas don't come easily or naturally, writers need techniques for finding or creating them. Writers need to invent new ideas or discover old ones at all phases of the writing process, from finding and developing a topic to narrowing an argument and searching for good evidence. And knowing how to invent and discover ideas when none seems apparent is also the best antidote for writer's block, helping you get going even when you think you have nothing to say.

The main premise behind the techniques discussed in this chapter is "the more you write, the more you think." Language begets more language, and more language begets ideas, and ideas beget still more ideas. Virtually all writers have had the experience of starting to write in one direction and ending up in another; as they wrote, their writing moved their thinking in new directions—a powerful, messy, but ultimately positive experience and a good demonstration that the act of writing itself generates and modifies ideas. This occurs because writing lets people see their own ideas, and doing that, in turn, allows them to change those ideas. This chapter suggests ways to harness the creative power of language and make it work for you.

> ✌ **WRITING 1**
>
> Describe the procedures you usually use to start writing a paper. Where do you get the ideas—from speaking? listening? reading? writing? Do you do anything special to help them come? What do you do when ideas don't come?

✧ BRAINSTORMING

Brainstorming is systematic list making. You ask yourself a question and then list as many answers as you can think of. The point is to get lots of possible ideas on paper to examine and review. Sometimes you can do this best by setting goals for yourself: What are five possible topics for a paper on campus issues?

1.   Overcrowding in campus dormitories
2.   Prohibiting cars for first-year students
3.   Date rape
4.   Multiculturalism and the curriculum
5.   Attitudes toward alcohol on campus

Sometimes you can brainstorm best by leaving the question open-ended: What do you already know about multiculturalism and the curriculum that interests you?

> Racial diversity high among campus students
> Racial diversity low among faculty
> Old curriculum dominated by white male agenda
> New curriculum dominated by young feminist agenda
> How to avoid simplistic stereotypes such as those I've just written?

In making such lists, jotting down one item often triggers the next, as is seen above. Each item becomes a possible direction for your paper. By challenging yourself to generate a long list, you force yourself to find and record even vague ideas in concrete language, where you can examine them and decide whether or not they're worth further development.

✧ FREEWRITING

Freewriting is fast writing. You write rapidly, depending on one word to trigger the next, one idea to lead to another, without worrying about conventions or correctness. Freewriting helps you find a focus by writing nonstop and not censoring the words and ideas before you have a chance to look at them. Try the following suggestions for freewriting:

1.   Write as fast as you can about an idea for a fixed period of time, say five or ten minutes.
2.   Do not stop writing until the time is up.
3.   Don't worry about what your writing looks like or how it's organized; the only audience for this writing is yourself.

If you digress in your freewriting, fine. If you misspell a word or write something silly, fine. If you catch a fleeting thought that's especially interesting, good. If you think of something you've never thought of before, wonderful. And if nothing interesting comes out—well, maybe next time. The following five-minute freewrite shows John's attempt to find a topic for a local research project:

> I can't think of anything special just now, nothing really comes to mind, well maybe something about the downtown mall would be good because I wouldn't

mind spending time down there. Something about the mall . . . maybe the street vendors, the hot dog guy or the pretzel guy or that woman selling T and sweat-shirts, they're always there, even in lousy weather—do they like it that much? Actually, all winter. Do they need the money that bad? Why do people become street vendors—like maybe they graduated from college and couldn't get jobs? Or were these the guys who never wanted anything to do with college?

John's freewrite is typical: He starts with no ideas, but his writing soon leads to some. This kind of writing needs to be free, unstructured, and digressive to allow the writer to find thoughts wherever they occur. For John, this exercise turned out to be a useful one, since he ultimately wrote a paper about "the hot dog man," a street vendor.

✌ LOOP WRITING

Loop writing is a sequenced set of freewrites. Each freewrite focuses on one idea from the previous freewrite and expands it. To loop, follow this procedure:

1. Freewrite for ten minutes to discover a topic or to advance the one you are working on.
2. Review your freewrite and select one sentence closest to what you want to continue developing. Copy this sentence, and take off from it, freewriting for another ten minutes. (John might have selected "Why do people become street vendors?" for further freewriting.)
3. Repeat step 2 for each successive freewrite to keep inventing and discovering.

✌ ASKING REPORTER'S QUESTIONS

Writers who train themselves to ask questions are training themselves to find information. Reporters ask six basic questions about every news story they write: Who? What? Where? When? Why? and How? Following this set of questions leads reporters to new information and more complete stories.

1. Who or what is involved? (a person, character, or thesis)
2. What happened? (an event, action, or assertion)
3. Where did this happen? (a place, text, or context)
4. When did it happen? (a date or relationship)
5. Why did it happen? (a reason, cause, or explanation)
6. How did it happen? (a method, procedure, or action)

While these questions seem especially appropriate for reporting an event, the questions can be modified to investigate any topic:

What is my central idea?
What happens to it?
Where do I make my main point? On what page?
Are my reasons ample and documented?
How does my strategy work?

∽ OUTLINING

Outlines are, essentially, organized lists. In fact, outlines grow out of lists, as writers determine which ideas go first, which later; which are main, which subordinate. Formal outlines use a system of Roman numerals, capital letters, Arabic numerals, and lowercase letters to create a hierarchy of ideas. Some writers prefer informal outlines, using indentations to indicate relationships between ideas.

When Carol set out to write a research essay on the effect of acid rain on the environment in New England, she first brainstormed a random list of areas that such an essay might cover.

> What is acid rain?
> What are its effects on the environment?
> What causes it?
> How can it be stopped?

After preliminary research, Carol produced this outline:

  I. Definition of acid rain
 II. The causes of acid rain
     A. Coal-burning power plants
     B. Automobile pollution
III. The effects of acid rain
     A. Deforestation in New England
        1. The White Mountain study
        2. Maple trees dying in Vermont
     B. Dead lakes

Note how Carol rearranged the second and third items in her original list to talk about causes before effects. The very act of making the outline encouraged her to invent a structure for her ideas. Moving entries around is especially easy if you are using a computer, because you can see many combinations before committing yourself to any one of them. The rules of formal outlining also cause you to search for ideas: If you have a Roman numeral I, you need a II; if you have an A, you need a B. Carol thought first of coal-burning power plants as a cause, then brainstormed to come up with an idea to pair with it.

Writing outlines is generative: In addition to recording your original thoughts, outlines actually generate new thoughts. Outlines are most useful if you modify them as you write in accordance with new thoughts or information.

∽ CLUSTERING

A *clustering diagram* is a method of listing ideas visually to reveal their relationships. Clustering is useful both for inventing and discovering a topic and for exploring a topic once you have done preliminary research. To use clustering, follow this procedure:

1. Write a word or phrase that seems to be the focus of what you want to write about. (Carol wrote down *acid rain*.)

2. Write ideas related to your focus in a circle around the central phrase and connect them to the focus phrase. If one of the ideas suggests others related to it, write those in a circle around it. (Carol did this with her idea *solutions*.)

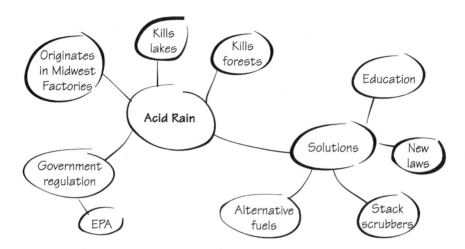

3. If one idea (such as *solutions*) begins to accumulate related ideas, start a second cluster with the new term in the center of your paper.

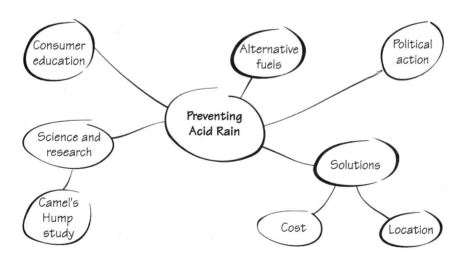

⅔ STARTING A DIALOGUE

One of the most powerful invention techniques is talking to a partner with the deliberate intention of helping each other find ideas. The directions are simple: Sit across from each other for five or ten minutes and talk about possible topics or approaches or ways of finding sources. It doesn't matter who starts or finishes, since the principle at work here is that oral language, like written language, begets ideas. At some point it will be helpful to write down what you are talking about so that you have a record to return to.

You can also start a deliberate dialogue by writing back and forth with a classmate. Each of you starts by posing a problem to solve; the other responds by offering a solution. Continue to exchange ideas, each time pushing harder to solve your classmate's problem as well as your own.

Suggestions for Invention and Discovery

1. Brainstorm a list of five possible topics to write about.
2. Freewrite for ten minutes about the most interesting topic on your list.
3. Loop back in your freewriting, selecting the most interesting or useful point, and freewrite again with that point as the focus.
4. Ask the reporter's questions about the topic.
5. Make an outline of a possible structure for your paper.
6. Cluster ideas about your topic; then cluster on a related idea that occurs during the initial clustering.
7. Talk with a partner for seven minutes; take turns helping each other find or advance one idea, seeing how fast you can go.

✍ WRITING 2

Pose a writing problem and try to solve it using at least three of the invention strategies described in this chapter. Which seems most helpful? If you still have a problem to solve, try several of the other techniques and see if they help.

✍ SUGGESTIONS FOR WRITING AND RESEARCH

Individual

Explain your own most useful invention technique for finding ideas. Explain your technique and support it with samples from your earlier papers. Give clear directions to teach other writers how to use it.

Collaborative

Find a common writing topic by having each person in the group or class select one of the invention and discovery techniques described in this chapter and practice using it for ten minutes. Make a collective list of the topic ideas generated this way. Then ask each individual to select one topic and write for another five minutes. Again make a list of topics and the important ideas generated about them. Discuss the ideas together and try to arrive at a consensus on a common writing topic.

PART TWO

❧ ❧ ❧

WRITING ESSAYS

Chapter

☙ **8** ☙

Recounting Experience

*Writing allows me to hold up a mirror to my life
and see what clear or distorted images stare back at me.*

RICK

Good stories occur everywhere and can be told about anything. They are as likely to occur in your own neighborhood as in some exotic locale. Potential stories happen daily; what makes potential stories actual stories is putting them into language, recounting them, orally or in writing. Good stories are entertaining, informative, lively, and believable; they will mean something to those who write them and to those who read them.

All stories, whether they're true (nonfiction) or imagined (fiction), are accounts of something that happened—an event or series of events, after which something or somebody is changed. Whether the story is "The Three Little Pigs" or *The Adventures of Huckleberry Finn* (both fiction), or Darwin's *The Voyage of the Beagle* or an account of your own canoe trip (nonfiction), it includes the following elements: a character (*who?*) to whom something happens (*what?*), in some place (*where?*), at some time (*when?*), for some reason (*why?*), told from a particular perspective (*how?*). In other words, any time you render a full account of a personal experience, you answer what we called "reporter's questions" in the previous chapter—the who, what, where, when, why, and how questions reporters ask themselves to make sure their reports of news stories are complete. Whether your story is engaging or not depends upon the subject, your interest in telling it, and the skill with which you weave together these story elements.

☙ CHARACTER (WHO)

In personal experience writing, your main character is yourself, so try to give your readers a sense of who you are through your voice, actions, level of awareness, and description. The characters in a good story are believable and interesting; they come alive for readers. Study the elements that give life to characters in your story: voice, actions, insight, and detail.

Voice

In your own story, your language reveals who you are—playful, serious, rigid, loose, stuffy, honest, warm, or whatever. In the following excerpt, in which Beth relates her experience playing oboe during a two-hour Saturday morning orchestra rehearsal, we learn she's serious, fun-loving, impish, and just a little lazy:

> I love that section. It sounds so cool when Sarah and I play together like that. Now I can put my reed back in the water and sit back and listen. I probably should be counting the rests. Counting would mean I'd have to pay attention and that's no fun. I'd rather look around and watch everyone else sweat.

Actions

Readers learn something about the kind of person you are from your actions. For example, when Karen recalls her thoughts playing in a basketball tournament, we learn something of her insecurity, fears, and skills all at once:

> This time, don't be so stupid Karen—if you don't take it up court, you'll never get the ball. Oh, God, here I go. Okay, they're in a twenty-one—just bring it up— Sarah's alone—fake up, bounce pass—yes, she hits it! I got the assist!

Insight

One of the best ways to reveal who you are is to show yourself becoming aware of something, gaining a new way of seeing the world, a new insight. While such awareness can occur for apparently unexplainable reasons, it most often happens when you encounter new ideas or have experiences that change you in some way. In writing a paper about why she goes to the library to write a paper, Judith clarifies first for herself—then for her readers—the relationship between feeling safe and being creative.

> Two weeks ago, a female student was assaulted not far from where I live—that's why I've taken to locking my door so carefully. I am beginning to understand the importance of feeling safe in order to be creative and productive. Here, in the library, I feel secure, protected from real violence and isolated from everyday distractions. There are just enough people for security's sake but not so many that I feel crowded. And besides, I'm surrounded by all these books, all these great minds who dwell in the hallowed space! I am comfortable, safe, and beginning to get an idea.

Telling Details

Describe yourself and other participants in your story in such a way that the details and facts help tell your story. A telling detail or fact is one that advances your characterization of someone without your having to render an obvious opinion. For example, you could characterize your little sister by pointing out the field hockey stick in the corner of her room, the photograph of the seventh-grade field hockey team on the wall, and the teddy bear next to her pillow. You could characterize her coach by pointing to the logo on her sweatshirt: "Winning isn't everything. It's the only thing."

↭ WRITING 1

Start to characterize yourself. Write four paragraphs, and in each one, emphasize one of these individualizing elements: voice, actions, awareness, and any telling details of your life. Select any or all that seem worthy of further exploration and write a few more paragraphs.

↭ SUBJECT (WHAT)

People write about their personal experiences to get to know and understand themselves better, to inform and entertain others, and to leave permanent records of their lives. Sometimes people recount their experiences casually, in forms never intended for wide circulation, such as journals, diaries, and letters. Sometimes they write in forms meant to be shared with others, such as memoirs, autobiographies, or personal essays. In college, the most common narrative forms are personal experience essays.

Subjects for good stories know no limits. You already have a lifetime of experiences from which to choose, and each experience is a potential story to help explain who you are, what you believe, and how you act today. Here are some of the topics selected by a single first-year writing class:

- Playing oboe in Saturday orchestra rehearsals
- Counseling disturbed children at summer camp
- A job picking strawberries on a farm
- Winning a championship tennis match
- Clerking at a drugstore
- Playing in a championship basketball game
- Solo camping as part of an Outward Bound experience
- Touring Graceland in Memphis, Tennessee
- Painting houses during the summer
- Working on an assembly line in a battery factory

When you write a paper based on personal experience, ask yourself: Which experience do I want to write about? Will anybody else want to read about it? Here are some suggestions.

Winning and Losing

Winning something—a race, a contest, a lottery—can be a good subject, since it features you in a unique position and allows you to explore or celebrate a special talent. At the same time, exciting, exceptional, or highly dramatic subjects, such as scoring the winning goal in a championship game or placing first in a creative writing contest, may be difficult to write because they've been used so often that readers have very high expectations.

The truth is that in most parts of life there are more losers than winners. While one team wins a championship, dozens do not. So there's a large, empathetic audience out there who will understand and identify with a narrative about losing. Although more common than winning, losing is less often explored in writing because it is more painful to recall. Therefore, there are fresher, deeper, more original stories to tell about losing.

Milestones

Perhaps the most interesting but also the most difficult experience to write about is one that you already recognize as a turning point in your life, whether it's winning a sports championship, being a camp counselor, or surviving a five-day solo camping trip in midwinter. People who explore such topics in writing often come to a better understanding of them. Also, their very significance challenges writers to make them equally significant for an audience that did not experience them. When you write about milestones, pay special attention to the physical details that will both advance your story and make it come alive for readers.

Daily Life

Commonplace experiences make fertile subjects for personal narratives. You might describe practicing for, rather than winning, the big game, or cleaning up after, rather than attending, the prom. If you are accurate, honest, and observant in exploring a subject from which readers expect little, you are apt to pleasantly surprise them and draw them into your story. Work experiences are especially fruitful subjects, since you may know inside details and routines of restaurants and retail shops that the rest of us can only guess. For example, how long is it before McDonald's tosses its unsold hamburgers? How do florists know which flowers to order when?

A Caution About Love, Death, and Divorce

Several subjects that you may need to write about may not make good topics for formal papers that will be shared with classmates and instructors. For example, you are probably too involved in a love relationship to portray it in any but the rosiest terms; too close to the recent death of someone you care about to render the event faithfully; too angry, confused, or miserable to write well of your parents' divorce. Writing about these and other close or painful experiences in your journal or diary can be immensely cathartic, but there is no need to share them with others.

ॐ **WRITING 2**

Make a list of a dozen experiences about which you could tell stories. Think of special insight you gained as well as commonplace events that were instructive or caused change. Share your list with classmates and find out which they would most like to hear about.

ॐ SETTING (WHERE)

Experiences happen in some place at some time, and good stories describe these settings. To describe a believable physical setting, you need to re-create on paper the sights, sounds, smells, and physical sensations that allow readers to experience it for themselves. In addition to telling details that support your plot or character development, try to include evocative details, colorful details of setting and character that will let your readers know you were really there. In the following example, Heather portrays details of the farm where she spent the summer picking strawberries:

> The sun is just barely rising over the treetops and there is still dew covering the ground. In the strawberry patch, the deep green leaves are filled with water droplets and the strawberries are big and red and ready to be picked. The patch is located in a field off the road near a small forest of Christmas trees. The white house, the red barn, and a checkerboard of fields can be seen in the distance. It is 5:30 A.M. and the day has begun.

The evocative details are those that appeal to your senses, such as sight, touch, and smell: *dew covering the ground, deep green leaves, strawberries . . . big and red, white house, red barn,* and *checkerboard of fields.*

The details of a setting reveal something essential about your story without your explaining them. For example, in telling a story about your sister, you might describe the physical objects in her room, which in turn describe important elements of her character: the hockey stick, soccer ball, gym bag, sweatpants, and baseball jersey, and the life-size posters of Michael Jordan and Jackie Joyner-Kersee. In other words, skillful description helps you "tell" the story without your telling it outright.

ॐ **WRITING 3**

Describe in detail one of the settings in which your experience took place. Appeal to at least three senses, and try to include details that "tell" some of your story without needing further explanation or overt value judgments on your part.

✄ SEQUENCE (WHEN)

In every story, events are ordered in some way. While you cannot alter the events that happened in your experience, as a writer you need to decide which events to portray and in what order to present them.

Selecting Events

You have dozens of places to start and end any story, and at each point along the way, many possible details and events are worth relating. Your final selection should support the theme of your story. To decide which events to portray, figure out how much detail you intend to devote to each one. In writing about her basketball career, Karen could have told about her four years playing in high school, her senior year alone, one game, or even less. Because she wanted to focus on a climactic point in great detail, she selected "even less"— she writes her entire six-page paper about the final six minutes of her final game.

In selecting events, consider using one of two strategies that writers commonly use to maintain reader interest: showing cause and effect and building suspense. When writers recount an experience to show cause and effect, they pair one event (having an accident, meeting a person, taking a journey) with another event or events that were caused as a result (undergoing physical therapy, making a friend, learning a new language).

In using suspense, writers raise questions or pose problems but delay answering or solving them. If the writer can make the question interesting enough, the problem pressing enough, readers will keep reading to learn the answers or solutions—in other words, to find out what happens. Karen's paper asks indirectly, "What is it like to play a championship game from the perspective of a substitute player?"

Ordering Events

The most common way to sequence events is to use chronological order, presenting events in the sequence in which they happened. Chronological order can be straightforward, following a day from morning to night as Heather does in her narrative about picking strawberries. Chronology also orders one student's twenty-two days of Outward Bound. And chronology orders Karen's six minutes at the end of one basketball game. Sometimes, however, the order is deliberately broken up, so that readers are first introduced to an event in the present and then, later in the story, are allowed to see events that happened earlier in time through flashbacks. For example, the Outward Bound journal could start with his first day of solo camping and then, in another entry, flash back to the early days to explain how he got there. Such a sequence has the advantage of stimulating readers' interest by opening with a point of exceptional drama or insight.

✄ WRITING 4

Outline the sequence of events of your story in the order that makes the most sense. Is the arrangement chronological? If not, what is it? How do you decide which event to begin with? Which one to end with?

✌ THEME (WHY)

We can talk about the "why" of a story on two levels: First, why did the events occur in the story? What motivated or caused them? Well-told stories will answer this question, directly or indirectly. But we can also ask the writer: "Of all the stories you could write, why did you write this one?" Or, more bluntly, every reader asks, at least tacitly: "So what? What's the meaning or significance of this story? What did I learn by reading it?" Well-told stories will also answer this question; readers will see and understand both why you wrote it and why they read it.

However, first drafts of personal experience narratives often do not reveal clear answers to these questions, even to the writers themselves. First drafts are for getting the events down on paper for writers to look at. In subsequent drafts, the meaning of these events—the theme—should become clearer. If it doesn't, the writer should drop the subject.

In experiential stories, the theme isn't usually explicitly stated in the first paragraph, as the thesis statement often is in expository or argumentative writing. Instead, storytellers may create a meaning that is not directly stated anywhere and that becomes clear only at the end of the narrative. Many themes fall into three broad categories: slices of life, insights, and turning points.

Slices of Life

Some stories simply let readers see what life is like for someone else. Such stories exist primarily to record the writer's memories and to convey information in an interesting way. Their primary theme is, "This is what my life is like." Beth's story of the Saturday orchestra rehearsal is a slice of life, as she chooses to focus on a common practice rather than a more dramatic performance. After using interior monologue for nine paragraphs, in the last paragraph she speaks to the readers directly, explaining what the meaning of music is in her everyday life.

> As hard as it is to get up every Saturday morning, and as hard as it is to put up with some people here, I always feel good as I leave rehearsal. A guest conductor once said, "Music sounds how feelings feel." It's really true. Music evokes emotions that can't be described on paper. Every human feeling can be expressed through music—sadness, love, hatred. Music is an international language. Once you learn it you can't forget it.

Insights

In contrast to the many but routine experiences that reveal slices of life is the single important experience that leads to a writer's new insight, change, or growth. Such an experience is deeply significant to the writer, and he or she makes sure that readers see the full value of the experience, usually by commenting explicitly on its meaning. The final paragraph of Karen's championship basketball game reveals just such an insight.

> It's over now, and I've stopped crying, and I'm very happy. In the end I have to thank—not my coach, not my team—but Walpole for beating us so badly that I got

to play. I can't get over it. I played. And my dream came true, I hit a three pointer in the Boston Garden.

Turning Points

Turning points are those moments in one's life when something happens that causes the writer to change or grow in some large or small way—more than routine, less than spectacular—perhaps somewhere in between slices of life and profound insights. In fact, many of the best personal experience stories have for themes a modest change or the beginning of growth. Although such themes may be implied throughout the story, they often become clear only in a single climactic moment or episode. Mary's camp counselor story shows her progress from insecurity to confidence in gaining the trust of a ten-year-old. The following excerpt takes place after she has rescued Josh from ridicule by other campers.

> He ran in and threw himself on my bed, crying. I held him, rubbing his head for over an hour. "I love you, Mary. You're the best big sister in the whole world and you're so pretty! I love you and don't ever want you to leave."

∾ **WRITING 5**

Freewrite for ten minutes about the meaning of your story as you have written it so far, addressing some of these questions: What have you discovered about yourself? Were there any surprises? Does your story interest you? Why or why not? What do you want readers to feel or know at the end?

∾ PERSPECTIVE (HOW)

When reporters ask themselves "How?" they mean "How did the news event happen?" However, in this chapter, "How?" means "How is the story told?"—from what vantage point, position, or perspective are you telling the story? Perspective addresses the question "How close—in time, distance, or spirit—are you to the experience?" Do you write as if it happened long ago or yesterday? Do you summarize what happened or put readers at the scene? Do you explain the experience or leave it mysterious? In other words, you can control, or at least influence, how readers respond to a story by controlling the perspective from which you tell it.

Point of View

Authorial perspective is established largely by point of view. Using the first person (*I*) puts the narrator right in the story as a participant. This point of view is usually the one used in personal experience writing, as Beth and Karen do in earlier examples.

The third person (*he* or *she*) establishes a distinction between the person narrating the events and the person experiencing them and thus tends to depersonalize the story.

This perspective is more common in fiction, but it has some uses in personal essays as well. In the following example, for instance, Karen opens her personal experience essay from the imagined perspective of the play-by-play announcer who broadcasts the championship game; the point of view is first person, but from the perspective of a third person:

> 2:15 Well folks, it looks as if Belmont has given up, the coach is preparing to send in his subs. It has been a rough game for Belmont. They stayed in it during the first quarter, but Walpole has run away with it since then. Down by twenty with only six minutes left, Belmont's first sub is now approaching the table.

Verb Tense

Verb tense establishes the time when the story happened or is happening. The tense used to relate most of the events in a story is called the *governing tense*. Personal experience stories are usually set in either the past or the present.

Once upon a Time: Past Tense

The most natural way to recount a personal experience is to write in the past tense; whatever you're narrating did happen sometime in the past. Lorraine uses the past tense to describe an automobile ride with her Native American grandfather to attend a tribal conference:

> I sat silently across from Grandfather and watched him slowly tear the thin white paper from the tip of the cigarette. He gathered the tobacco in one hand and drove the van with the other. I memorized his every move as he went through the motions of the prayer, which ended when he finally blew the tobacco out of the window and into the wind.

Even though the governing tense for your personal narrative may be the past tense, you may still want to use other tenses for special purposes.

Being There: Present Tense

The present tense provides the illusion that the experience is happening at the moment; it leaves no time for your reflection. This strategy invites readers to become involved with your story as it is happening and invites them to interpret it for themselves.

If you want to portray yourself thinking rather than talking—in what is called interior monologue—you may choose to use fragment sentences and made-up words since the flow of the mind doesn't obey conventional rules of language. For example, when Beth describes her thoughts during orchestra rehearsal, she writes an interior monologue; we hear her talking to herself while trying to blow her oboe. Note how she provides clues so that we understand what is going on around her.

> No you don't really mean that, do you? You do. Rats. Here we go. . . . Pfff . . . Pfff . . . Why isn't this playing? Maybe if I blow harder . . . HONK!! Great. I've just made a total fool of myself in front of everyone. Wonderful.

> ☙ **WRITING 6**
>
> Write one page of a possible story using the first person, past tense, and a second page using the first person, present tense. From which perspective do you prefer to tell the story? Why?

☙ SHAPING THE WHOLE PAPER

Read through the complete final draft of Karen Santosuosso's essay recounting her experience in the Eastern Massachusetts women's basketball championship. Note that she elects to play creatively with both time and perspective in telling her story: She opens her story by recounting the last six minutes of the game as told by the play-by-play announcer in the broadcast booth, letting us see the game from his imagined perspective, represented in italics. Second, she recounts the same six minutes from her own perspective on the floor, using normal type, but including occasional snatches of the announcer's voice in italics. As you read this account, notice how she characterizes herself through her voice and actions, uses present tense to create suspense, and lets the theme emerge indirectly only in the last paragraph.

Three Pointer in the Boston Garden

Karen Santosuosso

2:15 Well folks, it looks as if Belmont has given up, the coach is preparing to send in his subs. It has been a rough game for Belmont. They stayed in it during the first quarter, but Walpole has run away with it since then. Down by twenty with only six minutes left, Belmont's first sub is now approaching the table.

Meghan Sullivan with the ball goes coast to coast and lays it in for two. She has sparked Walpole from the start. The fans have livened up a bit, but oddly enough they aren't Walpole's fans, they're Belmont's.

"KAREN IS SMALL, SHE'S NOT TALL, WE LOVE KAREN!"

Meanwhile, Belmont's number eleven, 5'1" Karen Santosuosso, replaces Michelle Hayes. With three minutes left, Belmont's Kristin Sullivan brings up the ball, spin dribbles—but the ball is stripped away by Meghan Sullivan, who takes it down court. She throws on the fly and hits a three pointer! That girl has truly amazing talent.

Belmont with the ball. Santosuosso brings it up, passes to Jones, Jones shoots and hits it! Belmont may be out of the game, folks, but these reserves are having the time of their lives.

Walpole brings it up. Called for traveling.

Both Belmont and Walpole are now emptying their benches. As she sits down, Meghan Sullivan receives a standing ovation. An honor-roll student and all-star athlete, she'll be attending Holy Cross this fall. What a night she had, eighteen points.

One and a half minutes left, D'Andrea inbounds to Santosuosso, who brings it up. Walpole sitting back in a twenty-one zone, Karen, over half court, stops, pops, swish. Santosuosso for three! The Belmont fans are going wild—

"KAREN IS SMALL, SHE'S NOT TALL, WE LOVE KAREN!"

As the clock winds down, fifteen seconds left, Walpole has started celebrating. The Belmont bench is quiet, but they have nothing to be ashamed of. Smith misses the second free throw, Hathaway on the rebound, gives it off to Keohane—and that will do it, folks. Walpole is the Eastern Massachusetts women's basketball champion.

2:15 Where are they? Oh, there they are, Mom's waving and Dad has the camera. It's awesome, they come to every game, even to watch me bench.

"KAREN IS SMALL, SHE'S NOT TALL, WE LOVE KAREN!"

Oh, God, I'm so embarrassed, complete silence and those guys start yelling again. I can't believe them. Nothing is happening and no one else is cheering. . . . Well, at least maybe coach hears them now—not that it will make any difference. With six minutes to go, we're still down by eighteen. No matter how badly the starters play, he won't let the subs play. My senior year, six minutes left in the Boston Garden—in my career as the shortest player in the history of Belmont High—and he hasn't played me yet.

At least my friends are on my side—they're all sitting in the bleachers on the Walpole side with their faces painted maroon and blue and wearing Belmont clothes. I want to wave to them again. Oh no, he caught me waving. No big deal, three months ago I would have minded, but what difference does it make now?

Meghan Sullivan with the ball goes coast to coast and lays it in for two. She has sparked Walpole from the start.

"Girls you have got to keep your heads into the game. Don't let them get you down. You have worked so hard all season, and you are as good as they are. Look at our record, 18-2-0, there's no reason you can't play with them."

"Coach, they're killing us. They are making us look like fools. We're down by eighteen with six minutes left. It's hopeless."

"I don't want to hear any of you talk like that. You have worked too hard to get to this point only to give up now. Remember every sweat-dripping, suicide-sprinting, drill-conditioning, nine o'clock Saturday morning practice!"

"All right girls, now get out there, play hard, and have fun."

"KAREN IS SMALL, SHE'S NOT TALL, WE LOVE KAREN!"

I can't believe those guys, I love them. My own personal fan club, and they're even audible in the Boston Garden. Those guys won't stop until he plays me. Five and a half minutes left, I wonder if he has the guts. . . .

"Karen, come here. Go in for Michelle."

Am I imagining this? There's a whole five minutes left and he's letting me play—but I'll take it!

"Hi, sub, number eleven for twenty-four. Is that your fan club in the balcony screaming for you?"

Wow, even the timekeeper noticed these guys. He pays more attention than coach. "Yeah, they're my friends. At every game, I put up three-point-shot halftime shows just for them. We have a joke on the team that he only plays five and a half players, because the sixth hardly ever gets in."

"Tweet." Too bad the timekeeper isn't our coach—anyway, I'm playing—on the parquet floor of the Boston Garden—actually playing, not just warming up or putting on a halftime show.

"Karen take the ball up."

"No way, Kristin, you take it." What am I afraid of?

Kristin's dribbling up the court and, as usual, she's showing off. Here she goes, spin dribble—no wait, Sullivan strips the ball. . . .

This time, don't be so stupid Karen—if you don't take it up court, you'll never get the ball. Oh, God, here I go. Okay, they're in a twenty-one—just bring it up—Sarah's alone—fake up, bounce pass—yes, she hits it! An assist!

There goes Sullivan again, burning up the court, right by all of us and she hits the one on the fly. *She* is truly awesome!

Kristin's got two defenders on her and . . . "Kristin, I'm open." So, she won't pass and she misses her shot clean, bricked it off the backboard, what a heave.

Walpole brings it up. Called for traveling.

"Karen, coach wants you to bring it up."

If I don't do this now, I'll hate myself tomorrow. If I miss, big deal, I tried. I have to try it. Just think of Dad . . . think of the guys. Over half court, approaching the key, they aren't challenging me. Stop, pop, and there it goes. Oh please, please go in, please . . .

Karen, over half court, stops, pops, swish. Santosuosso for three!

Yes! I did it. I can't believe. . . . Oh my, oh my . . . I did it! Where's Dad? I can't find him, where is he? I can't believe I'm crying. There he is, Dad! And Mom's jumping up and down—we must look like twins dancing around, only she's in the stands and I'm on the court.

"KAREN IS SMALL, SHE'S NOT TALL, WE LOVE KAREN!"

Now if I could only stop crying so I could finish playing. Thirty seconds left, I hope I make it.

As the clock winds down, Walpole has started celebrating. The Belmont bench is quiet, but they have nothing to be ashamed of.

Walpole's jumping up and down—and so am I. . . .

Walpole is the Eastern Massachusetts women's basketball champion.

It's over now, and I've stopped crying, and I'm very happy. In the end I have to thank—not my coach, not my team—but Walpole for beating us so badly that I got to play. I can't get over it. I played. And my dream came true, I hit a three pointer in the Boston Garden.

↩ SUGGESTIONS FOR WRITING AND RESEARCH

Individual

Write a personal experience essay based on Writings 2–6 in this chapter. Find a subject that will let you show some change or learning on your part. Plan to write this narrative in several drafts, each one exploring a different aspect of your experience. See Chapters 22–23 for suggestions about revising this essay.

Collaborative

As a class, write the story of your writing class so far in the semester. Each class member contributes one chapter (one page) to this tale. Each member chooses any moment (funny, momentous, boring, routine) and describes it so that it stands on its own as a complete episode. Choose two class members to collect all the short narrative chapters and weave them into a larger narrative with a beginning, middle, and end.

Chapter

❧ **9** ❧

Exploring Identity

*I am the son of a man who does not know who Mick Jagger is, but
can recite the names of all the state capitals in alphabetical order.*

JEREMY

How do you write about your own past, present, and future? How much does your current identity depend upon your past or anticipate your future? The four assignments explored here let you construct in writing a limited but substantial version of yourself. An *autobiography* asks you to explore how you became who you are today by reconstructing and reflecting on your past, while a *language autobiography* focuses on your specific development as a user of language. A *self-portrait* asks you to present a version of your present self to share with the world. And a *career profile* asks you to build upon both past and present to anticipate an interesting future.

❧ **AUTOBIOGRAPHY**

It is presumptuous, of course, for most people to write an autobiography while still in college, with so much life still ahead. However, it may be useful to use your enrollment in college as a life marker—a point of reflection—to examine how your precollege years have shaped the person you are today.

In order to write even a partial autobiography, it helps to remember and then examine the important influences that come to mind: family, home life, neighborhood environment, school, favorite activities (hobbies, sports, clubs), political and religious experiences, and so on. It also helps to research your past by interviewing those who have watched you grow up: parents, siblings, relatives, teachers, neighbors, and friends. And it helps to dig, too, among saved artifacts that offer clues to your development: journals, diaries, letters, report cards, photographs, letters, posters, toys, scrapbooks, and the like. Following are a number of approaches to experiment with or help you get started.

Locating Artifacts

Explore a dimension of your growing up based on something you once valued—a present from someone, something you saved up money to purchase, or an object that played an important role in the past. When you recall this artifact, describe it in as much concrete detail as you can so that others who did not grow up with you can see it. In the following example, Sara writes about a journal she kept as a child:

> When I was in fifth grade, I used to keep a journal that I would write in every day. Its shiny red cover was decorated with yellow sunflowers and it had my initials embossed in gold in the corner—SAH. It also had a lock and key so my older sisters couldn't read it. I would write freely in it about being happy or sad, or about adventures with my girlfriend, Molly, or sometimes I just wrote when I felt lonely. I stopped writing in it when I was in sixth grade, because then I started having boyfriends.

Describing Interests

Write a version of yourself based on what you care about most passionately. Think about old interests that carry through to the present as well as more recent interests (playing guitar, collecting sports cards, running, ballet) and describe them as well as explore the reason they fascinate you. In the following example, Rene focuses on so common an activity as playing cards:

> I have always been a card player. I have played with family, friends, boyfriends, and strangers. I have played by myself and with as many as ten or twelve people. Card playing is an exercise in skill and chance, a social event, and a creative act. I say creative because on more than one occasion I've invented my own game—though you won't find them in *The Complete Hoyle*. When I meet someone who doesn't play cards, I look at them with suspicion—what's wrong with you I think, not liking to play cards is not natural.

Using Quotations

Write a version of yourself based on what somebody has said about you. Sometimes when you overhear others talking about you, you gain insight into how others in the world perceive you. Think about overheard conversations, the way your parents characterize you, what friends have said, or what a teacher has written on a report card. In the following example, Joy seems rather pleased that she was often described as a handful when she was young:

> I was what some people would call a problem child. When I talk to my aunt or grandmother about my younger years, they just shake their heads and say, "You were a pisser, that's for sure." But they also smile when they say it, so it couldn't have been that bad.

Writing Snapshots

One of the most daunting aspects of writing an autobiography or self profile of any kind is our own complexity as human beings. While some of us may be said to have one clearly dominant characteristic that stands out above others, most of us see ourselves as more complicated than that, made up of multiple and even contradictory selves, making it especially difficult to write unified and coherent portraits of ourselves. One way to solve the coherence problem is to write in a form that allows you to celebrate multiplicity without pretending to unity and to portray contradictions without resolving them.

Writing prose snapshots is just such a form, allowing writers to compose their autobiographies through a collection of short prose pictures or verbal snapshots. Each individual snapshot focuses on one dimension of self the way a single photograph captures one particular scene. At the same time, a collection of such verbal episodes, carefully juxtaposed against others, tells a larger, more complete story the way a collection of photographs in an album or exhibit tells a larger story.

Consider a single prose snapshot as an autonomous paragraph of a few sentences or a few hundred words, but separated from preceding and following snapshots by white space rather than transitional phrases. Each snapshot tells its own small story, and each may be followed by another snapshot telling either a related or a quite different story. An accumulation of such snapshots allows writers to present complex versions of self in a brief amount of space and time. Here, for example, are three of thirteen snapshots from Becky's personal portrait:

- My mother grew up in Darien, Connecticut, a Presbyterian. When she was little, she gave the Children's Sermon at her church.
- My father grew up in Cleveland, Ohio, a Jew. When I went away to college, he gave me the Hebrew Bible he received at his Bar Mitzvah.
- The only similarity between my parents' families is freckles. They both have them, which means I get a double dose. Lucky me. My mother once told me that freckles are "angels' kisses." Lucky me.

Note that Becky explores her conflicting religious heritage as well as her common freckle inheritance by briefly but carefully juxtaposing mother against father. She covers a lot of ground with a few sentences, allowing the reader to figure out the transitions without writing them in herself. (Additional versions of snapshot writing will be presented later in this chapter. For the complete version of Becky's snapshot essay, see the end of this chapter.)

✌ WRITING 1

Make a list of five episodes in your life, each of which reveals something about who you are or what you value. Arrange these episodes in two different ways: first, chronologically, with earliest episode first, most recent last. Then arrange according to significance, with least important first, most important last. Which way would you prefer telling your story?

⊰ **LANGUAGE AUTOBIOGRAPHY**

In many ways, getting a college education is getting a language education, learning course by course, new vocabulary, new meanings, and new ideas. Consequently, one of the most interesting types of autobiography to write is one that focuses on your growth and development as a user of language. Such "language autobiographies" focus on those moments when you learned something connected with language—reading, writing, speaking, listening. Such studies need not be limited to verbal languages, as other languages such as music, mathematics, visual, and body languages also have much to teach. You would go about constructing a language autobiography in the same way as any autobiography but now limiting your investigation to those experiences and episodes that shed light on you as a reader, writer, and speaker.

To construct a language autobiography, list all the possible places and episodes where you remember language making a special impact on you—home, neighborhood, school, hobbies, sports, clubs, and so forth. Also, list the books, films, television shows, and song lyrics that you remember most vividly as well as any stories, poems, and lyrics you may have written yourself.

Creating a Time Line

One way to begin a language autobiography is to create a time line. Draw a vertical line on a sheet of paper. On one side write a chronological list of all the books you remember reading in school, and on the other side, make a similar chronology of readings (including being read to) at home. Expand the entries on either side of the line by adding in other episodes of language learning—keeping a journal, writing letters, entering a short story contest, and so on, as Figure 9.1 illustrates. Include age or grade in parentheses next to each entry.

Adding Research

Research will add an element of persuasive authenticity to this project. First, plan to talk—in person, on the telephone, via e-mail—with your parents, aunts, uncles, and brothers or sisters in the event they remember something about the "young you" growing up that you do not. Next, browse among the books in your current bookshelf as well as among any childhood books still available, and make a list of the titles that most influenced you. Finally, dig—at home or in your own school files—among your own literate artifacts, and look again at old school reports, report cards, stories, poems, journals, or letters to the editor—any written documents that provide clues to early language influences or habits.

It makes good sense to write this paper in chronological order, beginning with earliest influences and working forward to watch yourself develop. However, consider reversing chronology or using flashbacks if that way of telling your story seems more interesting

Time Line

Personal	School
Read to: Dr. Seuss--lots (age 4?) *The Giving Tree* *Good Night, Moon*	
Winnie the Pooh	Learn to write a paragraph (2nd grade)
	Learn cursive (3rd)
The Jungle Book (9)	
Letters to Mike (10)	
	Where the Red Fern Grows (6th)
Write and perform play: *The Moon Man* (Summer Festival of the Arts)	
Write story about Mars (12)	
Catcher in the Rye (13)	*The Diary of Anne Frank* (8th)
	Presentation of moon study in biology class (10th)
Keep personal journal (15) *Story of Michael Jordan*	Paper on *Caged Bird* read aloud in class Keep basketball diary
Letter to editor published (17)	Two poems published in school paper (11th)
	Research paper on Ecuador (12th)
College entrance essay (successful!)	

Figure 9.1   Time line

to you. Megan, for example, began her language autobiography by recounting a trip home to rummage through the attic looking for artifacts:

> I took great pains to discover my first creative writing. Luckily, my mother has saved boxes of drawings, class pictures, and old stories from my kindergarten year on in our attic. I found two Big Chief writing tablets from first grade, a story about my summer vacation from second grade, and the complete manuscript of my first novel from fifth grade.

Including Artifacts

Once you've located a poem, letter, or journal entry from the past, you need to decide how to use it: If you simply mention it, you are suggesting it is notable but not very important. If you summarize it, you are suggesting the content is important, but there's no need to see the document itself. If you reproduce it, you are telling us the artifact is pretty interesting reading in and of itself. Keep in mind that the more space you devote to a single artifact in your essay, the more importance it assumes for your reader.

In the sample below, Susan includes a complete poem written in elementary school when she was seven years old.

My Special Tree

My special tree is very old.
It is large and green and in our yard.
It is a maple tree.
Birds fly in and out of its leaves.
It has a big curved branch.
I hang on tight and swing.

In fact, Susan's language autobiography is composed of fourteen artifacts solely from her elementary years, ages five through eleven. She prefaces the collection with this introduction:

> The truth and honesty of my childhood writing is the centerpiece of my language autobiography. In rummaging through my essays, research papers, stories, poems, and even journal entries from later years, I realized my true voice was already shaped by the time I entered seventh grade. I am now fourteen years older than the earliest sample included here, but it speaks for me as well now as it did then.

As a reader I infer that the simplicity of childhood dreams and pleasures is still very much alive in this now more experienced college student. Because Susan's paper is made up almost entirely of her own early writing, neither edited nor explained, I must do most of the interpretive work, a method which works so long as the material is self-explanatory. However, many times when a writer refers to an old poem or mentions a book, we don't understand its importance unless the writer tells us. Here, for example, when Kate writes about her first encounter with classical literature, she is explicit about why it was so important:

> When I was in sixth grade, we were assigned *The Odyssey*. It was the thickest book I'd ever tried to read, so it seemed like a big deal. How could I possibly read so many pages—here were at least 200? I trudged home with it in my backpack, feeling its weight, grumpily calculating how long it would take to read it. But I ended up reading it hungrily—Homer's beautiful writing drew me in to the most ghastly and majestic stories I'd ever experienced. Good literature had affected me for the first time.

Writing Snapshots

Snapshots prove to be an ideal way to represent a language autobiography, since each snapshot focuses on one specific literate event or idea, while the whole collection shows a broad range of influences. In the following series, Abbey gives a thematic title to each episode in her snapshot essay—four of twenty are included here.

SILENCE. They say that girls who have had a bad relationship with their fathers will have bad relations with men later in life. One day my father went on a business trip and never came back. I have not spoken to him since I was six years old. I have no relationship with my father. This does not bode well for me.

WRITING. "The Fire Bird," my dramatic sixth-grade entry into a national fable contest, tells the story of a pure white dove, a smoky gray pigeon, and their battle against the evil black raven and his destructive reign of fire. It is the first thing I have written that requires a title page. I do not win, but am honored to be nominated.

READING. BAM! BAM! BAM! My mom has learned very quickly how to rouse me from reading. Lost in words, I am deaf to the world outside. The only sound I hear is the clicking of Nancy Drew's pumps on the hidden staircase or the tapping of chalk as Laura Ingalls finishes her homework. Mom bangs a pot on the kitchen wall, and finally I hear.

SPEAKING. Twelve years old, dressed completely in white, I stand in front of a crowd of people, reciting Sylvia Plath from memory. I no longer remember the title or the words, just that she spoke of being broken, of being imprisoned in a white cast, and at twelve I knew how she felt.

᧤ WRITING 2

List five artifacts from five different years that reveal something significant about your life as a language user. Write a paragraph about each one, describing it and explaining its significance. When you are able, retrieve these artifacts, study them closely, and expand your paragraphs accordingly.

᧤ SELF-PORTRAIT

A different way to explore your identity is to focus primarily on the present, on who you are now rather than the way you were or even how you got here. Composing a profile of your current self is a hard assignment—maybe impossible, since full self-knowledge is difficult to obtain. But even attempting to profile yourself may prove to be instructive and rewarding, since the attempt asks you to reflect upon what you've learned to value and what next you hope to become.

To portray yourself in writing, you might start internally by constructing an idea you have of yourself. Conversely, you might start by looking around you, at your room or possessions, and jot down what they seem to say about you. You might start looking in your notebooks, journals, diaries, or letters to see how you represent yourself in writing. Or you might simply start by looking in a mirror and describing who or what you see there. Let's look at the opening paragraphs of several writers looking for ways to portray themselves.

Looking Inside

Alexis writes imaginatively, as if she were somewhere else, suggesting that there's an inner Alexis quite different and less wary than the person she normally appears to be.

> I linger beneath the surface, enjoying the prickly coldness of water against my forehead and eyelids. I am alone. I am naked. I cannot see, but I can feel, imagine, and remember. I am free to accept the moods of my conscience. They do not scare me here as they do in the real world. Down here, the solitude echoes as the water whirls around my body and I remember. . . .

Looking Outside

Doug opens his language autobiography by examining familiar objects in his room, looking for clues about how they might help define him:

> My life is plagued by half-completed projects. One glance at my room proves this. Nothing plays the role it was meant to because at some point I lost interest.
>
> Take my fish tank, for instance. A lovely fish tank. Twenty gallons of water-holding potential, not a leak in sight. The plan was to buy a few fish, toss them in the tank, and throw some food at the little guys every few days. That was last summer. For one reason or another, I never got around to buying the fish, never filled it with water, never made the aquarium I intended.
>
> As the months passed, the Plexiglas top of the tank became a place to put things—photos, newspapers, rolled-up Grateful Dead posters, guitar strings, loose change, dirty socks, poly-sci notebooks, more change . . . I never put anything in the tank, only on top, only because I had no place left in my cluttered room to store things.

Notice especially how casual specific details—guitar strings, posters, dirty socks—allow us to see other dimensions of Doug's life that he doesn't choose to dwell on but which, nevertheless, help fill our view of who he is.

Looking in the Mirror

Sandra starts her essay by methodically staring at herself in a mirror, describing precisely what she sees in the glass so we can see her too. She uses the mirror image to reflect inward on her origins, teaching us something of her past:

I go stare into a mirror which stares back at me—white skin, dark hair, and deep brown eyes. Do I still see the little girl whose daddy called her *muy bonita?* Am I the little girl who struggled to follow her father's step to the salsa beat? Am I the little girl who stole his lemon slices when he was drinking? Am I the little girl who hated the smell of alcohol on her daddy's breath?

I close and open my eyes, scrunch up my face, tug at my cheeks. Now I see images of a woman. I wear a black leather jacket, faded denim jeans, Steve Madden shoes with three-inch heels, and a shirt so short that when I move, you can see my belly button ring. I have hips and breasts, a feminine walk, and flirtatious eyes—I have filled out well.

In the first paragraph Sandra's questions teach us about growing up with an alcoholic father, suggesting a tough young life; in the second paragraph her clothing shows us that, mature now, she's become one tough woman.

Establishing Distance

Cathleen opens her portrait by writing from a third-person point of view, even calling herself by another name, creating distance between herself the writer and herself the subject:

At this time the subject would like to be known as Kitten Junkie, but first she wants to clarify her pseudonym. She chose the name for pop culture and aesthetic reasons only. Kitten was raised in a liberal activist atmosphere where marijuana was treated as an hors d'oeuvre. As a child she remembers finger-painting in a Rainbow Coalition office, her eyes watering from thick clouds of marijuana smoke. However, other than aspirin, she does not use drugs of any kind. She is not a junkie. She can't remember the last time she smoked weed and has no plans to smoke it in the future—as she reminds her mother, she is every parents' dream.

Cathleen's stance here is ironic, sassy, and analytical all at the same time, so that her tone teaches us as much about who she is as the details of her liberal activist youth. We know she is serious; at the same time, we know she doesn't take herself too seriously.

Writing Snapshots

Snapshots are also a useful alternative to conventional paragraphs for presenting views of your present self. Even if you *limit* a self-profile to the approximate present—to glimpses or versions of you during only the past year or so—the choices are many. Snapshots allow you to show rapid glimpses of who you are and what you value, suggesting in a brief space the many dimensions of a single individual. Following are three such snapshots from three different writers, each covering a lot of material in a limited space, each also using deliberate rhetorical repetition to create memorable rhythms:

Jeremy:

I am the son of a man who does not know who Mick Jagger is, but who can recite the names of all of the state capitals in alphabetical order. I am the son of a woman whose only work experience prior to motherhood was cashiering at Sears, but who now runs her own business.

Sarah:

Mine, mine, mine. I am greedy. I love my stuff. I love Elizabeth, my faded and fuzz-worn-off teddy bear with the pink dress knitted by my aunt. I love the address book that I bought at Harrod's in London, the only souvenir I could afford, now filled with the names of my favorite people. I love my puffy white comforter—have you ever wondered what it feels like to sleep on a cloud? When I look at my stuff, I see parts of myself—the endurance of Elizabeth, the irony of Harrod's, the calmness of my comforter.

Marcianna:

I prefer summer to winter. I prefer hot to cold. I prefer dark chocolate to milk chocolate. I prefer writing by myself to talking in front of the class. I prefer dressing up to dressing down. I prefer "trick-or-treat" to buying candy. I prefer friends to enemies and laughing to worrying.

✌ WRITING 3

Write three different openings for a possible self-portrait: first, define yourself by describing something in your room or apartment and speculating upon what it says about you. Second, describe yourself by standing in front of a full-length mirror and make inferences about what your clothes and facial expressions say about you. Third, give yourself a fictive name and then write about yourself from the third-person point of view (as he or she). Which of these portrait techniques captures you best?

✌ CAREER PROFILE

This identity assignment involves exploring who you intend to become next by exploring either potential college majors or out-of-college careers. If you know what you want to pursue as life work, it makes the most sense to investigate and report on that; if you are undecided about either career or major, it makes good sense to explore those toward which you are leaning or most attracted. If you are more uncertain than that, uncertain even about why you are in college or what to major in, this assignment could be your chance to give voice to that confusion and look more closely at possible resolutions.

Whatever your college goals, you might start with a close look at your own past work experiences as well as at any other firsthand experience with what seem to be attractive careers. You also might explore how you selected a college to attend as well as a discipline in which to major. Any of these initial investigations will be relevant to what you intend to do next with your life. For this paper, plan to include a substantial amount of research into possible careers or majors as well as to look closely at your own past experience and interests.

Mining the Past

Your past interests, hobbies, and successes often anticipate your college major as well as your life work. Even in those instances where you encounter whole new fields in college, from engineering to social work, the origins of your new interest are often evident in past activities. In the following example, Chris outlines a childhood dream.

> As far back as I can remember, I wanted to be a weather reporter. If I couldn't be a professional sailboat racer, I wanted to be a weatherman. I never had any desire to be the guy on TV, but I wanted to know what the guy on TV knows.

Early on in his paper, Chris also explores impediments to pursuing this dream.

> One day, when I was in eighth grade, my teacher told me I had to drop back from the regular math group to the slower math group. Math was not my strong suit. But when obtaining a degree in meteorology—the degree you need to be a weather reporter I got the same answer every time: "You need lots of math."

Researching the Present

If you're interested in finding out what certain jobs or professions are like, look in the phone book and find out where, in your community, people actually practice this work. If you want to research the global dimensions of this career, your local library or bookstore will have occupation guidebooks to read. And if you want to find out how your college or university could help prepare you, talk to professors in the disciplines that interest you as well as visit your career counselor or job placement service.

To find out firsthand whether or not even to attempt to major in meteorology, Chris set up an appointment with Paul Sisson of the National Weather Service in Burlington, Vermont, whom he interviewed in person:

> I asked several questions, eventually getting around to my fear of math. "Yes, you have to be pretty proficient in math to survive meteorology. If you can't survive calculus, you won't be able to survive the rest." I looked across at Paul in his blue oxford shirt, navy knit tie, pressed khakis, and boat shoes with a combination of respect, envy, and hopelessness.

But in spite of his difficulty with math, writing this paper became a way for Chris to explore his fascination with weather more deeply and resolve what next to do about it:

> In a way I'd never considered before, my desire to be a meteorologist stems from my desire to help people since the ability to predict severe weather can help everyone from airline pilots to farmers to flood victims. More than ever, I realize

that this is what I want to do with my education. I believe my desire to learn will help me overcome my difficulties with mathematics.

Writing Snapshots

Snapshots also work well to present your explorations of career choices. In the following excerpts from a ten-snapshot essay, Sonya explores her dual interest in the environment and in teaching, letting each snapshot reveal a stage on her way to selecting a college major:

1. What I care about is the environment, and what I want to do is teach younger children to care about it too. That's what brings me to college, and to this English class, writing about what I want to be when I grow up.

2. My first teaching was this past summer on the Caribbean island of South Caicos. In the classroom one morning, I tried to teach local teenagers about the fragility of their island environment, but they did not seem to hear me or attend to my lesson, and I left class very frustrated. Later on, we went to the beach, and they taught me back my morning lessons, and I felt so much better. I thought then I wanted to be a teacher. . . .

3. The School of Education scares me. A lady named Roberta and a professor named Merton gave me a list of classes I would need in order to major in education. "Environmental education is not a real field, yet," the professor said. I realized it would take four years and many courses and still I wouldn't be studying the environment or be sure of ever teaching about it in public schools.

4. The School of Natural Resources excites me. Professor Erickson is my advisor, and in one afternoon, she helped me plan a major in "Terrestrial Ecology." I now know, for the first time, exactly what I'm doing in college. I need to study natural resources first, later on decide whether I want to teach, work in the field or what.

Earlier drafts of Sonya's essay included a fair amount of complaining and editorializing, but when she switched to writing snapshots, she focused only on the highlights of her decision-making process and skipped most of the complaining that characterized earlier drafts.

♂ WRITING 4

Write about where you currently stand in terms of college major or career choice. Include the names of people you might talk to or places you might visit to gather more information. When you finish writing, plan to set up appointments with the most available prospects.

Shaping the Whole Essay

This short autobiographical essay (six hundred words) portrays Rebecca Rabin's mixed religious heritage, strong commitment to Protestantism, and current participation in Christian rituals. Each individual snapshot focuses on a single small event—a cross necklace, prayer, a church, and so forth. Each actually tells a small story, complete with beginning, middle, and punch line. At the same time, the cumulative effect of these thirteen snapshots reveals Becky's broad tolerance, education, and interest in a spirituality that goes well beyond separate religious creeds. This theme emerges as one experience is juxtaposed against another, past tense against present tense, without editorializing, allowing readers to supply the connective tissue by filling in the white spaces for themselves.

I Know Who I Am
Rebecca Rabin

- My mother grew up in Darien, Connecticut, a Presbyterian. When she was little, she gave the Children's Sermon at her church.

- My father grew up in Cleveland, Ohio, a Jew. When I went away to college, he gave me the Hebrew Bible he received at his Bar Mitzvah.

- The only similarity between my parents' families is freckles. They both have them, which means I get a double dose. Lucky me. My mother once told me that freckles are "angels' kisses." Lucky me.

- When my parents married and decided to have kids, they agreed to raise them Christian. Thus, I was brought up Presbyterian. According to Jewish tradition and law, Judaism is passed on via a person's mother. Thus I am not a Jew.

- I have attended First Presbyterian Church of Boulder, Colorado, for most of my life. When I was baptized, Reverend Allen said: "Becky is being baptized here today, brought by her believing mother and her unbelieving—but supportive—father."

- When I was little, I was terrified of the darkness. Sometimes, I would wake up in the night and scream. It was my mother who came in to comfort me, smoothing my hair, telling me to think of butterflies and angels.

- I once went through this stage in which I was quite certain I was meant to be Greek Orthodox. That was one my parents couldn't quite understand. My fascination came from watching a movie that had an Eastern Orthodox wedding with the bride wearing beautiful flowers.

- I have always said my prayers before going to bed. Lying silently in the dark, talking to God. Like the disciples in the Garden of Gethsemane, I have been known to fall asleep while praying. Now I pray on my knees, it is harder to doze off that way.

- My father's sister's family is Orthodox Jewish. Things in their house are very different from things in our house. They have two of everything: refrigerators, sinks, dish sets, and dishwashers. When I go there, though, I always feel comfortable in the most basic way—as if my heart says, Oh yes, this is where I belong.

- I am not afraid of the dark anymore. Except sometimes, when I wake up in a sweaty panic, certain there's an axe murderer in my room. My father, the Jewish psychiatrist, says that nightmares are from going to sleep before your brain has sufficiently wound down from a busy day. He says reading before sleeping will cure that.

- When I am in Vermont, I attend North Avenue Alliance Church. I chose it because it is big, like my church at home. The last two Sundays I have sung solos. The first time, I sang "Amazing Grace." The second time, I sang "El Shaddai," which is partly in Hebrew. In church, my heart says, Oh yes, this is where I belong.

- I wear a cross around my neck. It is nothing spectacular to look at, but I love it because I bought it at the Vatican. Even though I am not a Catholic, I am glad I bought it at the Pope's home town. Sometimes, when I sit in Hebrew class, I wonder if people wonder, "What religion is she, anyway?"

- From a very early age I have always known exactly what I wanted: To live in a safe world. To worship freely. To sing. I know what I believe. I know who I am. I know where I'm going.

✌ SUGGESTIONS FOR WRITING AND RESEARCH

Individual

1. Explore your identity by selecting one of the four types of autobiographical writing discussed in this chapter—autobiography, language autobiography, self-portrait, or career profile. Write an impressionistic first draft based on remembered experiences. Write a researched second draft based on artifacts or interviews. Write a snapshot-shaped third draft, mixing memory and research in a structure that pleases you.

2. If you keep a journal or diary, compile an edited, annotated edition of it so that it profiles or portrays you in a way you are willing to share with the world. Edit or otherwise rewrite entries to protect privacy or improve clarity. Construct an introduction that explains the time period this document covers, and add notations as needed for those who do not know you to understand you better.

Collaborative

Compile a class book of autobiographical sketches that includes a contribution from each class member. Elect a team of editors who will collect, compile, introduce, and produce these essays. Ask your instructor to write an afterword explaining the assignment. (For more information on publishing class books, see Chapter 28.)

Chapter

༄ 10 ༄

Profiling People

*When I met Lisa for coffee, it was like the meeting of two total strangers. But we're
both from the South, so when she told me about growing up in Charleston, we got
a lot closer. I still need to find what makes Lisa uniquely Lisa. We'll meet again.*

REBECCA

Reading and writing profiles about people teach us not only about others, but also about ourselves. Newspaper and magazine writers commonly profile the rich, famous, and powerful people of the world. Some popular magazines, such as *Vanity Fair* and *Rolling Stone,* feature well-researched profiles in virtually every issue. Weekly newsmagazines, such as *Time, Newsweek,* and *Sports Illustrated,* use short profiles as regular features. And literary magazines, such as *The New Yorker, Atlantic,* and *Esquire,* are well-known for their lengthy, in-depth profiles. Profiles are not easy to write, but they are rewarding because the process of writing them inevitably brings the writer and subject closer together.

༄ WRITING A PROFILE

The purpose of a profile is to capture a person's essence on paper. Good profiles generally focus on a single aspect of the subject's life or personality and make some sort of comment on it. In short, a good profile tells a story about its subject.

Profiles are usually about people other than the writer, but you, the writer, are ultimately in control. You decide what to include, what to omit, how to describe the subject and his or her surroundings, where to begin, and where to end. However, the ultimate purpose of a profile is to convey a sense of who your subject is. You must develop a portrait that is essentially true.

Profiles lie on a spectrum between two related forms, interviews and biographies. Interviews are conversations between a writer and a living person. They are commonly the result of a single visit, though some may be based on multiple visits. Published interviews often transcribe the interviewer's questions followed by the subject's responses.

Biographies are usually book-length studies of people, dead as well as living. A biographer's sources include letters from and to the subject; diaries and notebooks; stories by relatives and acquaintances; newspaper and magazine reports; previously published interviews; legal and medical records; and the subject's published writing or other work—all available resources that shed light on the life and character of the subject.

Like interviews, profiles include direct conversations with living people. Like biographies, they make use of other sources of information about the subject. Profiles are usually longer than interviews, but they are considerably shorter than biographies. Profiles are more tightly focused than both interviews, which may contain questions on a wide variety of subjects, and biographies, which attempt to convey information on all aspects of a subject's life. In contrast, a profile selectively presents information to create a unified portrait.

A common college assignment will ask you to profile a professor or staff member who works at your college or university, people who work in the local community, or other students. Profile writing requires, first, a willing subject; second, time to collect information about the subject; and third, the skill to focus on one aspect of the subject and develop a clear theme.

✌ WRITING 1

Describe any profiles that you recall reading in magazines or newspapers. What details do you remember about the person profiled? Why do you think these details remain in your memory?

✌ FINDING A SUBJECT

Profiles can be written about virtually any person willing to hold still long enough to reveal something about himself or herself. While a list of people you might profile is unlimited, some subjects are more accessible than others for students in a college writing class. The profile examples in this chapter are all taken from assignments to profile classmates. But you can select for your own profile subject anyone in whom you are especially interested.

Relatives make good profile subjects because they are usually more than willing to cooperate, and the knowledge resulting from the profile will contribute to your family history. However, some family members are not easily accessible for more than one visit. And family members such as parents, who are emotionally close to you, may be difficult to profile because you may lack the objectivity necessary to portray them realistically.

Members of the campus community are usually willing subjects: professors, librarians, cafeteria workers, resident assistants, alumni staff, and coaches, to name some of the obvious ones. In addition, the local community contains other potential subjects: shopkeepers, street vendors, police officers, city administrators, and various local characters of good and ill repute. The advantage of profiling members of your community is that you can learn about people in various occupations and social circumstances. The disadvantages include the unavailability of busy people for extended interviews and the extra time it takes to conduct off-campus interviews.

Your own classmates can also provide a wealth of characters with varied backgrounds and interests. The advantages of profiling your peers includes their willingness to be profiled, their availability, and the chance to get to know what they are like. The only disadvantage is that writing about a classmate may not expose you to a wider range of people, though this certainly depends on who is in your class. Note, too, that it is very difficult to profile objectively students with whom you're romantically involved.

> ### ⌁ WRITING 2
>
> Make a list of people in your family, campus community, and local community whom you might be interested in profiling. Make a similar list of classmates who might make good profile subjects. Talk with your instructor about which subjects would be best for a profile assignment.

⌁ PROVIDING BACKGROUND INFORMATION

To write an effective profile, you need to learn as much about your subject as possible. You will eventually use some of this background information when you draft the profile itself, to provide a context for the subject's words and actions. However, much of it is useful primarily during your planning and invention stages, as you decide on a focus for the profile and a direction for more research.

Finding background information requires good research skills. You must take advantage of all the available sources of information and follow up on new leads wherever you find them.

Preliminary Interviews

Talking with your subject often is the best place to start. In addition to providing valuable background information, he or she can give you leads to further information. If you are profiling a classmate, interview him or her for ten or fifteen minutes to get started. For other subjects, call ahead and set up a time to meet.

Public Information

If your subject has a résumé, ask to see it. If your subject has published something—whether a letter to the editor or a book—get a copy. If your subject has made a speech or taken a public stand, find a record of it. If the person has been the subject of an interview or biography, read it to see what previous writers have found out.

Writing Portfolios

A rich source of recent background information is your subject's writings—especially if your subject is a student. Assigned papers or essay tests will tell you something of the person's intellectual interests, but journals or letters may reveal more personal information.

Friends and Acquaintances

An obvious source of background information is the people who know, live, and work with your subject. Each conversation you have with a friend or acquaintance of your subject is itself a small interview. It is a good idea to begin such a conversation with specific questions and to take notes. Good questions include the following: What is this person like? How did you come to know each other? What do you most often do together? Whom does this person admire? What does this person want to do or become next?

✑ WRITING 3

Locate a profile subject and find out as much about his or her background as you can. Take good notes. What further questions are suggested by the subject's responses? Set a time to meet again and probe your subject's history further.

✑ DESCRIBING PHYSICAL APPEARANCE

Often the first thing we notice about people we've just met is their appearance. Profiles, too, commonly introduce the subject through descriptions of how the person looks and acts. In the early stages of your writing process, you may want to take notes on every aspect of your subject's appearance. Later, you should select details that capture your subject's individuality and reveal his or her personality.

When describing your subject, use words that appeal to the senses, that express size, shape, color, texture, and sound. Be as specific as possible. You may want to describe physical appearance, clothing, and habits or gestures. Pam can capture Mari Anne's appearance by employing all three types of information:

> While sitting at her desk, Mari Anne keeps twirling her hair to help her think and relax. She is dressed in solid colors, black and red, and has a dozen bracelets on her right arm. She smiles as if nervous, but as I got to know her, I found that smile always on her face. She is five feet two inches tall, has naturally curly brown hair, a dark complexion, and dark brown eyes, and is always smiling. She is a second-generation American since both sets of grandparents came from Greece.

Pam's description tells readers much about Mari Anne's tastes (solid colors, lots of jewelry) and personality (friendly but perhaps a little high-strung). Pam also uses physical description (dark complexion, dark brown eyes) to lead directly to background information (her Greek heritage). It is important to realize how much you, the writer—either consciously or unconsciously—can shape such descriptions. By mentioning Mari Anne's nervous hair twisting at the beginning of the description and her nervous smile a bit later, Pam makes readers see the subject as nervous.

✍ WRITING 4

As an exercise, sit with a classmate and spend ten minutes taking detailed notes on what he or she looks like; pay attention to face, body, height, clothing, gestures, and expressions. Write one to two pages organizing these physical details to convey a dominant impression.

✍ DESCRIBING THE SETTING

Effective description of the setting contributes to realism in a profile and advances readers' understanding of your subject. As in all descriptions, you should note specific details of the physical environment, using sensory words that help readers experience what it was like to be there.

The setting you describe should be the one in which you and your subject met and talked. If this is where your subject lives, you have the opportunity to observe an environment that he or she created and that no doubt reflects much of his or her personality. Do your best to record the details that tell the most about your subject's special interests. If the interview takes place elsewhere, your description may be primarily a way of creating a realistic backdrop, but you may also reveal a great deal about your subject by the way he or she reacts to a less personal environment. Settings can be described on their own—usually near the beginning of the profile—or subtly and indirectly along with the action of the interview.

In the following example, Caleb meets Charles at Charles's favorite off-campus hangout in order to get to know him better. Before turning to his subject, Caleb describes the ambiance of their meeting place:

> Charles suggested we meet in the Other Place, a downtown bar commonly referred to as OP. Charles had basketball practice until seven or eight, so he'd meet me there around 9:30. The OP had a dive bar ambiance to it. In the far corner of the bar stood two outdated pinball machines being hugged respectively by a barefoot girl and a portly man wearing jeans and a leather vest. Both were shaking their hips to the noises of their machines. Above the actual bar a mute hockey game was taking place on television.

Caleb describes the "dive" with a good eye for detail, implying in his descriptions of its people that it's not a place he feels comfortable in. But he lets readers know that Charles is comfortable there.

Beth indirectly includes a description of the setting in her profile:

> Becky sits cross-legged at the foot of the bottom bunk on her pink and green homemade quilt. She leans up against the wall and runs her fingers through her brown shoulder-length hair. The sounds of James Taylor's "Carolina on My Mind" softly fill the room. Posters of John Lennon, James Dean, and Cher look down on us from the walls. Becky stares at the floor and scrunches her face as if she is thinking hard.

Beth includes a rich number of sight words about Becky's home away from home—describing bed, posture, person, posters, and room—letting the detail contribute to the portrait of the person. By including the James Taylor song as well as the three posters, she allows readers to make inferences about what these say about Becky's tastes. Beth slips in the setting details quietly so as not to detract attention from the subject herself.

When including details of setting, think about whether they should strongly reinforce your verbal portrait of the subject, provide a colorful platform for your interaction to take place, or stay quietly in the background.

↪ WRITING 5

If possible, arrange to visit the place where your profile subject lives, and record as much sensory information about that place as you can. Capture what is on the walls and floors, out the windows, on the desk, and on the bed and under it. Also note the brand names of things; their sizes, colors, smells; and the sounds from the CD player and from down the hall as well.

↪ GIVING SUBJECTS A VOICE

Interviews allow people to reveal information about themselves that contributes to a portrait. Interviews are the primary source of information about many profile subjects, especially classmates. You can conduct interviews in three ways.

First, you and your subject can have an informal conversation in which you get to know the subject, usually without notes. In an informal interview, ten prepared questions may be more intrusive than helpful. If you don't take notes, be sure to capture your recollections of the conversation within twenty-four hours, or you'll forget most of them.

The second method of interviewing is to take notes from the subject's responses to prepared questions. You'll need to write fast in a small notebook, catching the essence of your subject's responses and filling in the details later. Note taking is especially helpful during an interview because it lets you see what information you've got as you go along and sometimes helps you decide where to go next with your questions.

As a third method, tape-recording captures your subject's language *exactly* as it was spoken. If you plan to bring a tape recorder to the interview, be sure to secure your subject's permission in advance. While reconstructing dialogue from tape may sound easier than reconstructing it from notes, transcribing the conversation accurately and selecting which passages to use in the profile demand a great deal of time and patience. Plan to use especially those ways of speaking that seem most characteristic of your subject.

In all but the most informal interviews, you should come with prepared questions. In your first interview, ask a coherent set of questions to provide you with an overview of your subject. While questions might vary depending on the age, status, and occupation of your subject, the following is a good starting list:

1. Where did you grow up? What was it like?
2. What do your parents do for a living?
3. Do you have any brothers or sisters? What are they like? How are you similar to or different from them?
4. How do you spend your free time?
5. What kinds of jobs have you held?
6. What are your favorite books, movies, or recordings?
7. How did you come to be where you are today?
8. What do you intend to do next?

During subsequent interviews, narrow your questions to a more limited range of interests. For example, if your first interview revealed that photography is your subject's favorite hobby, in your second interview focus several questions on his or her involvement with photography. Your profile will succeed according to the amount of detailed information you get your subject to reveal through interviews. After the interview with Becky, Beth wrote the following narrative from her tape transcript:

> Finally, after minutes of silence she says, "I don't ever remember my father ever living in my house, really. He left when I was three and my sister was just a baby, about a year old. My mom took care of us all. Forever, it was just Mom, Kate, and me. I loved it, you know? Just the three of us together."

Beth is aware that if she can capture the small details of Becky's childhood along with her teenage conversational style (*like, you know*), Becky's story will be all the more plausible—which in turn will make Beth's profile more readable.

✒ WRITING 6

Interview your profile subject in a place that is convenient and comfortable to you both. If this is a first interview, start with informal, conversational questions like those presented in the text.

✒ SELECTING A POINT OF VIEW

Your profile will be written from either the first-person or the third-person point of view. The point of view you choose will do much to establish the tone and style of your profile. In the first-person point of view, the narrator (*I*) has a presence in the story. You can use this point of view to let readers know that you are presenting the subject through the filter and perhaps the bias of your own eyes, as Caleb does in the following passage:

> I walked in past the overfilled coat rack and scanned the room for Charles. The smell of smoke was overwhelming and it was hard to see. It surprised me that Charles would hang out in a joint like this, especially since he seemed to be such a disciplined athlete. I spotted him with a group of tall guys near the pinball machines. We made eye contact and he rose to greet me.

In the third-person point of view, *I* is never used, only proper names (*Joan, Sara*), nouns for persons (*athlete, grocer*), and the pronouns *he, she,* and *they.* You can use this point of view to keep yourself out of the narrative, focusing instead on the subject and on the words and actions that any observer might witness. Caleb does this in the following passage as he describes an exchange that takes place where he and Charles are sitting:

> Again the girls came over, accosting Charles. One of them gave him a Budweiser, which he gladly accepted.
> "You want to play tennis tomorrow?" the shorter girl asked.
> "Sure, what time?" Charles asked in a soft voice.
> "Whenever you're free."
> "How about three? I have a Saturday practice tomorrow morning."

Many profiles are actually written, as Caleb's is, in a combination of first person and third person. Although most of the focus is naturally on the subject, many writers find that since they select what to report and what to ignore, some acknowledgment of their presence is the most honest approach.

Write your first draft from whatever points of view seem comfortable to you. In the revising stages, you can experiment with increasing or decreasing your presence in the narrative, and in editing you will check that the point of view is consistent.

↬ WRITING 7

Write one page about your profile subject in the first-person point of view and then one page in the third person, using essentially the same information in each version. Which do you prefer? Why?

↬ DEVELOPING A THEME

Ultimately, profiles tell stories about people. Though you have read only a few examples of student profiles in this chapter, already you may have drawn some conclusions about what these people are like. Profile writers select the details and dialogue and background information that tell the story they want to tell about their subject. Most often, that story builds as the profile progresses so that the last page or paragraph focuses on the most important point the writer wants to make about the subject. This central theme can be revealed *explicitly*, with the writer telling readers what to think, or *implicitly*, with the evolution of the profile making the theme clear.

Explicit Theme

Pam concludes her profile, which focuses on Mari Anne's continued love of gymnastics, with her own summary, a small judgment on what the realities of college life have done to Mari Anne's passion.

> [Mari Anne says,] "I want to be able to judge [gymnastics] here at college, so I will still have to pass part two of the exam. I would especially like to judge creative matches that show each gymnast's unique ability." As of now Mari Anne has not had time to fit gymnastics into her schedule. She is too busy trying to keep up with her studies. But you can bet that next year will find her on the floor or behind the judge's table; Mari Anne has too much passion for the sport to stay out for very long.

Implicit Theme

Beth never suggests what readers should think of Becky. She lets her subject's words end the profile, allowing readers to make their own inferences about who Becky is and what she stands for.

> "I think that because I didn't have my dad, we're closer to my grandparents. Because Mom was so young and they helped us out all the time. They gave us property to build a house and everything. So we're a lot closer because she could always count on them. That's the most important thing, you know, being able to count on people."

ᢛ WRITING 8

In the profile you are developing, would you like your theme to emerge explicitly or implicitly? If implicitly, how would you conclude the profile so that readers would be most likely to understand the theme you intend?

ᢛ PROFILE OF A CLASSMATE AS A WRITER

An interesting variation on the profile assignment is studying, then profiling a classmate—or anybody else for that matter—as a writer. Such an assignment makes use of all the profiling strategies outlined above, but holds the focus on how a person developed as a writer. Such projects usually begin with classmates interviewing each other and asking questions such as the following:

- How did you learn to write?
- What kinds of writing have you done on your own?

- What was your experience as a writer in school?
- What's the best writing experience you ever had, in school or out?
- What's the worst writing experience you ever had, in school or out?
- What do you most enjoy about writing?
- Describe your writing habits.
- What causes you the most problems when you write?
- Describe your experience writing in college.
- Describe your experience writing in this class.

In the following example, Megan quotes her profile subject, Nicole, talking about the writing she has done on her own and plans to do in the future:

> "I've kept a journal—more of a diary—since I was twelve or thirteen. But that's private writing. In high school I used to write stories and poems occasionally, but I never showed these to anyone. I never felt confident in my creative writing." Nicole studies the edges of the quilt on her dormitory bed and rubs her chin. "I'm trying to change that. I think I have a lot of West Virginia stories to tell—that's where I grew up. My dad worked in a steel mill for thirty years. If you don't come from a town with a steel mill, you don't know what that's like. Those towns are dying off now—I'd like to capture that life in words."

Nicole, in turn, profiles Megan through a combination of close observation and interview, which she weaves together in the following paragraph:

> Megan tries to keep her own desk clean, but often finds it muddled with glasses and coffee mugs. She'll sit in front of her secondhand AT&T computer and compose on the screen, stopping every half an hour to get up and move around. Being a fast typist allows her to write fast, letting her thoughts flow rapidly, not worrying much about punctuation or capitalization—at least on early drafts. Her favorite pen is a black fine point Uniball Micro, which she uses to catch stray thoughts away from the computer. "I prefer to write almost everything on my computer, but I do love to write letters, lists, and journal entries with my Uniball—it's very inky and smooth and spidery."

In addition to paraphrasing and directly quoting interview subjects, plan to share with each other samples of past and current writing, including journals, letters, school assignments, and self-initiated writing. When you compose your final profile draft, weave together the several types of information you've collected—observations, interview material, and original sources. Before concluding this paper, share a version with your subject and ask for reactions, including corrections in fact or adjustments in tone.

Strategies for Writing Profiles

1. When interviewing a classmate, agree on guidelines beforehand: whether or not to use tape recorders; where, when, and for how long to exchange visits; where else to meet; what sources of background to share; and so on.
2. Use your own narration to summarize, to provide background and context, and to interpret. Try to strike a balance between writing about your subject and letting him or her speak.

3. Quote interview material directly to reveal your subject's personality and beliefs. Subjects who talk directly to readers characterize themselves and provide living proof to support your inferences about them.

4. Share drafts with your profile subject. When subjects see early drafts, they may tell you important information that will improve your representations of them.

5. You are the author of the profile you write: Take your subject's comments into account as much as you can, but maintain ownership of your interpretation and characterization.

ᔕ WRITING 9

Write a response to any of the questions above; then give this response to the classmate who is writing your profile to help him or her get started.

ᔕ SHAPING THE WHOLE PAPER

Beth Devino's finished draft of her profile "Becky" is presented here. Notice that Beth chose to shape the profile as if it took place at one sitting in Becky's dormitory room; in fact, she interviewed Becky at several different times and places over several weeks. Beth removes herself almost completely from the profile, letting Becky's own words do most of the characterizing—though at times she also presents information from the dormitory setting and her own summary of background information.

Notice the focus on trust and dependability, which develops early in the profile and carries through to the concluding words that express Becky's attitudes toward men. The final line makes a very strong ending, but be aware that the writer, not her subject, created this ending. Over the course of several interviews, Becky talked about a wide range of subjects, including sports, teacher education, and college life. It was Beth who shaped these many conversations into a coherent essay with a beginning, a middle, and an end.

Becky

Beth Devino

Becky sits cross-legged at the foot of the bottom bunk on her pink and green homemade quilt. She leans up against the wall and runs her fingers through her brown shoulder-length hair. The sound of James Taylor's "Carolina on My Mind" softly fills the room. Posters of John Lennon, James Dean, and Cher look down on us from the walls. Becky stares at the floor and scrunches her face as if she is thinking hard.

Finally, after minutes of silence she says, "I don't ever remember my father ever living in my house, really. He left when I was three and my sister was just a baby, about a year old. My mom took care of us all. Forever, it was just Mom, Kate, and me. I loved it, you know? Just the three of us together."

Becky smiles and continues, "And I remember little things, you know, like we would all sleep together in Mom's bed. We'd all climb in. Little things like, I remember one night there was a bat in the house and Mom is afraid of bats and I was only, like, five, and Mom climbed under the covers with my little sister and I had to go down and call my grandmother to get the bat out of the house.

"But I'm really proud of my mother for bringing up my sister and me on her own. She had to work, sometimes two jobs, and she worked really hard. I don't remember a sad time then, ever. I had the happiest childhood. You know, some of my friends who have whole families complain about fights with their parents, but I have no complaints about anything. I never felt like I needed anything or that I lacked anything."

Becky pauses, hugging her knees close to her chest, rocking slightly. "Hmmm," she mumbles. She traces her lips with the back of her fingernail. "Oh, I always do this when I'm thinking or I'm upset—now I'm just thinking."

Becky Harris grew up in West Granville, a small town where people knew and supported each other. She came to the university to major in elementary education: "I really think those early years are so crucial, when children are first learning how to live in and trust the world."

On campus, she lives in Connors Hall, with Trish, a roommate from Maine who is fast becoming a best friend. Trish agrees, saying, "Outside of classes, we do everything together, share tapes, run on the weekends, borrow clothes, and talk late almost every night. The posters? She put those up. I really like them."

Becky offers me a cup of herbal tea, then makes a cup for herself and resumes her place on the bed. "Anyway, Mom and I have the strangest relationship. It's like we're friends—she's my mom, but we're friends more than she's my mom because, when my dad left, I kinda had to grow up overnight and take care of my sister 'cause my mom was working so much. I mean, she never left us alone or anything, but I had to do things. I had to learn to dress myself and all those little things really fast, earlier than lots of kids because she didn't have time for both of us.

"Oh! I have this watch bear. I put it over my bed, somewhere where it can watch me all the time." She gets up from the bed and slides across the linoleum floor to her desk to pick up the little white stuffed animal. "Two years ago, I lost Watch Bear. For almost three whole days, and I didn't have anybody watching over me. But I found him, he was under the bed. I brought him with me to college to sit on my desk to watch over me and to make sure I'm safe like he did when I was little."

Becky carefully places Watch Bear against the wall near her and continues, "Anyway, I hope that if I were ever in the same situation as my mom, I could be as strong as she. 'Cause that would scare me to have a car and this brand new house that they just built and have to take care of everything. 'Cause my dad never paid any money, never a cent of child support, ever in his life to us.

"I've seen him maybe three or four times in ten or twelve years. Once two years ago, at Thanksgiving, I saw him, and that was when he had just gotten remarried, and I met his wife. I really liked her a lot and I really liked their kids. I got along with them, but you know, I don't think of him as part of my family. I don't even really think of him as my father, really. I mean biologically, but that's all. I used to get really sad sometimes that I didn't have a dad. But I don't feel like I've missed anything in my life, ever. I'd rather have my mother happy than to have her live with someone just to make a whole family 'cause I think we had a whole family.

"My mother has never said a bad thing about my father, ever, in her life. And if it was me who got dumped with two children, I would just—would always be bad-talking, I'm sure. She never wanted us to hate him and wanted us to have the opportunity to get to know him if we wanted to when we could choose to. When he called and asked us to go to Thanksgiving with him a few years ago, I didn't want to, but Mom really encouraged us to 'cause she said maybe he's going to reach out and try to change his ways and be your dad. So we went. Kate, my sister, who was just a year when he left, never knew him at all. She was very uncomfortable there, but I talked to him a little bit."

Becky pauses and traces her lips again. "I used to have really bad feelings towards men in general. Like, I didn't trust them at all 'cause I thought that, you know, they were all sort of like him; you couldn't count on them for anything. I just don't think there's. . . . I get so mad that there's people that would just leave someone with children—especially their own, you know? I'm better now. I have a boyfriend and I trust him a lot, but I question everything he does. When he makes commitments I don't really think he's going to come through, you know? I wonder about that a lot because—I don't really have a reason to distrust all men but, you know?

"I think that because I didn't have my dad, we're closer to my grandparents. Because Mom was so young and they helped us out all the time. They gave us property to build a house and everything. So we're a lot closer because she could always count on them. That's the most important thing, you know, being able to count on people."

∼ SUGGESTIONS FOR WRITING AND RESEARCH

Individual

1. Write a profile of a classmate using the information you have collected in Writings 2–7. In doing this assignment, write several drafts. Share these with your subject, listen to his or her response to your profile, and take those comments into account when writing your final draft. Keep in mind the golden rule of profile writing: Do unto your subject as you would want him or her to do unto you.

2. Write a profile of somebody in the university community who is not a student: a professor, counselor, security officer, cafeteria staffperson. Be considerate in arranging interview times; focus on the work this person does; plan to share your resulting profile with the subject.

3. Write a profile of a family member. Interview this person and collect as much information about him or her as you can: letters, yearbooks, photographs. Plan to contribute your final draft to whoever in your family collects such records. (If nobody does, would you want to start collecting yourself?)

Collaborative

1. Pair up all students in the class—including the instructor if there is an uneven number of students. Within each pair, members will take turns being subject and writer, with each profiling the other to find out the details of his or her writing life. Each student should then compose a profile of the pair partner *as a writer.* As a class, plan to publish all such profiles in a class book. Elect a team of editors to collect, collate, introduce, and produce the class book. (See Chapter 28 for more information on publishing class books.)

2. Write a collaborative profile of your class. First discuss what a class profile might be like: Would it be a collection of individual profiles arranged in some order? Or would it consist of written bits and pieces about people, places, and events, arranged as a verbal collage? Would there be a place for visual components in this class profile? Would you want to challenge other writing classes to develop similar profiles and share them with one another?

Chapter

❧ 11 ❧

Explaining Things

When I give directions to somebody about how to find some place, it's much easier to draw a map than write it out in words. I mean, with words you have to be so precise but with a map you just draw lines.

JANE

To explain something is to make it clear to somebody else who wants to understand it. Explaining is fundamental to most acts of communication and to nearly every type of writing, from personal and reflective to argumentative and research writing. At the same time, explanatory writing is also a genre unto itself; for example, see a newspaper feature on baseball card collecting, a magazine article on why dinosaurs are extinct, a textbook on the French Revolution, a recipe for chili, or a laboratory report.

❧ WRITING TO EXPLAIN

Whether you write to explain something as part of a larger intention or write to report information, explanatory writing (also called expository or informational writing) answers questions such as these:

- What is it?
- What does it mean?
- What are the consequences?
- How does it work?
- How is it related to other things?
- How is it put together?
- How do you get there?
- Why did it happen?
- Why did it fail?

To write a successful explanation, you need to find out first what your readers want to know, then what they already know and what they don't know. If you are able to determine—or at least make educated guesses about—these audience conditions, your writing task becomes clear. When you begin to write, keep in mind three general principles that typify much explanatory writing: (1) It focuses on the idea or object being explained rather than on the writer's beliefs and feelings; (2) it often—not always—states its objective early in what might be called an informational thesis; and (3) it presents information systematically and logically.

In writing classes, explanation usually takes the form of research essays and reports that emphasize informing rather than arguing, interpreting, or reflecting. The assignment may be to describe how something works or to explain the causes and effects of a particular phenomenon. This chapter explains how to develop a topic, articulate your purpose, and use strategies appropriate for your audience.

᪥ WRITING 1

How good are you at explaining things to people? What things do you most commonly find yourself explaining? What is the last thing you explained in writing? How did your audience receive your explanation?

᪥ FINDING A TOPIC

Topics with a limited, or specific, scope are easier to explain carefully and in detail than topics that are vague, amorphous, or very broad. For example, it's hard to know where to start with subjects such as mountains, cities, automobiles, or sound systems: What, exactly, are you interested in writing about? EVERYTHING? If so, the task is daunting for even the world's foremost expert. However, a specific mountain range, city, or automobile would be a better place to start. But even then you would still need to know what it is about this subject that sparks your curiosity. For example, general subjects such as mountains, automobiles, or music sound systems are so broad that it's hard to know where to begin. However, a specific aspect of sound systems, such as compact discs (CD's), is easier. Within the subject of CD's, of course, there are several topics as well (design, manufacturing process, cost, marketing, sound quality, comparison to tape and vinyl recordings, and so on). If your central question focuses on how CD's are manufactured, you might address some of these other issues (cost, marketing, comparison) as well, but only insofar as they illuminate and advance your focus on manufacturing.

Once you have a focused topic on a central question, you need to assemble information. If you're not an expert yourself, you'll need to consult authorities on the topic. Even if you are already an expert, finding supporting information from other experts will help make your explanation clear and authoritative. Keep your audience in mind as you begin your research. You don't want to waste time researching and writing about things your audience already knows.

✂ DEVELOPING A THESIS

A thesis is simply a writer's declaration of what the paper is about. Stating a thesis early in an explanatory work lets readers know what to expect and guides their understanding of the information to be presented. In explanatory writing, the thesis states the answer to the implied question your paper sets out to address: What is it? How does this work? Why is this so?

QUESTION   **Why do compact discs cost so much?**

THESIS   **CD's cost more than records because the laser technology required to manufacture them is so expensive.**

The advantage of stating a thesis in a single sentence is that it sums up the purpose of your paper in a single idea that lets readers predict what's ahead. Another way to state a single-sentence thesis is to convey an image, an analogy, or a metaphor that provides an ongoing reference point throughout the paper and gives unity and coherence to your explanation—a good image keeps both you and your readers focused.

QUESTION   **How are the offices of the city government connected?**

THESIS   **City government offices are like an octopus, with eight fairly independent bureaus as arms and a central brain in the mayor's office.**

The thesis you start with may evolve as you work on your paper—and that's okay. For example, suppose the more you learn about city government, the less like an octopus and the more like a centipede it seems. So, your first thesis is really a working thesis, and it needs to be tentative, flexible, and subject to change; its primary function is to keep your paper focused to guide further research.

✂ WRITING 2

Find a topic that needs explaining to someone for some reason, and write out a working thesis. If you are addressing a *when?* or *how?* question, find an analogy to something familiar that will help your reader understand your explanation better.

✂ USING STRATEGIES TO EXPLAIN

Good strategies that can be used to explain things include: defining, describing, classifying and dividing, analyzing causes and effects, and comparing and contrasting. Which strategy you select depends on the question you are answering as well as the audience to whom you are explaining. You could offer two very different explanations to the same question depending on who asked it: For example, if asked "Where is Westport Drive?" you would respond differently to a neighbor familiar with local reference points ("one block north of Burger King") than to a stranger, who would not know where Burger King was either. With this caution in mind on considering who is the receiver of the explanation, here is a brief overview of possible strategies, using CD's as a sample topic:

QUESTION   STRATEGY

What is it?   Define
A CD is a small plastic disc containing recorded music.

What does it mean?   Define
Today, when you talk about a "recording," you mean a cassette tape or a CD, not a grooved vinyl disc.

What are its characteristics?   Describe
A vinyl disc is round with tiny grooves covering its surface in which a needle travels to play sound.

What will its consequences be?   Analyze cause and effect
If you scratch a vinyl record with a knife, it will skip.

How does it work?   Describe process
On a CD, a laser beam reads the tiny dots on the spinning disc and sends back sound.

How is it related to other things?   Compare and contrast
A CD transmits clearer sound than a cassette tape.

How is it put together?   Classify and divide
A basic sound system includes an input (to generate sound), a processor (to amplify and transmit sound), and an output (to make sound audible).

To what group does it belong?   Classify and divide
CD players, along with tuners and tape cassette decks, are sources of music in a sound system.

Why did it happen?   Analyze cause and effect
A CD skips either because its surface is dirty or because the player is broken.

If your paper is on a tightly focused topic and answers a narrow, simple question, you may need to use only one strategy. More often, however, you will have one primary strategy that shapes the paper as a whole and several secondary strategies that can vary from paragraph to paragraph or even sentence to sentence. For example, to explain why the government has raised income taxes, your primary strategy would be to analyze cause and effect, but you may also need to define terms such as *income tax*, to classify the various types of taxes, and to compare and contrast raising income taxes to other budgetary options. In fact, almost every explanatory strategy makes use of other strategies: How, for example, do you describe a process without first dividing it into steps? How can you compare and contrast without describing the things compared and contrasted?

Defining

To *define* something is to identify it, to set it apart so that it can be distinguished from similar things. Writers need to define any terms central for reader understanding in order to make points clearly, forcefully, and with authority.

Formal definitions are what you find in a dictionary. They usually combine a general term with specific characteristics.

> A computer is a programmable electronic device [general term] that can store, retrieve, and process data [specific characteristics].

Usually, defining something is a brief, preliminary step accomplished before you move on to a more important part of the explanation. When you need to define something complex or difficult or when your primary explanatory strategy is definition, you will need an extended definition consisting of a paragraph or more. This was the case with Mark's paper explaining computers, in which he defined each part of a typical computer system. After defining the central processing unit (CPU), he then defined computer memory:

> Computer storage space is measured in units called "kilobytes" (K). Each K equals 1,024 "bytes" or approximately 1,000 single typewriter characters. So one K equals about 180 English words, or a little less than half of a single-spaced typed page, or maybe three minutes of fast typing.
>
> Personal computers generally have their memories measured in "megabytes" (MB). One MB equals 1,048,567 bytes (or 1,000 K), which translates into approximately 400 pages of single-spaced type. One gigabyte (GB) equals 1,000 MB or 400,000 pages of single-spaced type!

Describing

To *describe* a person, place, or thing means to create a verbal image so that readers can see what you see; hear what you hear; or taste, smell, and feel what you taste, smell, and feel. In other words, effective descriptions appeal to the senses. Furthermore, good description contains enough sensory detail for readers to understand the subject, but not so much as to distract or bore them. Your job, then, is to include just the right amount of detail so that you put readers in your shoes.

To describe how processes work is more complicated than describing what something looks like: In addition to showing objects at rest, you need to show them in sequence and motion. You need to divide the process into discrete steps and present the steps in a logical order. This is easier to do with simple processes, such as making a peanut butter and jelly sandwich, than for complex processes, such as manufacturing an automobile.

Whenever you describe a process, show the steps in a logical sequence that will be easy for readers to follow. To help orient your readers, you may also want to number the steps, using transition words such as *first, second,* and *third.* In the following example, taken from an early draft of his paper, Keith describes the process of manufacturing compact discs:

> CD's start out as a refrigerator-sized box full of little plastic beads that you could sift your hands through. They are fed into a giant tapered corkscrew—a blown-up version of an old-fashioned meat grinder. As the beads pass down the corkscrew, they are slowly melted by the heated walls.

At the bottom of their descent is a "master recording plate" onto which the molten plastic is pressed. The plastic now resembles a vinyl record, except that the disc is transparent. The master now imprints "pits," rather than grooves, around the disc, the surface resembling a ball of Play-Doh after being thrown against a stucco wall—magnified 5,000 times.

Comparing and Contrasting

To *compare* two things is to find similarities between them; to *contrast* is to find differences. Comparing and contrasting at the same time helps people understand something two ways: first, by showing how it is related to similar things, and second, by showing how it differs. College assignments frequently ask you to compare and contrast one author, book, idea, and so on with another.

People usually compare and contrast things when they want to make a choice or judgment about them: books, food, bicycles, presidential candidates, political philosophies. For this reason, the two things compared and contrasted should be similar: You'll learn more to help you vote for president by comparing two presidential candidates than one presidential candidate and a Senate candidate; you'll learn more about which orange to buy by comparing it with other types of oranges (navel, mandarin) than with apples, plums, or pears. Likewise, it's easiest to see similarities and differences when you compare and contrast the same elements of each thing. If you describe one political candidate's stand on gun control, describe the other's as well; this way, voters will have a basis for choosing one over the other.

Comparison-and-contrast analysis can be organized in one of three ways. (1) A point-to-point analysis examines one feature at a time for both similarities and differences; (2) a whole-to-whole analysis presents first one object as a whole and then the other as a whole; (3) a similarity-and-difference analysis presents first the similarities between two things, then the differences, or vice versa.

Use a point-to-point or similarity-and-difference analysis for long explanations of complex things, such as manufacturing an automobile, in which you need to cover everything from materials and labor to assembly and inspection processes. But use a whole-to-whole analysis for simple objects that readers can more easily comprehend. In the following whole-to-whole example, a student explains the difference between Democrats and Republicans:

> Like most Americans, both Democrats and Republicans believe in the twin values of equality and freedom. However, Democrats place a greater emphasis on equality, believing equal opportunity for all people to be more important than the freedom of any single individual. Consequently, they stand for government intervention to guarantee equal treatment in matters of environmental protection, minimum wages, racial policies, and educational opportunities.
>
> In contrast, Republicans place greater emphasis on freedom, believing the specific rights of the individual to be more important than the vague collective rights of the masses. Consequently, they stand for less government control in matters of property ownership, wages and the right to work based strictly on merit and hard work, and local control of schools.

Note how the writer devotes equal space to each political party, uses neutral language to lend academic authority to his explanation, and emphasizes the differences by using parallel examples as well as parallel sentence structure. The careful use of several comparison-and-contrast strategies makes it difficult for readers to miss his point.

An *analogy* is an extended comparison that shows the extent to which one thing is similar in structure and/or process to another. Analogies are effective ways of explaining something new to readers, because you can compare something they are unfamiliar with to something they already know about. For example, most of us have learned to understand how a heart functions by comparing it to a water pump. Be sure to use objects and images in analogies that will be familiar to your readers.

Classifying and Dividing

People generally understand short more easily than long, simple more easily than complex. To help readers understand a complicated topic, it helps to classify and divide it into simpler pieces and to put the pieces in context.

To *classify* something, you put it in a category or class with other things that are like it:

Like whales and dolphins, sea lions are aquatic mammals.

To *divide* something, you break it into smaller parts or subcategories:

An insect's body is composed of a head, a thorax, and an abdomen.

Many complex systems need both classification and division to be clear. To explain a music sound system, for example, you might divide the whole into parts: headphones, graphic equalizer, tape deck, CD player, preamplifier, amplifier, tuner, and speakers. To better understand how these parts function, you might classify them into categories:

Inputs	Tuner
	Tape deck
	Compact disc player
Processors	Preamplifiers
	Amplifiers
	Graphic equalizers
Outputs	Speakers
	Headphones

Most readers have a difficult time remembering more than six or seven items at a time, so explaining is easier when you organize a long list into fewer logical groups, as in the preceding example. Also be sure that the categories you use are meaningful to your readers, not simply convenient for you as a writer.

Analyzing Causes and Effects

Few things happen all by themselves. Usually, one thing happens because something else happened; then it, in turn, makes something else happen. You sleep because you're tired, and once you've slept, you wake up because you're rested, and so on. In other words, you already know about cause and effect because it's a regular part of your daily life. A *cause* is something that makes something else happen; an *effect* is the thing that happens.

Cause-and-effect analyses are most often assigned for college papers to answer *why* questions: Why are the fish dying in the river? Why do CD's cost more than records? The most direct answer is a *because* statement.

> Fish are dying because of low oxygen levels in the lake.

Each answer, in other words, is a thesis, which the rest of the paper must both defend and support:

> There are three reasons for low oxygen levels . . . .

Cause-and-effect analyses also try to describe possible future effects:

> If nitrogen fertilizers were banned from farmland that drains into the lake, oxygen levels would rise, and fish populations would be restored.

Unless there is sound, widely accepted evidence to support the thesis, however, this sort of analysis may lead to more argumentative writing. In this example, for instance, farmers or fertilizer manufacturers might complicate the matter by pointing to other sources of lake pollution—outboard motors, paper mill effluents, urban sewage runoff—making comprehensive solutions harder to reach. Keep in mind that most complex situations have multiple causes. If you try to reduce a complex situation to an overly simple cause, you are making the logical mistake known as *oversimplification*.

✍ WRITING 3

Decide which of the five strategies described in this section best suits the primary purpose of the explanatory paper you are drafting. Which additional or secondary strategies will you also use?

✍ ORGANIZING WITH LOGIC

If you explain to your readers where you're taking them, they will follow more willingly; if you lead carefully, step by step, using a good road map, they will know where they are and will trust you.

Your method of organization should be simple, straightforward, and logical, and it should be appropriate for your subject and audience. For example, to explain how a sound system works, you have a number of logical options: (1) you could start by putting a CD in a player and end with the music coming out of the speakers; (2) you could describe the system technically, starting with the power source to explain how sound is made in the speakers; (3) you could describe it historically, starting with components that were developed earliest and work toward the most recent inventions. All of these options follow a clear logic that, once explained, will make sense to readers.

～ **WRITING 4**

Outline three possible means of organizing the explanatory paper you are writing. List the advantages and disadvantages of each. Select the one that best suits your purpose and the needs of your audience.

～ MAINTAINING A NEUTRAL PERSPECTIVE

First, you need to understand that absolute neutrality or objectivity is impossible when you write about anything. All writers bring with them assumptions and biases that cause them to view the world—including this explanatory project—in a particular way. Nevertheless, your explanations will usually be clearer and more accessible to others when you present them as fairly as possible, with as little bias as possible—even though doing this, too, will depend upon who your readers are and whether they agree or disagree with your biases. In general, it's more effective to emphasize the thing explained (the object) rather than your personal beliefs and feelings. This perspective allows you to get information to readers as quickly and efficiently as possible without you, the writer, getting in the way.

To adopt a neutral perspective, write from the third-person point of view, using the pronouns *he, she,* and *it.* Keep yourself in the background unless you have a good reason not to, such as explaining your personal experience with the subject. In some instances, adopting the second-person *you* adds a friendly, familiar tone that keeps readers interested.

Be fair; present all the relevant information about the topic, both things you like about it and things you dislike. Avoid emotional or biased language. Remember that your goal is not to win an argument, but to convey information.

～ **WRITING 5**

For practice, write a one-paragraph explanation of your topic from a deliberately biased perspective (political, class, gender, etc.); then compare your version with those written by your classmates and decide on the strategies needed to avoid biases in your writing.

～ SHAPING THE WHOLE PAPER (STUDENT ESSAY)

In the following essay, Keith Jordan asks the question "How is the music that our generation listens to and takes for granted actually made?" He says, in effect, read my essay ("Let me fill you in.") and I'll explain how CD players operate and how the discs are manufactured. His organization is simple as he starts with the playing of the disc and backtracks to how they are made. Keith's voice throughout is that of a knowledgeable tour guide. Although his personality is clear ("How much can you screw up a yes or no?"), his biases do not affect the report. Although the primary explanatory strategy in Keith's essay is cause and effect, he uses most of the other strategies discussed in this chapter as well: definition, process description, and comparison and contrast. His essay is most remarkable for its effective use of analogy. At various points, he asks his readers to think of radar, the game of telephone, jimmies on an ice cream cone, corkscrews, meat grinders, player pianos, and Play-Doh.

The Sound Is Better than the Music: The Making of Compact Discs
Keith Jordan

Our generation is the music generation. We buy and listen to more music more often than any generation before us, but few of us actually understand how this music is made. The purpose of your whole sound system, from recording to speaker, is to reproduce music that sounds as much like the original source as possible. The compact disc (CD) technology that we take for granted reproduces music better than previous recording systems because it's both simpler and more complicated than they were. Let me fill you in.

A CD player operates by sending out a laser beam of light that bounces off an object, like radar, and returns with a message, which becomes the music. On one CD there are hundreds of thousands of tiny pits that resemble those of a player piano scroll, telling the piano which keys to hit. The CD player reads either a simple "yes" or a "no"—a pit or no pit—from the disc. The laser in your CD player detects the distance to the disc to determine whether there's a pit, which will be farther away, or not.

What's the difference, you ask, in receiving music from tiny pits versus the grooves on vinyl records or magnetic deposits on tapes? The result is less interference between the message sent and the message received. Do you remember playing the "telephone game" in fifth grade? You know, the one where someone on one side of class whispers something in your ear and it gets passed along until it gets to the last person, who says what he or she was told? This is much the way in which your sound system works: a recording—either a disc or a tape—is like the first person, and your speakers are like the last. In the case of the vinyl records or magnetic tapes, I whisper some line in your ear and you pass it on, but by the time it reaches the last person, it's been touched and twisted and has a few more words attached. However, in the case of the CD, I whisper either a "yes" or a "no," and by the time it reaches the last person it should be exactly the same—this is where the term "digital" comes from, meaning either there is a signal ("yes") or there isn't one ("no"). How much can you screw up a yes or a no?

The CD manufacturing process, however, is not so simple. To guarantee that almost nothing will interfere with (scratch or break) the digital message encoded on the plastic, the disc is metallized, a process that deposits a thin film of metal, usually aluminum, on the surface; you see it as a rainbow under a light. Since light won't bounce back from transparent plastic, the coating acts as a mirror to bounce back the laser beam. The disc is mirrored by a precise spray-painting process called "sputtering." You couldn't just dip the thing because then the pits would fill in or melt. The clear disc is inserted into a chamber and placed opposite a piece of pure aluminum called a "target," which is bombarded with electricity, causing the aluminum atoms to jump off and embed themselves into the surface of the disc, like jimmies on an ice cream cone. Then the disc is "spin-coated" yet again with a fine film of resin, which becomes the outer coating on the CD.

Once the resin is cured by a brief exposure to ultraviolet light, your CD is pretty much idiot-proof. As long as you don't interrupt the light path in the film, your CD will perform perfectly, even with small scratches, so long as they don't diffract the laser beam—and even then you may be able to rub them smooth with a finger. No object, other than a ray of light, comes in contact with the recorded surface of a compact disc.

The CD is finished when it is stamped with the appropriate logo and allowed to dry. In other words, the way CD's are played and made eliminates the interference that caused distortion in earlier music systems. These steps, which have taken me several hours to explain on paper, take a mere seven seconds on the assembly line from materials that cost no more than a pack of gum. The technology behind CD systems guarantees a faithful sound recording and a disc that will last virtually forever—longer, perhaps, than some of the music we listen to.

✌ SUGGESTIONS FOR WRITING AND RESEARCH

Individual

1. Write a paper explaining any thing, process, or concept. Use as a starting point an idea you discovered in Writing 2. When you have finished one draft of this essay, look back and see if there are places where your explanation could be improved through use of one of the explanatory strategies described in this chapter.
2. Select a writer of your choice, fiction or nonfiction, who explains things especially well. Read or reread his or her work and write an essay in which you analyze and explain the effectiveness of the explanation you find there.
3. Revise by directing your last draft to a less knowledgeable audience or one with more specialized expertise. (For help with this revision, see "Switching" in Chapter 23.)

Collaborative

Form writing groups based on mutual interests; agree as a group to explain the same thing, process, or concept. Write your explanations separately and then share drafts, comparing and contrasting your different ways of explaining. For a final draft, either (1) rewrite your individual drafts, borrowing good ideas from others in your group, or (2) compose a collaborative single paper with contributions from each group member.

Chapter

ॐ 12 ॐ

Arguing For and Against

When I argued against federal gun control laws with my roommates, it was pretty easy to convince them I was right, but when I wrote the same arguments in my English paper, my writing group challenged every point I made and kept asking for more evidence, more proof. Do you have to have evidence for everything you write?

WOODY

No, you don't *have* to have evidence for every paper you write, but if you want reasonable people—for example, the instructors and students of a college or university—to believe your assertions, interpretations, and arguments, you'd better have evidence—substantial, plentiful, convincing evidence.

Argument is deeply rooted in the American political and social system, in which free and open debate is the essence of the democratic process. Argument is also at the heart of the academic process, in which scholars investigate scientific, social, and cultural issues, hoping through the give-and-take of debate to find reasonable answers to complex questions. Argument in the academic world, however, is less likely to be about winning or losing—as it is in political and legal systems—than about changing minds or altering perceptions about knowledge and ideas.

Argument as rational disagreement, rather than as quarrels and contests, most often occurs in areas of genuine uncertainty about what is right, best, or most reasonable. In disciplines such as English, history, and philosophy, written argument commonly takes the form of interpretation, in which the meaning of an idea or text is disputed. In disciplines such as political science, engineering, and business, arguments commonly appear as position papers, in which a problem is examined and a solution proposed.

ॐ WRITING TO CHANGE PEOPLE'S MINDS

The reason for writing an argumentative paper in the first place is to persuade other people to agree with a particular point of view. Arguments focus on issues about which there is some debate; if there's no debate, there's no argument. College assignments commonly ask you to argue one side of an issue and defend your argument against attacks from skeptics.

In a basic position-paper assignment, you are asked to choose an issue, argue a position, and support it with evidence. Sometimes your investigation of the issue will lead you beyond polar positions toward compromise—a common result of real argument and debate in both the academic and political worlds. In other words, such a paper may reveal that the result of supporting one position (*thesis*) against another (*antithesis*) is to arrive at yet a third position (*synthesis*), which is possible now because both sides have been fully explored and a reasonable compromise presents itself. This chapter explains the elements that constitute a basic position paper: an arguable issue, a claim and counterclaim, a thesis, and evidence.

Issue

An *issue* is a controversy, something about which there is disagreement. For instance, mountain bikes and cultural diversity are things or concepts, not in themselves issues. However, they become the foundation for issues when questions are raised about them and controversy ensues.

ISSUE   **Do American colleges adequately represent the cultural diversity of the United States?**

ISSUE   **Should mountain bikes be allowed on wilderness hiking trails?**

These questions are issues because reasonable people could answer them in different ways; they can be argued about because more than one answer is plausible, possible, or realistic.

Position

Virtually all issues can be formulated, at least initially, as yes/no questions about which you will take one position or the other: pro (if the answer is yes) or con (if the answer is no).

ISSUE   Should mountain bikes be allowed on trails in Riverside Park?

PRO   Yes, they should be allowed to share pedestrian trails.

CON   No, they should not be allowed to share trails with pedestrians.

Claims and Counterclaims

A *claim* is a statement or assertion that something is true or should be done. In arguing one side of an issue, you make one or more claims in the hope of convincing an audience to believe you. For example, you could make a claim that calls into question the educational experience at Northfield College:

CLAIM   **Northfield College fails to provide good education because the faculty is not culturally diverse.**

Counterclaims are statements that oppose or refute claims. You need to examine an opponent's counterclaim carefully in order to refute it or, if you agree with the counterclaim, to argue that your claim is more important to making a decision. For example, the following counterclaim might be offered against your claim about the quality of Northfield College education:

COUNTERCLAIM **The Northfield faculty are good scholars and teachers; therefore, their race is irrelevant.**

You might agree that "Northfield faculty are good scholars and teachers" but still argue that the education is not as good as it would be with more diversity. In other words, the best arguments provide not only good reasons for accepting a position, but also good reasons for doubting the opposition. They are made by writers who know both sides of an issue and are prepared for the arguments of the opposition.

Thesis

The primary claim made in an argument is called a *thesis*. The thesis in an argumentative paper is the major claim the paper makes and defends.

THESIS **Northfield College should enact a policy to make the faculty more culturally diverse by the year 2005.**

In taking a position, you may make other claims as well, but they should all work to support this major claim or thesis.

CLAIM **The faculty is not culturally diverse.**

CLAIM **A culturally diverse faculty is necessary to provide a good education for today's student.**

CLAIM **The goal of increased cultural diversity by the year 2005 is achievable and practical.**

In arguing a position, you may state your thesis up front, with the remainder of the paper supporting it (thesis first), or you may state it later in the paper after weighing the pros and cons with your reader (delayed thesis). As a writer, you can decide which approach is the stronger rhetorical strategy after you fully examine each claim and the supporting evidence. Each strategy, thesis first or delayed, has its advantages and disadvantages.

Evidence

Evidence makes a claim believable. Evidence consists of facts, examples, or testimony that supports a claim. For example, to support a claim that Northfield College's faculty lacks cultural diversity, you might introduce the following evidence:

EVIDENCE **According to the names in the college catalog, 69 of 79 faculty members are male.**

EVIDENCE **According to a recent faculty survey, 75 of 79 faculty members are Caucasian or white.**

EVIDENCE **According to Carmen Lopez, an unsuccessful candidate for a position in the English Department, 100 percent of the faculty hired in the last ten years have been white males.**

Most arguments become more effective when they include documentable source material; however, shorter and more modest argument papers can be written without research and can profitably follow a process similar to that described here.

❧ WRITING 1

An issue debated by college faculty is whether or not a first-year writing course should be required of all college students. Make three claims and three counterclaims about this issue. Then select the claim you most believe in and write an argumentative thesis that could form the basis for a whole essay.

❧ FINDING AN ISSUE

You'll write better and have a more interesting time if you select an issue that interests you and about which you still have real questions. A good issue around which to write a position paper will meet the following criteria:

1. It is a real issue about which there is controversy and uncertainty.
2. It has at least two distinct and arguable positions.
3. Resources are available to support both sides.
4. Writing a position paper on the issue fits the time frame and scope of the assignment.

In selecting an issue to research and write about, consider both national and local issues. The advantage of selecting national issues is that you are likely to see them explained and argued on the national news programs—*CBS Evening News* or *All Things Considered* on NPR—or in national news publications such as *Time, Newsweek,* and the *New York Times.* In addition, you can count on your audience having some familiarity with the subject. The disadvantage is that it may be difficult to find local experts or a site where some dimension of the issue can be witnessed. These are examples of national issues:

> Are SATs a fair measure of academic potential?
> Should handgun ownership be outlawed in the United States?
> Does acid rain kill forests?

Local issues are derived from the community in which you live. You will find issues like these argued about in local newspapers and on local news broadcasts:

> Should a new mall be built on the beltway?
> Should mountain bikes be allowed in Riverside Park?
> Should Northfield College require a one-semester course introducing students to diverse American cultures?

The advantage of local issues is that you can often visit a place where the controversy occurs, interview people who are affected by it, and find generous coverage in local news media. The disadvantage is that the subject won't be covered in the national news.

Perhaps the best issue is a national issue (hikers versus mountain bikers) with a strong local dimension (this controversy in a local park). Such an issue will enable you to find both national press coverage and local experts, and you can be reasonably sure that both your instructor and your classmates will know something about it and will be interested in your position on it.

✑ WRITING 2

Make a list of three national and three local issues about which you are concerned. Next, select the three issues that seem most important to you and write each as a question with a yes or no answer. Finally, note whether each issue meets the criteria for a good position paper topic.

✑ ANALYZING AN ISSUE

The most demanding work in writing a position paper takes place after you have selected an issue but before you actually write the paper. To analyze an issue, you need to conduct enough research to explain it and identify the arguments of each side.

In this data-collecting stage, treat each side fairly, framing the opposition as positively as you frame the position. Research as if you are in an honest debate with yourself; doing so may even cause you to switch sides—one of the best indications of open-minded research. Furthermore, empathy for the opposition leads to the selection of qualified assertions and heads off overly simplistic right-versus-wrong arguments. Undecided readers who see merit in the opposing side respect writers who acknowledge an issue's complexity.

Establishing Context

Provide full context for the issue you are writing about, as if readers know virtually nothing about it. Providing context means answering these questions: What is this issue about? Where did the controversy begin? How long has it been debated? Who are the people involved? What is at stake? Use a neutral tone, as Issa does in discussing the mountain bike trail controversy:

> With all these new riders, there is a need for places to ride, and this is where the wilderness trail controversy begins. The mountain bike is designed to be ridden on dirt trails, logging roads, and fire trails in backwoods country. However, other trail users who have been around much longer than mountain bikers prefer to enjoy the woods at a slow, leisurely pace. They find the rapid and sometimes noisy two-wheel intruders unacceptable.

Claims For (Pro)

List the claims supporting the pro side of the issue. Make each claim a distinctly strong and separate point, and make the best possible case for this position, identifying by name the most important people or organizations that hold this view. Issa makes the following claims for opening up wilderness trails to mountain bikes:

1. All people should have the right to explore the wilderness so long as they do not damage it.
2. Knobby mountain bike tires do no more damage to hiking trails than Vibram-soled hiking boots.
3. Most mountain bike riders are respectful of the wilderness and courteous to other trail users.

Claims Against (Con)

List the claims supporting the con side of the issue—the counterclaims. It is not necessary to have an equal number of reasons for and against, but you do want an approximate balance.

1. Mountain bike riders ride fast, are sometimes reckless, and pose a threat to slower-moving hikers.
2. Mountain bike tires damage trails and cause erosion.

Annotated References

Make an alphabetical list on note cards or computer files of the references you consulted during research, briefly identifying each according to the kind of information it contains. The same article may present claims from both sides as well as provide context. Following are three of Issa's annotated references:

> Buchanan, Rob. "Birth of the Gearhead Nation." Rolling Stone 9 July 1992: 80-85. Marin Co. CA movement advocates more trails open to mountain bike use. Includes history. (pro)
>
> "Fearing for Desert, A City Restricts Mountain Bikes." New York Times 4 June 1995, A24. Controversy in Moab, Utah, over conservation damage by mountain bikes. (con)
>
> Schwartz, David M. "Over Hill, Over Dale on a Bicycle Built for . . . Goo." Smithsonian 25.3 (June 1992): 74-84. Discusses the hiker vs. biker issue, promotes peaceful coexistence, includes history. (pro/con)

Annotating your list of references allows you to check and rearrange your claims at any time during the writing process. In addition, if you write and organize your references now, your reference page will be ready to go when you've finished writing your paper.

✌ WRITING 3

Select one of the issues you are interested in, establish the necessary context, and make pro and con lists similar to those described in this section, including supporters of each position. Make the best possible case for each position.

ᴥ TAKING A POSITION

Once you have spread out the two positions fairly, determine which side is stronger. Select the position that you find most convincing and then write out the reasons that support this position, most compelling reasons last. This will be the position you will most likely defend; you need to state it as a thesis.

Start with a Thesis

Formulate your initial position as a *working thesis* early in your paper-writing process. A working thesis asserts your major claim; it is merely something to start with, not necessarily to stick with—which is why we're calling it "working": it's still in the process of being developed and made final. Even though it's tentative, it serves to focus your initial efforts in one direction, and it helps you articulate claims and assemble evidence to support it.

WORKING THESIS	Hikers and mountain bikers should cooperate and support each other in using, preserving, and maintaining wilderness trails.

Writers often revise their initial positions as they reshape their paper or find new evidence; however, do your best to make each assertion of a thesis as strong as possible—even if you're pretty sure it's not final. Your working thesis should meet the following criteria:

1. It can be managed within your confines of time and space.
2. It asserts something specific.
3. It proposes a plan of action.

ᴥ WRITING 4

Take a position on the issue you have identified. Formulate a working thesis that you would like to support. Test your thesis against the criteria listed for good theses.

ᴥ DEVELOPING AN ARGUMENT

Your *argument* is the case you will make for your position; it is the means by which you will try to persuade your readers that your position is correct. Good arguments, as Woody discovered at the beginning of this chapter, need solid and credible evidence and clear and logical reasoning.

Assembling Evidence

A claim is meaningless without evidence to support it, and good evidence can come from a variety of sources: *facts, examples, inferences, informed opinion,* and *personal experience* all provide believable evidence.

Facts and Examples

Facts are verifiable and agreed upon by everyone involved regardless of personal beliefs or values. Facts are often numerical or statistical, and they are recorded in some place where anybody can look them up—a dictionary, an almanac, a public report, or a college catalog.

Water boils at 212 degrees Fahrenheit.

Northfield College employed 79 full-time faculty and enrolled 1,143 full-time students in 1996.

Five hundred Japanese-made "Stumpjumper" mountain bikes were sold in the United States in 1981.

Examples can be used to illustrate a claim or clarify an issue. If you claim that many wilderness trails have been closed to mountain biking, you can mention examples you know about:

The New Jersey trails at South Mountain, Eagle Rock, and Mills Park have all been closed to mountain bikes.

Facts and examples can, of course, be misleading and even wrong. For hundreds of years malaria was believed to be caused by "bad air" rather than, as we know today, by a parasite transmitted through mosquito bites; however, for the people who believed the bad-air theory, it was fact.

The accumulation of a certain number of facts and examples should lead to an interpretation of what those facts mean—an *inference* or a generalization. For example, if you attend five different classes at Northfield College and in each class you find no minority students, you may infer that there are no minority students on campus. However, while your inference is reasonable, it is not a fact, since your experience does not allow for your meeting all the possible students at the college.

Facts are not necessarily better or more important than inferences; they serve different purposes. Facts provide information, and inferences give that information meaning.

Sometimes inference is all that's available. For example, statistics describing what "Americans" believe or do are only inferences about these groups based on information collected from a relatively small number of individuals. To be credible, however, inferences must be reasonable and based on factual evidence.

Expert Opinion

Expert opinion makes powerful evidence. When a forest ranger testifies about trail damage caused by mountain bikes or lug-soled hiking boots, his training and experience make him an expert. A casual hiker making the same observation is less believable. To use expert opinion in writing arguments, be sure to cite the credentials or training that makes this person's testimony "expert."

Personal Testimony

A useful kind of evidence is testimony based on *personal experience.* When someone has experienced something firsthand, his or her knowledge cannot easily be discounted. If you have been present at the mistreatment of a minority student, your eyewitness testimony will carry weight, even though you are not a certified expert of any kind. To use personal testimony effectively, provide details that confirm for readers that you were there and know what you are talking about.

Demonstrating Reasoning

To build an effective argument, consider the audience you must persuade. In writing about the mountain bike controversy, for example, ask yourself these questions:

- Who will read this paper: members of an environmentally conscious hiking club, members of a mountain bike club, or your instructor?
- Where do I think they stand on the issue? (Hikers are often opposed to mountain bikes, mountain bikers are not, but you would need more information to predict your instructor's position.)
- How are their personal interests involved? (Hikers want the trails quiet and peaceful, bikers want to ride in the wilderness, and your instructor may or may not care.)
- What evidence would they consider convincing? (A hiker would need to see convincing examples of trails being improved by mountain bike use, bikers would accept anecdotal testimony of good intentions, and you're still not sure about your instructor.)

The more you know about the audience you're trying to sway, the easier it will be to present your case. If your audience is your instructor, you'll need to make inferences about his or her beliefs based on syllabus language, class discussion, assigned readings, or personal habits. For example, if your instructor rides a mountain bike to work, you may begin to infer one thing; if he or she assigns Sierra Club readings in the course, you infer something else; and if the instructor rides a mountain bike and reads *Sierra Club,* well, you've got more homework ahead. Remember that inferences based on a single piece of evidence are often wrong; find out more before you make simple assumptions about your audience. And sometimes audience analysis doesn't work very well when an instructor assumes a deliberately skeptical role in reading a set of papers. It's best to assume you will have a critical reader and to use the best logic and evidence available. Following are some ways to marshal careful and substantial evidence.

First, establish your credibility. Demonstrate to your audience that you are fair and can be trusted. Do this by writing in neutral, not obviously biased language—avoid name-calling. Also do this by citing current sources by respected experts, and don't quote them out of context. Identify elements that serve as common ground between you and the audience; be up front and admit when the opposite side makes a good point.

CREDIBLE   Northfield College offers excellent instruction in many areas; however, its offerings in multicultural education would be enhanced by a more diverse faculty.

LESS CREDIBLE   Education at Northfield College sucks.

Second, use logic. Demonstrate that you understand the principles of reasoning that operate in the academic world: Make each claim clearly, carefully, and in neutral language. Make sure you have substantial, credible evidence to support each claim. Make inferences from your evidence with care; don't exaggerate or argue positions that are not supported by the evidence.

LOGICAL   Since 75 of 79 faculty members are white or Caucasian, and 69 of 79 are male, it would make good sense to seek to hire more black, Hispanic, and Native American women faculty when they are available.

ILLOGICAL   Since all Northfield faculty members are racists, they should all be fired.

Third, appeal to your audience's emotions. It's fair to use means of persuasion other than logic to win arguments. Write with vivid details, concrete language, and compelling examples to show your audience a situation that needs addressing. It is often helpful, as well, to adopt a personal tone and write in friendly language to reach readers' hearts as well as minds.

EMOTIONAL   When Bridget Jones, the only black student in Philosophy 1, sits down,
APPEAL   the desks on either side of her remain empty. When her classmates choose partners for debate, Bridget is always the last one chosen.

꒰ WRITING 5

Develop an informal profile of the audience for your position paper by answering the questions posed in this section. Make a list of the kinds of evidence most likely to persuade this audience.

꒰ ORGANIZING THE PAPER

To organize your paper, you need to know your position on the issue: What is the main point of your argument? In other words, move from a working thesis to a final thesis: Confirm the working thesis that's been guiding your research so far, or modify it, or scrap it altogether and assert a different one. You should be able to articulate this thesis in a single sentence as the answer to the yes/no question you've been investigating.

THESIS   Wilderness trails should be open to both mountain bikers and hikers.

THESIS   Wilderness trails should be closed to mountain bikes.

Your next decision is where in this paper should you reveal your thesis to the reader—openly up front or strategically delayed until later? Neither strategy is necessarily right or wrong, but the decision is important because each one has a different psychological effect on your reader.

Thesis-First Organization

When you lead with a thesis, you tell readers from the beginning where you stand on the issue. The remainder of the essay supports your claim and defends it against counterclaims. Following is one good way to organize a thesis-first argument:

1. Introduce and explain the issue. Make sure there are at least two debatable sides. Pose the question that you see arising from this issue; if you can frame it as a yes/no, for/against construction, both you and your reader will have the advantage throughout your answer of knowing where you stand.

    Minority students, supported by many majority students at Northfield College, have staged a weeklong sit-in to urge the hiring of more minority faculty across the curriculum. Is this a reasonable position? Should Northfield hire more minority faculty members?

2. Assert your thesis. Your thesis states the answer to the question you have posed and establishes the position from which you will argue. Think of your thesis as the major claim the paper will make.

    Northfield College should enact a policy to make the faculty more culturally diverse as soon as is reasonably possible.

    Writers commonly state their thesis early in the paper, at the conclusion of the paragraph that introduces the issue.

3. Summarize the counterclaims. Explain the opposition's "counterclaims" before elaborating upon your own claims, because doing that gives your own argument something to focus on—and refute—throughout the rest of the paper. Squeezing the counterclaims between the thesis (2) and the evidence (5) reserves the strongest places—the opening and closing—for your position.

    COUNTERCLAIM 1 Northfield college is located in a white middle-class community, so its faculty should be white and middle class also.

    COUNTERCLAIM 2 The Northfield faculty are good scholars and teachers; therefore, their race is irrelevant.

4. Refute the counterclaims. Look for weak spots in the opposition's argument, and point them out. Use your opponent's language to show you have read closely but still find problems with the claim. To refute counterclaim 1:

    If the community in which the college is located is "white middle class," there is all the more reason to offer that diversity in the college.

    Your reputation is often stronger when you acknowledge the truth of some of the opposition's claims (demonstrating your fairness) but point out the limitations as well. To refute counterclaim 2:

Although Northfield College offers excellent instruction in many areas, its instruction in multicultural education would be enhanced by a more diverse faculty.

5.   Support your claims with evidence. Spell out your own claims clearly and precisely, enumerating them or being sure to give each its own full-paragraph explanation, and citing supporting evidence. This section will constitute the longest and most carefully documented part of your essay. The following evidence supports the thesis that Northfield needs more cultural diversity.

According to the names in the college catalog, 69 of 79 faculty members are male.

According to a recent faculty survey, 75 of 79 faculty members are Caucasian or white.

According to Carmen Lopez, an unsuccessful job candidate for a position in the English Department, all faculty hired in the last ten years have been white males.

6.   Restate your position as a conclusion. Near the end of your paper, synthesize your accumulated evidence into a broad general position, and restate your original thesis in slightly different language.

While Northfield College offers a strong liberal arts education, the addition of more culturally diverse faculty members will make it even stronger.

There are several advantages to leading with a thesis. First, your audience knows where you stand from the first paragraph. Second, your thesis occupies both the first and last position in the essay. In addition, this is the most common form of academic argument.

Delayed-Thesis Organization

Using the delayed-thesis type of organization, you introduce the issue, discuss the arguments for and against, but do not obviously take a side until late in the essay. Near the end of the paper, you explain that after listening carefully to both pros and cons, you have now arrived at the most reasonable position. Concluding with your own position gives it more emphasis. The following delayed-thesis argument is derived from the sample student essay at the end of this chapter:

1.   Introduce the issue and pose a question. Both thesis-first and delayed-thesis papers begin by establishing context and posing a question. Following is the question for the mountain bike position paper:

Should mountain bikes be allowed on wilderness trails?

2.   Summarize the claims for one position. Before stating which side you support, explain how the opposition views the issue:

To traditional trail users, the new breed of bicycle [is] alien and dangerous, esthetically offensive, and physically menacing.

3.   Refute these claims. Still not stating your own position, point out your difficulties with believing this side:

> Whether a bicycle—or a car or horse for that matter—is "alien . . . and esthetically offensive" depends on your personal taste, judgment, and familiarity. And whether it is "dangerous" depends on how you use it.

In addition, you can actually strengthen your position by admitting that in some cases the counterclaims might be true.

> While it's true that some mountain bikers—like some hikers—are too loud, mountain biking at its best respects the environment and promotes peace and conservation, not noise and destruction.

4. Summarize the counterclaims. You are supporting these claims, and so they should occupy the most emphatic position in your essay, the last:

> Most mountain bikers respect the wilderness and should be allowed to use wilderness trails.

5. Support your counterclaims. Now give your best evidence; this should be the longest and most carefully documented part of the paper:

> Studies show that bicycle tires cause no more erosion or trail damage than the boots of hikers, and far less than horses' hooves.

6. State your thesis as your conclusion. Your rhetorical stance or strategy is this: You have listened carefully to both the claims and counterclaims, and after giving each side a fair hearing, you have arrived at the most reasonable conclusion.

> It's clear that mountain bikers don't want to destroy trails any more than hikers do. The surest way to preserve America's wilderness areas is to establish strong cooperative bonds among the hikers and bikers, as well as those who fish, hunt, camp, canoe, and bird-watch, and encourage all to maintain the trails and respect the environment.

There are many advantages to delayed-thesis arguments. First, the audience is drawn into your struggle by being asked to weigh the evidence and arrive at a thesis. Second, readers are kept in suspense about your position; their curiosity is aroused. Finally, readers understand your difficulty in making a decision.

WRITING 6

Make two outlines for organizing your position paper, one with the thesis first, the other with a delayed thesis. Share your outlines with your classmates and discuss which seems more appropriate for the issue you have chosen.

SHAPING THE WHOLE PAPER

In the following paper, Issa Sawabini explores whether or not mountain bikers should be allowed to share wilderness trails with hikers. In the first part of the paper he establishes the context and background of the conflict; then he introduces the question his paper will address: "Is any resolution in sight?" Note his substantial use of sources, including the Internet and interviews, cited in the MLA documentation style. Issa selects a delayed-thesis strategy, which allows him to air both sides of the argument fully before revealing his solution, a compromise position: So long as mountain bikers follow environmentally sound guidelines, they should be allowed to use the trails.

On the Trail: Can the Hikers Share with the Bikers?

Issa Sawabini

The narrow, hard-packed dirt trail winding up the mountain under the spreading oaks and maples doesn't look like the source of a major environmental conflict, but it is. On the one side are hikers, environmentalists, and horseback riders who have traditionally used these wilderness trails. On the other side, looking back, are the mountain bike riders sitting atop their modern steeds wanting to use them too. But the hikers don't want the bikers, so trouble is brewing.

The debate over mountain bike use has gained momentum recently because of the increased popularity of this form of bicycling. Technology has made it easier for everyone to ride these go-anywhere bikes. These high-tech wonders incorporate exotic components, including quick gear-shifting derailleurs, good brakes, and a more comfortable upright seating position—and they can cost up to $2,000 each (Kelly 104). Mountain bikes have turned what were once grueling hill climbs into casual trips, and more people are taking notice.

Mountain bikes have taken over the bicycle industry, and with more bikes come more people wanting to ride in the mountains. The first mass-produced mountain bikes date to 1981, when five hundred Japanese "Stumpjumpers" were sold; by 1983 annual sales reached 200,000; today the figure is 8.5 million. In fact, mountain biking is second only to in-line skating as the fastest growing sport in the nation: "For a sport to go from zero to warp speed so quickly is unprecedented," says Brian Stickel, director of competition for the National Off Road Bicycle Association (Schwartz 75).

With all these new riders, there is a need for places to ride, and this is where the wilderness trail controversy begins. The mountain bike is designed to be ridden on dirt trails, logging roads, and fire trails in backwoods country. However, other trail users who have been around much longer than mountain bikers prefer to enjoy the woods at a slow, leisurely pace. They find the rapid and sometimes noisy two-wheel intruders unacceptable: "To traditional trail users, the new breed of bicycle [is] alien and dangerous, esthetically offensive and physically menacing" (Schwartz 74).

"The problem arises when people want to use an area of public land for their own personal purpose," says Carl Newton, forestry professor at the University of Vermont. "Eventually, after everyone has taken their small bit of the area, the results can be devastating. People believe that because they pay taxes for the land, they can use it as they please. This makes sense to the individual, but not to the whole community." Newton is both a hiker and a mountain biker.

When mountain bikes first came on the scene, hikers and environmentalists convinced state and local officials to ban the bikes from wilderness trails (Buchanan 81; Kelly 104). The result was the closing of many trails to mountain bike use: "Many state park systems have banned bicycles from narrow trails. National Parks prohibit them, in most cases, from leaving the pavement" (Schwartz 81). These trail closings have separated the outdoor community into the hikers and the bikers. Each group is well organized, and each group believes it is right. Is any resolution in sight?

The hikers and other passive trail users have a number of organizations, from conservation groups to public park planning committees, who argue against allowing mountain bikes onto narrow trails traditionally traveled only by foot and horse in the past. They believe that the wide, deeply treaded tires of the mountain bikes cause erosion and that the high speeds of the bikers startle and upset both hikers and horses (Hanley B4; Schwartz 76).

The arrival of mountain bikes during the 1980s was resisted by established hiker groups, such as the Sierra Club, which won debate after debate in favor of closing wilderness trails to mountain bike activities. The younger and less well organized biking groups proposed compromise, offering to help repair and maintain trails in return for riding rights, but their offers were ignored. "Peace was not given a chance. Foes of the bicycle onslaught, older and better connected, won most of the battles, and signs picturing a bicycle crossed with a red slash began to appear on trail heads all over the country" (Schwartz 74).

In Millburn, New Jersey, trails at South Mountain, Eagle Rock, and Mills Park have all been closed. Anyone caught riding a bike on the trails can be arrested and fined up to $100. Local riders offered an amendment calling for trails to be open Thursday through Sunday, with the riders helping maintain the trails on the other days. The amendment was rejected. According to hiker Donald Meserlain, the bikes "ruin the tranquillity of the woodlands and drive out hikers, bird watchers, and strollers. It's like weeds taking over the grass. Pretty soon we'll have all weeds" (Hanley).

Many areas in western New York, such as Hunter's Creek, have also been closed to mountain bike use. Anti-biking signs posted on trails frequently used by bicyclists caused a loud public debate as bike riding was again blamed for trail erosion.

Until more public lands are opened to trail riding, mountain bikers must pay fees to ride on private land, a situation beneficial to ski resorts in the off season: "Ski areas are happy to open trails to cyclists for a little summer and fall income" (Sneyd). For example, in Vermont, bike trails can be found at the Catamount Family Center in Williston, Vermont, as well as at Mount Snow, Killington, Stratton, and Bolton Valley. At major resorts, such as Mount Snow and Killington, ski lifts have actually been modified to the top of the mountains, and each resort offers a full-service bike shop at its base.

However, the real solution to the conflict between hikers and bikers is education, not separation. In response to the bad publicity and many trail closings, mountain bikers have banded together at local and national levels to educate both their own member bike riders and the non-riding public about the potential alliance between these two groups (Buchanan 81).

The largest group, the International Mountain Bike Association (IMBA), sponsors supervised rides and trail conservation classes and stresses that mountain bikers are friends, not enemies of the natural environment. "The IMBA wants to change the attitude of both the young gonzo rider bombing downhill on knobby tires, and the mature outdoorsman bristling at the thought of tire tracks where boot soles alone did tread" (Schwartz 76). IMBA published guidelines it hopes all mountain bikers will learn to follow:

1. Ride on open trails only.
2. Leave no trace.
3. Control your bicycle.
4. Always yield trail.
5. Never spook animals.
6. Plan ahead. (JTYL)

The New England Mountain Bike Association (NEMBA), one of the largest East Coast organizations, publishes a home page on the Internet outlining goals: "NEMBA is a not-for-profit organization dedicated to promoting land access, maintaining trails that are open to mountain bicyclists, and educating riders to use those trails sensitively and responsibly. We are also devoted to having fun" (Koellner).

At the local level, the Western New York Mountain Bike Association (WNYMBA) educates members on proper trail maintenance and urges its members to cooperate with local environmentalists whenever possible. For instance, when angry cyclists continued to use the closed trail at Hunter's Creek, New York, WNYMBA used the Internet to warn cyclists against continued trail use: "As WNYMBA wishes to cooperate with Erie County Parks Department to the greatest extent possible on the use of trails in open parks, WNYMBA cannot recommend ignoring posted signs. The first IMBA rule of trail is 'ride on open trails only'" (JTYL).

Educated mountain biking, like hiking and horseback riding, respects the environment and promotes peace and conservation, not noise and destruction. Making this case has begun to pay off, and the battle over who walks and who rides the trails should now shift in favor of peaceful coexistence. "Buoyed by studies showing that bicycle tires cause no more erosion or trail damage than the boots of hikers, and far less than horses' hooves, mountain bike advocates are starting to find receptive ears among environmental organizations" (Schwartz 78).

Even in the Millburn, New Jersey, area, bikers have begun to win some battles, as new trails have recently been funded specifically for mountain bike use: "After all," according to an unnamed legislator, "the bikers or their parents are taxpayers" (Hanley).

The Wilderness Society now officially supports limited use of mountain bikes, while the Sierra Club also supports careful use of trails by riders so long as no damage to the land results and riders ride responsibly on the path. "In pursuit of happy trails, bicycling organizations around the country are bending backward over their chain stays to dispel the hell-on-wheels view of them" (Schwartz 83).

Education and compromise are the sensible solutions to the hiker/biker standoff. Increased public awareness as well as increasingly responsible riding will open still more wilderness trails to bikers in the future. It's clear that mountain bikers don't want to destroy trails any more than hikers do. The surest way to preserve America's wilderness areas is to establish strong cooperative bonds among the hikers and bikers, as well as those who fish, hunt, camp, canoe, and bird-watch, and to encourage all to maintain the trails and respect the environment.

Works Cited

Buchanan, Rob. "Birth of the Gearhead Nation." Rolling Stone 9 July 1992: 80-85.

"Fearing for Desert, A City Restricts Mountain Bikes." New York Times 4 June 1995: A24.

Hanley, Robert. "Essex County Mountain Bike Troubles." New York Times 30 May 1995: B4.

JTYL (editor). Western New York Mountain Bike Association Home Page. Western New York Mountain Bike Association. <http://128.205.166.43/public/wnymba/wnymba.html>. (4 Oct. 1995).

Kelly, Charles. "Evolution of an Issue." Bicycling 31 (May 1990): 104-105.

Koellner, Ken (editor). New England Mountain Bike Association Home Page. 19 Aug. 1995. New England Mountain Bike Association. <http://www.ultranet.com/~kvk/nemba.html>. (30 Sep. 1995).

Newton, Carlton. Personal interview. 13 Nov. 1995.

Schwartz, David M. "Over Hill, Over Dale on a Bicycle Built for . . . Goo." Smithsonian 25.3 (June 1992): 74-84.

Sneyd, Ross. "Mount Snow Teaching Mountain Biking." Burlington Free Press 4 Oct. 1992: E1.

ぺ SUGGESTIONS FOR WRITING AND RESEARCH

Individual

1. Write a position paper on the issue you have been working with in Writings 2–6. Follow the guidelines suggested in this chapter, using as much research as you deem appropriate.
2. Write a position paper on an issue of particular interest to your writing class. Consider topics such as (a) student voice in writing topics, (b) the seating plan, (c) the value of writing groups versus instructor conferences, or (d) the number of writing assignments.
3. Revise your argumentative paper to submit to your student newspaper, either as a special feature or as an opinion piece for the editorial page. Consider carefully any special appeal, argument, or word choice that would make this paper more convincing to your college audience.

Collaborative

1. In teams of two or three, select an issue; divide up the work so that each group member contributes some work to (a) the context, (b) the pro argument, and (c) the con argument (to guarantee that you do not take sides prematurely). Share your analysis of the issue with another group and receive feedback. Finally, write your position papers individually.
2. Follow the procedure for the first collaborative assignment, but write your final position paper collaboratively.

Interpreting Texts

When I read, I've learned to ask a lot of questions, such as who's telling the story?
What's the character like? Why does a certain action happen? What do symbols mean?
Things like that. But I never find as much meaning in the stories as my teachers do.

DIANE

To interpret a text is to explain what it means. To interpret a text also implies that the text can be read in more than one way—your interpretation is your reading; others may read it differently. The word *text* implies words, writing, books; however, virtually all works created by human beings can be considered as texts open to interpretation—films, music and dance performances, exhibits, paintings, photographs, sculptures, advertisements, artifacts, buildings, and even whole cultures. Perhaps the most popular forms of interpretive writing are published reviews of movies, music, books, and the like.

In college, the most common form of interpretive essay assignment is to write analytical essays about reading assignments in humanities and social science courses. Since words can mean more than one thing, texts composed of written words have multiple meanings: They can mean different things depending on who is reading them, and there is no one right answer.

To find out what a poem, essay, play, or story means, you need to hear it, look at its language, examine how it is put together, compare it with similar things, notice how it affects you, and keep asking why.

ॐ WRITING TO INTERPRET

The best texts to select for an interpretive assignment are those that are most problematic—texts whose meaning seems to you somewhat slippery and elusive—since these give you, the interpreter, the most room to argue one meaning against another. Your job is to make the best possible case that your interpretation is reasonable and deserves attention.

A typical assignment may be to interpret a poem, story, essay, or historical document—a complex task that draws upon all of your reasoning and writing skills: You may have to describe people and situations, retell events, and define key terms, analyze passages, and explain how they work, perhaps by comparing or contrasting the text with others. Finally you will argue for one meaning rather than another—in other words, develop a thesis, and defend this thesis with sound reasoning and convincing evidence.

This chapter explores numerous ways of developing textual interpretations, using for illustrative purposes Gwendolyn Brooks's poem "We Real Cool." Brooks's poem is especially useful because it is short, quickly read, yet full of potential meanings. Our questions about this text, as well as the strategies for finding answers, are virtually the same as we would use with any text—fiction, nonfiction, or poetry.

Read, now, the following poem by Gwendolyn Brooks, and follow along as we examine different ways of determining what it means.

We Real Cool

> THE POOL PLAYERS.
> SEVEN AT THE GOLDEN SHOVEL.

We real cool. We
Left school. We

Lurk late. We
Strike straight. We

Sing sin. We
Thin gin. We

Jazz June. We
Die soon.

✐ WRITING 1

After reading "We Real Cool," freewrite for ten minutes to capture your initial reaction. Ask yourself questions such as these: What did it remind me of? Did I like it? Do I think I understand it? What emotions did I feel?

✐ EXPLORING A TOPIC

A good topic for an interpretive essay addresses a question that has several possible answers. If you think the text is overly simple, you will have no real need to interpret it. In addition, choose a topic that interests or intrigues you; if it doesn't, chances are it won't interest or intrigue your readers either.

No matter what text you are interpreting, however, you need to figure out what it means to you before you can explain it well to someone else. Plan to read it more than once, first to understand what it's like, where it goes, what happens literally. As you read, mark passages that interest or puzzle you. Read the text a second time, more slowly, making marginal notes or journal entries about the interesting, questionable, or problematic passages. As you do this, look for answers and solutions to your previous concerns, rereading as many times as necessary to further your understanding. In selecting a text to interpret, ask yourself these questions:

Can this text be read in more than one way?
What are some of the different ways of reading it?
With which reading do I most agree?
Where are the passages in the text that support this reading?
Whom does my interpretation need to convince? (Who is my audience?)

ᴄᴏ **WRITING 2**

Select a text that you are interested in interpreting. Choose a work of fiction, nonfiction, poetry, or drama that you find interesting and enjoyable, yet that has about it elements you do not fully understand, and then write out the answers to the questions above. Do not at this time worry about developing any of these answers thoroughly.

ᴄᴏ **IDENTIFYING INTERPRETIVE COMMUNITIES**

How you read and interpret a text depends on who you are. Who you are depends on the influences that have shaped you—the communities to which you belong. All of us belong to many communities: families, social and economic groups (students or teachers, middle or working class), organizations (Brownies, Boy Scouts, Democrats, Masons), geographic locales (rural or urban, North or South), and institutions (school, church, fraternity). Your membership in one or more communities determines how you see and respond to the world.

The communities that influence you most strongly are called *interpretive communities;* they influence the meaning you make of the world. People who belong to the same community that you do are likely to have similar assumptions and therefore likely to interpret things as you would. If you live in an urban black community, jazz and rap music may be a natural and constant presence in your life; if you live in a rural white community, country and western music may be the norm; at the same time, as a member of either group you might also belong to a larger community that surrounds itself with classical music. All this means is that people who belong to different communities are likely to have different—not better or worse—perspectives from yours.

Before writing an interpretive essay, it is helpful to ask, "Who am I when writing this piece?" You ask this to examine the biases you bring to your work, for each of us sees the world—and consequently texts—from our own particular vantage point. Be aware of your age, gender, race, ethnic identity, economic class, geographic location, educational level, political or religious persuasion. Ask to what extent any of these identities emerges in your writing.

College is, of course, a large interpretive community. Various smaller communities exist within it called disciplines—English, history, business, art, and so on. Within any discipline there are established ways of interpreting texts. Often when you write an interpretive essay, you will do so from the perspective of a traditional academic interpretive community. Take care to follow the conventions of that community, whether you are asked to write a personal interpretation or an analytical interpretation.

Whether or not you deliberately identify yourself and your biases in your essay depends on the assignment you are given. Some assignments ask you to remove your personal perspective as much as possible from your writing; others ask that you acknowledge and explain it, while still others fall somewhere in between.

↪ WRITING 3

Make a list of all the interpretive communities to which you belong. Annotate your list, putting an asterisk next to those that seem to have the greatest influence on how you think or act and note how this influence manifests itself.

Personal Interpretation

In writing from a personal or subjective perspective, the interpreter and his or her beliefs and experiences are part of the story and need to be both expressed and examined. In examining "We Real Cool," for instance, you may bring your background into your writing to help your reader understand why you view the poem as you do. You may compare or contrast your situation to that of the author or characters in the text. Or you might draw upon particular experiences that cause you to see the poem in a particular way. For example, Mitzi Fowler's response (reprinted at the end of this chapter) begins with memories inspired by the poem:

> Gwendolyn Brooks's "We Real Cool" is a sad poem. It reminds me of the gang in high school who used to skip classes and come back smelling of cigarette smoke and cheap liquor—not that I knew it was cheap back then.

It is increasingly common for good interpretive essays to include both analytical and personal discussions, allowing you to demonstrate your skill at closely reading texts while acknowledging your awareness of the subjective nature of virtually all interpretive acts. To move in a more analytic direction, Mitzi would need to quote and discuss more lines directly from the poem, as she does later in her essay:

> These "cool" dropouts paid for their rebellion in drug overdoses, jail terms, police shootouts, and short lives. They "Die soon," so we never know where else their adventurous spirits might have taken them.

Analytical Interpretation

In analyzing a text, writers often focus on the content and deliberately leave themselves, the interpreters, in the background, minimizing personal presence and bias. If you are asked to write this way—to avoid first-person pronouns or value judgments—do your best to focus on the text and avoid language that appears biased. In reality, of course, authors reveal their presence by the choices they make: what they include, what they exclude, what they emphasize, and so on. But when you are aware of your inescapable subjectivity, aware that your own situation affects the inferences and judgments you make about others, this awareness will help you keep your focus on the subject and off yourself.

In writing about "We Real Cool," for example, your first reaction may be more personal than analytical, focusing on your own emotions by calling it a sad poem, as Mitzi does, or by expressing value judgments about the poem's characters:

> I think these guys are stupid, cutting their lives short drinking, stealing, and fighting.

However, a more analytical response would be to drop the first person ("I think") and the judgment ("these guys are stupid") and to focus more closely on the text itself, perhaps quoting parts of it to show you are paying close attention:

> The speakers in the poem, "The Pool Players," cut their lives short by hanging out at "The Golden Shovel," fighting, drinking, stealing, and perhaps worse.

There is no formula for arriving at or presenting an interpretation in essay form, but readers, especially English instructors, will expect you to address and explain various elements of the text that usually contribute substantially to what it means.

✌ WRITING 4

Examine more closely the text you have selected to interpret. In one paragraph, write out your initial and personal impressions of what this text means to you—why it interests you. In another paragraph, list those elements or features of the text that pose problems or raise questions that you'd like to answer.

✌ DEVELOPING AN INTERPRETATION

Convincing interpretive essays commonly, but not always, include the following information:

1. An overview of the text, identifying author, title, and genre and briefly summarizing the whole text
2. A description of form and structure
3. A description of the author's point of view
4. A summary of the social, historical, or cultural context in which the work was written
5. An assertion or thesis about what the text means—your main business as an interpreter

Your thesis is a clear, concise statement that identifies your interpretation, on which the readers then expect you to elaborate. While some interpretive essays include all of the above information, essays more commonly will emphasize some aspects while downplaying others.

Identify and Summarize the Text

All interpretive essays should begin by answering basic questions: What genre is this text—poem, play, story, or essay? What is its title? Who is the author? When was it published? In addition, all such essays should provide a brief summary of the text's story, idea, or information. Summarize briefly, logically, and objectively to provide a background for what else you plan to say about the text, as in this example:

> "We Real Cool," a poem by Gwendolyn Brooks, condenses the life story of pool-playing high school dropouts to eight short lines and foreshadows an early death on the city streets.

Explain the Form and Organization

To examine the organizational structure of a text, ask: How is it put together? Why start here and end there? What connects it from start to finish? For example, by repeating words, ideas, and images, writers call attention to them and indicate that they are important to the meaning of the text. No matter what the text, some principle or plan holds it together and gives it structure. Texts that tell stories are often organized as a sequence of events in chronological order. Other texts may alternate between explanations and examples or between first-person and third-person narrative. You will have to decide which aspects of the text's form and organization are most important for your interpretation. The following example pays close attention to Brooks's overall poetic structure:

> The poem consists of a series of eight three-word sentences, each beginning with the word "We." The opening lines "We real cool. We / left school" explain the characters' situation. The closing lines "We / Jazz June. We / Die soon" suggest their lives will be over soon.

Describe the Author's Perspective

Authorial perspective is the point of view from which the text is presented. In an article, essay, textbook, or other work of nonfiction, you can expect the author to write about truth as he or she sees it—just as we are doing in this textbook, trying to explain writing according to our own beliefs about writing. However, in a work of poetry, fiction, or drama the author's point of view may be quite different from that of the character(s) who narrate or act in the story. If you can describe or explain the author's perspective in your interpretive essay, you provide readers with clues about the author's purpose. For example, Kelly Sachs's essay (reprinted at the end of this chapter) opens by making a distinction between Brooks, the poet, and her characters, "Seven at the Golden Shovel":

> Gwendolyn Brooks writes "We Real Cool" from the point of view of members of a street gang who have dropped out of school to live their lives hanging around pool halls—in this case "The Golden Shovel." These guys are semiliterate and speak in slangy street lingo that reveals their need for mutual support in their mutually rebellious attitude toward life.

What he doesn't say, but clearly implies, is that Brooks herself is a mature and highly skilled user of formal English, and that in the poem she adopts the persona or mask of semiliterate teenagers in order to tell their story more effectively.

Place the Work in Context

What circumstances (historical, social, political, biographical) produced this text? How does this text compare or contrast with another by the same author or a similar work by a different author? No text exists in isolation. Each was created by a particular author in a particular place at a particular time. Describing this context provides readers with important background information and indicates which conditions you think were most influential. "We Real Cool" could be contextualized this way:

> "We Real Cool" was published in 1963, a time when the Civil Rights Movement was strong and about the time that African Americans coined the phrase "Black is beautiful." The poem may have been written to remind people that just because they were black did not mean they necessarily led beautiful lives.

Explain the Theme of the Text

In fiction, poetry, and reflective essays, the main point usually takes the form of an implicit theme, which in academic writing you might call a thesis, either stated or unstated. A main reason for writing an interpretive essay is to point out the text's theme. Examine what you think is the theme in the text. Ask yourself: So what? What is this really about? What do I think the author meant by writing this? What problems, puzzles, or ideas seem interesting? Good topics arise from material in which the meaning is not obviously stated.

When you write an interpretation from an objective or analytical perspective, you make the best case possible that, according to the evidence in the text itself, this is what the text means. In analytical writing, you generally state your thesis (about the text's theme) early in the essay, as Kelly does here about "We Real Cool" at the end of his first paragraph:

> The speakers in the poem, "We," celebrate what adults would call adolescent hedonism but they make a conscious choice for a short intense life over a long, safe, and dull existence.

When you write from a subjective or personal perspective, you make it clear that your interpretation is based on your own emotional reactions and memories as much as on the content of the text, and that other readers will necessarily read it differently. The controlling idea or theme of Mitzi's subjective interpretation of "We Real Cool" is revealed in her first sentence, which the rest of the essay supports.

> "We Real Cool" . . . reminds me of the gang in high school who used to skip classes and come back smelling of cigarette smoke and cheap liquor—not that I knew it was cheap back then.

Support Your Interpretation

Analytical interpretations are usually built around evidence from the text itself: Summarize larger ideas in your own language to conserve space; paraphrase more specific ideas also in your own words; and quote directly to feature especially colorful or precise language. If you include outside information for supportive, comparative, or contrastive reasons, document carefully where it came from. Most of the preceding examples referred to specific lines in the poem, as does this passage from Kelly's essay:

> Instead of attending school or finding employment, these seven "Lurk late," "Strike straight," "Sing sin," "Thin gin," and "Jazz June"—actions that are illegal, frowned upon by society, or harmful to other people. This is a bunch of kids to watch out for. If you see them coming, cross the street.

Subjective interpretations also include textual evidence, but often passages from the text are cited as prompts to introduce the writer's own memories, associations, or personal ideas. The more specific and concrete your examples, the better.

> I think everybody who ever went to a public high school knows these guys—at least most of them were boys—who eventually "Left school" altogether and failed to graduate. They dressed a little differently from the rest of us—baggier pants, heavier boots, dirtier shirts, and too long hair never washed. And if there were girls—too much makeup or none at all.

ᴈ **WRITING 5**

Look once more at the text you plan to interpret, and make brief notes about each of the elements described in this section. Which elements do you already know something about? Which ones require further research?

Interpreting Different Genres

The brief examples we've been looking at in this chapter are all based on a single short poem. Many of the texts you may choose to interpret will be longer and in genres other than poetry. If you are interpreting a work of fiction or nonfiction, or something else altogether such as an art exhibit, a concert, or a film, the basic elements of interpretation discussed here still apply, but there will be important differences.

Poetry

Of all language genres and forms, poetry exhibits the most intensive, deliberate, and careful use of language. With certain exceptions, poetic texts are far shorter than even short stories or one-act plays. Consequently, when interpreting poems, pay special attention to specific words, phrases, and lines; quote lines, phrases, and words exactly to support your points; and familiarize yourself with the basic poetic terms you learned in high school: line, stanza, rhyme, rhythm, meter, metaphor, and image.

Fiction

To write about a novel or short story, explain how the main elements function: the narrator (who tells the story), plot (what happens in the story), one or more characters (who are acting or being acted upon), setting (where things are happening), and theme (the meaning of the story). Be sure to keep in mind that the author who writes the story is different from the characters in the story, and that what happens in the story is different from the meaning of the story.

Drama

Plays are a special kind of fiction meant to be acted out upon a stage; consequently, all the elements of fiction apply, except that since the characters are acting out the story, a narrator is seldom present. In drama, the setting is limited to what is contained on the stage. You usually do not have access to what the characters are thinking—the actors need to speak out loud or with body language to reveal themselves, so the actors who play the parts are crucial to the play's success; consequently, if you interpret a play that you see performed, you have more information to work with than a play listened to or read silently. Remember, too, that plays are structured according to acts, which are, in turn, divided into scenes. When you quote from drama, identify first the act (V), then the scene (iii), finally the line or lines, for example, *Hamlet* V.iii, 24–29.

Movies

Fictional films are discussed in terms similar to those used for written fiction and drama (plot, character, setting, theme). However, additional elements also come into play: camera angles, special effects, and unlimited settings. Because of the complexity of orchestrating all of these elements, the film director rather than the screenwriter is often considered the "author" of a film.

Nonfiction

The elements of a nonfiction essay are similar to those of a fictional story, except that everything in the text is supposed to have really happened. For this reason, the author and the narrator of the story are one and the same. Informational nonfiction—essays, reports, and textbook chapters—is also meant to be believed; here you might say that "ideas" and "arguments" must be strong and well supported to be believed. When interpreting nonfiction, pay special attention to the author's theme or thesis, and to whether it is well supported or not. Note, too, any interesting developments in tone, style, form, or voice.

Other Media

In interpreting other kinds of texts—paintings, photographs, sculptures, quilts, concerts, buildings, and so on—always explain to your reader the basic identifying features, be they verbal, visual, musical, or something else: What is it? What is its name or title? Who created it? What are its main features? Where is it? When did it take place?

❧ GLOSSARY OF LITERARY TERMS

In writing interpretations of literary texts, the following terminology is commonly used:

Alliteration   The repetition of initial consonant sounds. ("On the bald street breaks the blank day.")

Antagonist   A character of force opposing the main character (the *protagonist*) in a story.

Climax   A moment of emotional or intellectual intensity or a point in the plot where one opposing force overcomes another and the conflict is resolved.

Epiphany   A flash of intuitive understanding by the narrator or a character in a story.

Figurative language   Language that suggests special meanings or effects such as metaphors or similes; not literal language. ("She stands like a tree, solid and rooted.")

Imagery   Language that appeals to one of the five senses; especially language that reproduces something the reader can see. ("His heart was an open book, pages aflutter and crackling in the wind.")

Metaphor   A direct comparison between two things. ("He is a fox.")

Narrator   Someone who tells a story; a character narrator is a part of the story (such as Huckleberry Finn) while an omniscient narrator tells a story about other people.

Persona   A mask, not the author's real self, worn by the author to present a story or poem.

Plot   The sequence of events in a story or play.

Point of view   The vantage point from which a story or event is perceived and told.

Protagonist   The main character or hero of a plot.

Rhyme   The repetition of sounds, usually at the ends of lines in poems, but also occurring at other intervals in a line (*moon, June, noon*).

Rhythm   The rise and fall of stressed sounds within sentences, paragraphs, and stanzas.

Simile   An indirect comparison using the words *as* or *like*. ("He is like a fox.")

Symbol   An object that represents itself and something else at the same time. A black rose is both a rose of a certain color and the suggestion of something evil or deathlike.

Theme   The meaning or thesis of a text.

❧ SHAPING THE WHOLE PAPER

This chapter concludes with two sample interpretive papers from which some of the foregoing illustrations have been taken. The first one might be called an *objective* or *critical* essay, while the other might be called a *subjective* or *personal* essay.

Analytic Response to "We Real Cool"

Kelly Sachs writes a brief analytical interpretation of "We Real Cool," called "High Stakes, Short Life," keeping himself in the background, writing in the third-person point of view. He presents his thesis early and supports it afterward with frequent quotations from the text, amplifying and explaining it most fully in his last paragraph.

High Stakes, Short Life
Kelly Sachs

Gwendolyn Brooks writes "We Real Cool" (1963) from the point of view of the members of a street gang who have dropped out of school to spend their lives hanging around pool halls—in this case "The Golden Shovel." These guys are semiliterate and speak in slangy street lingo that reveals their need for mutual support in their mutually rebellious attitude toward life. The speakers in the poem, "We," celebrate what adults would call adolescent hedonism—but they make a conscious choice for a short intense life over a long, safe, and dull existence.

For the "Seven at the Golden Shovel," companionship is everything. For many teenagers, fitting in or conforming to a group identity is more important than developing an individual identity. But for these kids, none of whom excelled at school or had happy home lives, their group *is* their life. They even speak as a group, from the plural point of view, "We," repeated at the end of each line; these seven are bonded and will stick together through boredom, excitement, and death.

From society's point of view, they are nothing but misfits—refusing to work, leading violent lives, breaking laws, and confronting polite society whenever they cross paths. Instead of attending school, planning for their future, or finding work, these seven "Lurk late," "Strike straight," "Sing sin," "Thin gin," and "Jazz June." Watch out for this bunch. If you see them coming, cross the street.

However, the most important element of their lives is being "cool." They live and love to be cool. Part of being cool is playing pool, singing, drinking, fighting, and messing around with women whenever they can. Being cool is the code of action that unites them, that they celebrate, for which they are willing to die.

The poet reveals their fate in the poem's last line. Brooks shows that the price of coolness and companionship is higher than most people are willing to pay. In this culture, the fate of rebels who violate social norms is an early death ("We / Die soon"), but to these seven, it is better to live life to the fullest than hold back and plan for a future that may never come. They choose, accept, and celebrate their lives, and "Die soon."

Personal Response to "We Real Cool"

In the following essay, "Staying Put," Mitzi Fowler writes about her personal reaction to Brooks's poem, describing how it reminds her of her own high school experience. Mitzi's theme, the sad lives of ghetto gangs, opens and closes the essay and provides the necessary coherence to hold it together. While she quotes the text several times, her primary supportive examples come from her own memories.

Staying Put

Mitzi Fowler

Gwendolyn Brooks's "We Real Cool" is a sad poem. It reminds me of the gang in high school who used to skip classes and come back smelling of cigarette smoke and cheap liquor—not that I knew it was cheap back then. I think everybody who ever went to a public high school knows these guys—at least most of them were boys—who eventually "Left school" altogether and failed to graduate. They dressed a little differently from the rest of us—baggier pants, heavier boots, dirtier shirts, and too long hair never washed—and if there were girls, too much makeup or none at all.

They had their fun, however, because they stayed in their group. They came late to assemblies, slouched in their seats, made wisecracks, and often ended up like the characters in *The Breakfast Club,* in detention after school or Saturday morning.

And no matter how straight-laced and clean-cut the rest of us were, we always felt just a twinge of envy at these careless, jaunty rebels who refused to follow rules, who didn't care if they got detentions, who didn't do homework, and whose parents didn't care if they stayed out all night. I didn't admit it very often—at least not to my friends—but some part of me wanted to have their pool hall or whatever adventures, these adult freedoms they claimed for themselves. However, I was always afraid—chicken, they would have said—of the consequences, so I practiced piano, did my algebra, and stayed put.

Then I think of the poem's last line and know why I obeyed my parents (well, most of the time), listened to my teachers (at least some of them), and stayed put (if you don't count senior cut day). These "cool" dropouts paid for their rebellion in drug overdoses, jail terms, police shootouts, and short lives. They "Die soon," so we never know where else their adventurous spirits might have taken them. In the end, this poem just makes me sad.

✌ SUGGESTIONS FOR WRITING AND RESEARCH

Individual

1. Write an interpretive essay about a short text of your choice. Write the first draft from an analytic stance, withholding all personal judgments. Then write the second draft from a personal stance, including all relevant private judgments. Write your final draft by carefully blending elements of your first and second drafts.

2. Locate at least two reviews of a text (book, recording, exhibit) with which you are familiar, and analyze each to determine the reviewer's critical perspective. Write your own review of the text, and agree or disagree with the approach of the reviewers you analyzed. If you have a campus newspaper, consider offering your review to the editor for publication.

Collaborative

As a class or small group, attend a local concert, play, or exhibition. Have each student take good notes and, when he or she returns home, write a review of this event that includes both an interpretation and a recommendation that readers attend it (or not). Share these variations on the same theme with others in your class or group and explore the different judgments that arise as a result of different perspectives.

Chapter

ᎧᎧ 14 ᎧᎧ

Reflecting On the World

Imagine and create. Never be content with just retelling something. And never be content with your first telling. Dive deeply into your mind and give the reader something to dream about.

CHRISSIE

Reflective writing raises questions about any and all subjects, reflecting these subjects back to readers, allowing a clearer view and exposing new dimensions, as if a mirror had been held up before them. Reflective writing allows both writer and reader to consider things thoughtfully and seriously, but it demands neither resolution nor definitive answers.

What I am calling reflective essays are, perhaps, the closest modern form to the kind of writing that began the essay tradition some four hundred years ago when French author Michel de Montaigne first published his *Essais* in 1580. An *essai* (from the French, meaning "an attempt") was a short piece of writing meant to be read at a single sitting on a subject of general interest to a broad spectrum of citizens. Montaigne's essays explored education, truth, friendship, cruelty, conversation, coaches, cannibals, and many other subjects. Essays were never intended to be the last word on a subject, but rather, perhaps, the first thoughtful and speculative word. E. B. White, Virginia Woolf, and George Orwell are among the best-known essayists in the twentieth century; among the best-known living essayists are Joan Didion, Russell Baker, and Ellen Goodman.

ᎧᎧ WRITING TO REFLECT

When you write to reflect, part of your motivation is to figure out something for yourself in addition to sharing your reflections with somebody else. A piece of reflective writing is both *the result of* and *an account of* the act of speculation, and it invites readers into the game of reflection, speculation, and wonder as well. Such writing is commonly characterized by a slight sense of indirection—as if the writer were in the actual process of examining a subject closely for the first time. This appearance of spontaneity is, of course, an illusion, as most good reflective writing—unlike journal writing—has been thoroughly rewritten, revised, and edited to achieve just the right tentative and spontaneous tone.

It is tone along with purpose that distinguishes reflective writing from other types of writing. In reflective essays, writers ask *why*? Why do people live and behave the way they do? Why does society develop this way rather than that? Why does one thing happen rather than another? Why do I value what I value? In asking such questions, and offering possible answers, writers try to make sense of the world.

Reflective writing can be on any subject under the sun (or moon, for that matter). However, unlike in explanatory or argumentative writing, the subject that causes the reflection in the first place is seldom the actual focus or topic of the essay. The topic of reflective writing is the meaning the writer finds in the subject. In other words, the writer observes or experiences something specific, then wonders about possible meanings or implications. For example, visiting the library (subject) could stimulate a reflection on knowledge and creativity (topic). The everyday experience comes to stand for larger issues and ideas.

Unlike many college writing assignments, reflective essay assignments ask for your opinion. Assignments that contain a direction word such as *imagine, speculate,* or *discuss* invite a kind of writing that features your ability to see a given subject from several sides and to offer tentative answers to profound questions. For these assignments, it may be important to include factual information; however, such information is background, not foreground, material. While reflecting, writers may narrate a story, explain, interpret, or argue while reflecting on something; however, it is not the story, explanation, interpretation, or argument that is foremost on their minds.

Reflective essays are as varied as the thought processes of the people who write them, but a typical pattern does exist. Many reflective essays describe a concrete subject or actual situation, then move on to deeper or more abstract matters that reveal what's really on the writer's mind.

Reflective essays *are* about something that's on your mind, perhaps something bothersome, distracting, or intriguing to you—but you may not always be conscious of what that is. No matter where or with what you start, try to bring that concern into focus—mirror it—and allow both yourself and your reader to look at it in a steady light.

Before writing a reflective essay of your own, you might read a few examples of good reflective writing, such as Montaigne's early essays or Stephen J. Gould's contemporary ones. Read newspaper columnists such as Ellen Goodman and Russell Baker. Browse among the *New York Times* best-selling nonfiction in a bookstore, looking especially at titles that pique your curiosity.

➷ WRITING 1

Describe your past experiences writing reflective or exploratory essays. Did you enjoy writing them? How did readers respond to them?

➷ FINDING AND DESCRIBING A SUBJECT

The subject of a reflective essay is often something concrete, observable, or even superficial that prompts you to further, deeper, more profound thought. It can be any *person,* any *place,* any *thing*—in other words, everything imaginable is a possibility. To find a subject, you may want to start by remembering something that has always interested or puzzled you, even if you don't know why—finding out is part of the process. You may want to review old journal entries, do some freewriting, or use other discovery techniques to find a possible subject; looping and clustering may help you decide whether a subject is a fertile one for reflection.

Often, reflective essays begin slowly or casually, the writer paying specific attention to something immediate or commonplace—a recent rock concert, a specific case of plagiarism, an idea raised in a book or lecture. Writers try to interest us in the concrete or particular before moving toward the more abstract or general. When beginning a reflective essay, record as much detail as you can, appealing to all five senses. Try to make your subject visible and tangible before asking us to consider its more abstract dimensions and associations.

In your final draft, you will want details that foreshadow your ultimate point without giving it away too soon: a mother's anxious expression leads into an essay on domestic violence, the spring fashions in a department store window anticipate a later reflection on annual rebirth, and so on. You will continue the careful layering of details throughout the revising and editing stages. As you start composing a reflective essay, however, concentrate on collecting impressions that seem particularly interesting or meaningful, even if you don't know exactly what they mean.

People

Chapter 10 discusses writing profiles of people, which requires getting to know subjects by interviewing, doing background research, visiting their homes, and so on. However, for subjects of reflective essays, brief and casual encounters with strangers may work as well as more developed relationships. In fact, reflecting on chance encounters almost guarantees speculative and inconclusive thoughts: Why is she doing that? Where does he live? How is their life like mine? Unlike profile writing, where the writer seeks real answers to such questions from the subject, the reflective author turns the encounter into a prompt to ponder other subjects. Reflective writing about people often leads to comparisons with the writer's own circumstances, character, or behavior. What makes such writing especially interesting, of course, is that no two writers who encountered or witnessed the same person would speculate, muse, or wonder in the same way. In the following passage, Mari writes about a chance encounter one Saturday morning:

> The sign on the front door says the store opens at eight o'clock. It's 7:56, so I put the bag of bottles on the ground.
>
> An old woman with a shopping cart full of bottles stands in front of Pearl Street Beverage Mart. Most of her long gray hair is tucked into a Red Sox baseball cap, but some of it hangs in twisted strands about her face. She wears an oversized yellow slicker, a striking contrast against the crystal blue morning sky. She looks at the bottles and her lips are moving as if she's telling them a story.
>
> Then the bottle lady turns and speaks to me: "These bottles in my cart here, you see which I mean? Well, I'm gonna get five hundred dollars for them and get me a fine stylin' dinner tonight. It's a good thing there's bottles, yes?"

In this early part of her essay, Mari first focuses on locating the circumstance in a particular time and place and on describing the appearance of the woman. She provides enough detail so that the readers can see the woman and, perhaps, join the writer in asking questions: Why is she wearing a raincoat when the sky is blue? Why is she speaking to the bottles?

✒ WRITING 2

Make a list of people whom you have encountered or observed within the past few weeks and who have, for one reason or another, made a sharp impression on you. Did any of them trigger reflective thoughts? Do they now? What are those thoughts?

Places

Places make good reflective subjects; writers can generally return to a place again and again for further information and ideas. You can, of course, reflect on just about any place under the sun, including those in your memories to which you cannot return. But you can also turn to places right under your nose and make something interesting and reflective from those too. Reflective essays that begin by describing a physical place often end up focusing on places of the heart, mind, and spirit.

In the following passage, Judith turned the university library into an object for reflective thought. The first part of the essay is narrative. We follow Judith as she locks her apartment door and hurriedly walks to the library. This description locates her place in a particular time and circumstance. Once at the library, she slows her self and her essay down.

> Inside the smoke-colored doors, the loud and busy atmosphere vanishes, replaced by the soft soothing hum of air conditioning and the hushed sound of whispering voices. The repetitive sound of the copy machine has a calming effect as I look for a comfortable place to begin my work.
>
> I want just the right chair, with a soft cushion and a low sturdy table for a leg rest. The chairs are strategically positioned with comfortable personal space around each one, so you can stretch your arms fully without touching a neighbor. . . . People seem to respect each other's need for personal space.

Like her search for the right chair, Judith's description is detailed and concrete. At the same time, she introduces the importance of quiet space to her peace of mind. She concludes with a generalization triggered by the space she describes, hinting at but not stating the topic of her reflection.

✒ WRITING 3

Make a list of places that would provide you with material for writing a reflective essay. Think about places you've visited in the recent past and those you still have access to. What associations do these places conjure up? What questions do they raise? Write a paragraph about each of three places on your list and see which offers the best prospect for further writing.

Things

Any thing (object, idea, occurrence) can be a subject for reflection, so long as you treat it accordingly. Perhaps the first things that come to mind are physical objects. But a very different kind of a thing is an event, whether public (a political rally, a basketball game) or private (drinking a cup of coffee, fishing for trout). And a still more abstract kind of thing is a concept (plagiarism, campus parking). In other words, to reflect on a thing allows you to reflect on virtually anything! Scott opens his essay with a generalized statement about water fountains:

> There is something serene about the sound of a water fountain. The constant patter of water splashing into water. The sound of endless repetition, the feeling of endless cycle. . . .

Two paragraphs later, he introduces us to the campus water fountain:

> The fountain pipe has eighty-two nozzles. Two weeks ago, only fifty-five of them were spraying water vigorously, eleven sprayed weakly, and sixteen didn't spray at all. This week eighty-one are spraying; someone has fixed the fountain.

Even though Scott starts with a preliminary generalization, he quickly gives the reader a careful, even technical description. The implication that Scott is writing from direct observation at a particular time and place roots the description in a specific circumstance. Readers are likely to believe Scott's reflection because he has taken pains to be accurate in his observation, and so they are likely to believe the rest of his account as well. While Scott has made it clear that listening to fountains is peaceful for him, in the early paragraphs of his essay readers still do not know why he chooses to write about them.

↪ WRITING 4

Make a list of things that interest you, cause you to pause, or raise questions. Do so either from memory or by walking around your room, house, neighborhood, or campus. Write a paragraph or so on three of the things you encounter and see which offers the best prospect for further writing.

↪ STEPPING BACK

The three reflective examples we have seen so far have taken everyday people, places, and things—a bottle woman, a library, a water fountain—as the subjects on which to reflect. The advantage of common over spectacular subjects is that everyone has some experience with them and is automatically curious to see what you make of them. At the same time, you need to make sure your reflection itself goes beyond the commonplace and introduces readers to new dimensions or perspectives. Your subject alone will carry interest just so far: the rest is up to your originality, creativity, and skill as you present the topic of your essay, the deeper and more speculative meaning you have found.

After introducing a subject, essayists often step back and pause, saying to the reader, in effect, "Wait a minute, there's something else going on here—stop and consider." Such pauses are a sign that connections are being made—between the present and the past, the concrete and the abstract, the literal and the symbolic. It signals to the reader that the essay is about to move in a new and less predictable direction. Sometimes, pauses are accompanied by slight shifts in voice or tone as the writer moves in closer to what's really on his or her mind. The exact nature of these shifts will, of course, be determined by the point the writer wants to make.

After describing the bottle woman in some detail, but without judgment, Mari steps back and shifts from the woman to the common objective that has brought them both to the store, returning bottles for a refund.

> I guess I had never thought about it before, but the world is full of bottles. They're everywhere, on shelves, in trash cans, on park benches, behind bushes, in street gutters. In the modern city, bottles are more common than grass.

By calling attention to her own new thoughts about bottles, Mari causes her readers, too, to think about a world full of bottles and about what, if anything, that may mean.

In the following passage, Scott stops observing the campus water fountain and begins wondering and remembering:

> I'm not sure where my love of fountains comes from. Perhaps it's from my father. When I was very young he bought a small cement fountain for our backyard. Its basin was an upturned shell and there was a little cherubic boy who peed into it.

Scott steps back and reveals that it is not, in fact, water fountains that are most on his mind, but memories of his father. He now moves away from observing the campus fountain in the present toward memories of fountains in the past.

✌ WRITING 5

Select one of the topics you have explored in Writings 2–4. Once you have firmly and concretely described or defined it, create a reflective pause by writing one or two pages about what, besides itself, this person, place, or thing makes you think about.

✌ SO WHAT?

In explanatory, argumentative, or interpretive writing, the point of the paper is commonly stated as an explicit thesis, often in the first paragraph. However, in successful narrative or reflective writing, the point may be conveyed only indirectly and nearly always emerges at the end rather than the beginning of the essay. That way, readers are drawn into the act of reflecting themselves and become more and more curious to find out what the writer thinks. In other words, reflective writers are musing, exploring, or wondering rather than arguing. In fact, reflective essays are most persuasive when least obviously instructive or assertive.

Mari includes her observation of the bottle woman first by re-creating the shopkeeper's sarcastic response to the woman after he totals the deposit her bottles have earned her:

> "Eight dollars and thirty-five cents, Alice. You must have new competition or else bottles are getting scarce. You haven't broken ten dollars in days."

As readers perhaps suspect, Alice has grossly overestimated the value of the bottles in her cart. However, this transaction appears to have a transforming effect on the observer/writer, who now adds her own bottles to the old woman's pile.

> I realized then that my bottles were more nuisance than necessity. They were cluttering up our back hall—the landlord had already complained once—and tonight my roommates and I were making chicken curry for our boyfriends. That wasn't the case with Alice, who seemed to have neither roommate, dinner, nor boyfriend.
>
> "Here you go, Alice. Maybe you can break ten bucks with these." I gave them to her freely, at the same time realizing their true value for the first time.

Though Mari's essay begins with a narrative about a trip to the bottle store and an encounter with an old woman, in the end, she takes us somewhere else. When her essay concludes, Mari is still ignorant of any real details about the woman's life, history, or prospects—though she, like us, has made educated guesses. Instead, Alice remains frozen in the essay for all time as the "bottle woman." It is what Mari makes of this encounter, and in turn asks us to make of it, that provides the real topic of this essay. It makes us reflect rather than directs us to act, not hitting us over the head with an editorial against poverty or for state-mandated bottle refund policies. Rather, Mari reconsiders, if ever so briefly, her own relatively comfortable life as a college student with her whole future ahead of her, raising subtle questions for the rest of us about class, privilege, and well-being in a throwaway society.

Scott's essay about water fountains has progressed from fountains in general to the campus fountain in particular, then to the memory of fountains, which remind him of his father. Only in the last paragraph does he reveal even more:

> My father died just after my eighteenth birthday. It was very sudden and very surprising and everything felt very unfinished. They say I am a lot like him in many ways, but I'm not sure. What I do know is that like him, I love the sound of water.

Even at the end, Scott's point emerges by implication rather than explication—he never states outright why he began this reflection or what he hopes readers will take from it. Some readers will infer that Scott's reflection is on the nature of reflection itself, how observation stimulates memory and how memory itself becomes part of people's everyday reality. In other words, even very personal reflective writing can open new insights for readers.

His father's death proved to be an especially difficult topic for Scott to write about; in fact, he might not have chosen "water fountains" had he known in advance exactly where they were to lead him. Scott's early drafts focused on water and fountains and people sitting around them—but did not mention his father. The deliberate attention to mental processes that writing demands actually caused this particular reflection to emerge. A writer may select to write about an object with only a vague idea of why it's attractive, interesting, or compelling. Commonly it's only through the act of reflective writing that the writer finds out the nature of the attraction, interest, or compulsion.

Remember that reflective essays raise issues but do not necessarily resolve them. The tradition of essay writing is the tradition of *trying* and *attempting* rather than *solving* or *concluding*. Ending by acknowledging that you see still other questions invites readers to join in the search for still more answers.

At the very end of a reflective essay, writers often return to the specific person, place, or thing that prompted the reflection in the first place. This brings about a sense of closure— if not for the topic itself, at least for this particular essay.

> ꙮ **WRITING 6**
>
> Write two conclusions to the reflective essay you have been working on. In one, state your point openly near the end of your paper; in the other, let it emerge gradually without spelling it out.

ꙮ SUGGESTIONS FOR WRITING REFLECTIVE ESSAYS

1.   Begin your reflection with something concrete that interests you, such as a specific person, place, or thing.
2.   Describe your subject carefully and locate it in a specific circumstance, using sensory details where appropriate.
3.   At some point, pause and move deliberately away from the actuality of your subject to the larger or different issue it brings to mind—the topic of your reflection.
4.   Organize your essay to move from the concrete and specific toward the abstract and general.
5.   Use a reflective voice—tentative, questioning, gentle—to invite readers to share in your thinking.
6.   Let your point emerge, explicitly or implicitly, near the end of your essay.
7.   Conclude by reminding your readers where your reflective journey began.

ꙮ SHAPING THE WHOLE PAPER

The finished draft of Judith Woods's short reflective essay, "Writing in Safety," generally follows the structure described in this chapter. It has a loosely narrative pattern and is written in the present tense to convey a sense of the event unfolding as it is read. It is structured from start to finish by her walk to and from the library. Her essay does not, however, actually tell a story since nothing, at least in a physical sense, happens—unless you count walking and sitting down. The journey emerges as a mental, almost spiritual, quest for safety—safety in which to think and create without fear. At the same time, the physical dimensions of her journey and the attention to descriptive detail contribute to make her journey believable and to ground readers in a reality they, too, have most likely experienced.

The point of Judith's reflection is articulated in the paragraph before the concluding paragraph: "I am beginning to understand the importance of feeling safe in order to be creative and productive." The final paragraph, retracing her path home, returns us to the place where the essay began, suggesting that this reflection is over, at least for the time being.

Writing in Safety

Judith Woods

It is already afternoon. I fiddle with the key to lock the apartment door after me. I am not accustomed to locking doors. Except for the six months I spent in Boston, I have never lived in a place where I did not trust my neighbors. When I was little, we couldn't lock our farmhouse door; the wood had swollen and the bolt no longer lined up properly with the hole, and nobody ever bothered to fix it. I still remember the time our babysitter, Rosie, hammered the bolt closed and we had to take the door off the hinges to get it open.

I heft the book bag on my shoulder and walk up College Street toward the library. The morning is clear and cold and the colorful leaves look like confetti on the sidewalk. As I pass and am passed by other students, I scrutinize everything around me, hoping to be struck with a creative idea for a topic for my English paper. Instead, my mind fills with a jumble of disconnected images, like a bowl of alphabet soup: the letters are there, but they don't form any words. Campus side-walks are not the best places for creativity to strike.

Approaching the library, I see skateboarders and bikers weaving through students who talk in clusters on the library steps. A friendly dog is tied to a bench watching for its owner to return. Subjects to write about? Nothing strikes me as especially interesting, and besides, my heart is still pounding from the walk up the hill. I wipe my damp forehead and go inside.

Inside the smoke-colored doors, the loud and busy atmosphere vanishes, replaced by the soft soothing hum of air conditioning and the hushed sound of whispering voices. The repetitive sound of the copy machine has a calming effect as I look for a comfortable place to begin my work.

I want just the right chair, with a soft cushion and a low sturdy table for a leg rest. The chairs are strategically positioned with comfortable personal space around each one, so you can stretch your arms fully without touching a neighbor. I notice that if there are three chairs in a row, the middle one is always empty. If people are seated at a table, they sit staggered so they are not directly across from one another. People seem to respect each other's need for personal space.

Like a dog who circles her bed three times before lying down, I circle the reading room looking for the right place to sit. I need to feel safe and comfortable so I can concentrate on mental activity. Some students, however, are too comfortable. One boy has moved two chairs together and covered himself with his coat, and now he is asleep in a fetal position. A girl sits at a table, head down, dozing like we used to do in first grade.

I find my place, an empty chair near a window, and slouch down into it, propping my legs on the low table in front. If my mother could see me, she'd reprimand me for not sitting up straight. I breathe deeply, close my eyes for a moment, and let myself become centered, forgetting both last night's pizza and tomorrow's philosophy exam. I need a few minutes to acclimate to this space, relax, and feel safe before starting my work.

Two weeks ago, a female student was assaulted not far from where I live—that's why I've taken to locking my door so carefully. I am beginning to understand the importance of feeling safe in order to be creative and productive. Here, in the library, I feel secure, protected from real violence and isolated from everyday distractions. There are just enough people for security's sake but not so many that I feel crowded. And besides, I'm surrounded by all these books, all these great minds who dwell in the hallowed space! I am comfortable, safe, and beginning to get an idea.

Hours later—my paper started, my exam studied for, my eyes tired—I retrace the path to my apartment. It is dark now, and I listen closely when I hear footsteps behind, stepping to the sidewalk's edge to let a man walk briskly past. At my door, I again fumble for the now familiar key, insert it in the lock, open the door, turn on the hall light, and step inside. Here, too, I am safe, ready to eat, read a bit, and finish my reflective essay.

✤ SUGGESTIONS FOR WRITING AND RESEARCH

Individual

1. Review your responses to Writings 2–6. Select the reflective possibility that interests you most; then compose a whole essay on the subject. When you have finished a draft, follow the suggestions at the end of the chapter to help you write your second draft.
2. Find an object in your room, dormitory, house, or neighborhood that seems especially commonplace or routine or that you otherwise take for granted (a chair, rug, window, comb, glass, plant, newspaper, fire hydrant). Describe it carefully and fully and draft an essay in which you reflect on its possible meaning or value.
3. Look up one of the authors mentioned in the chapter—Michel de Montaigne, Joan Didion, Russell Baker, Ellen Goodman—or a similar writer of your choice. Read several of this writer's essays; analyze the work in terms of the characteristics of reflective essays described in this chapter. Write a review of this author as an essayist.

Collaborative

1. As a class, select a place or event that has reflective possibilities. Write your own reflection on the subject, limiting each reflection to two single-spaced pages. Elect two class members to edit the collection of essays and publish them in bound volumes for the whole class, complete with table of contents, an editor's introduction, and an afterword by the course instructor. (For more on publishing class books, see Chapter 28.)
2. Break into small groups. In each group select an essayist, living or dead, and research his or her writing (you can use writers mentioned in this chapter or a local newspaper columnist). Share findings with other group members and together prepare a report to teach the class about the writer. Select and hand out readings in advance of your presentation.

PART THREE

CONDUCTING RESEARCH

Chapter

❧ 15 ❧

Strategies for Research

This isn't my best writing. To write this paper we had to have at least fifteen references
from ten different sources, and I was really struggling to find that many, and in the
end I just sort of stuck them in to get it done with. You really shouldn't read this paper.

MEGAN

Before agonizing too long over your next research project, stop to consider what, exactly, research entails. Keep in mind that in your nonacademic life you conduct practical research of one kind or another every time you search the want ads for a used car, browse through a library in search of a book, or read a movie review. You may not make note cards or report the results in writing, but whenever you ask questions and then systematically look for answers, you are conducting research.

In college, the research you conduct is academic rather than practical. In other words, it's designed to result in a convincing paper rather than a purchase or an action. Academic research is part of the writing process for most of your papers.

You rarely begin writing an explanation, argument, or interpretation already knowing all the facts and having all the information you'll need: to fill in the gaps in your knowledge, you conduct research. In fact, one type of assignment, the research essay, is specially designed to introduce you to the process of conducting research and writing a paper based on your findings. Research essays are generally longer, require more extensive research, use a more formal style and format, and take more time than other papers.

Meaningful research results from real questions that you care to answer. It's exciting work. And if you're not now excited about something, once you start engaging in research activities you may surprise yourself: Take a tour of your library and find out what's there; get on the Internet and explore the resources of the World Wide Web; above all, start thinking, writing, and talking to people about ideas, and exciting things will happen.

❧ A LOOK AT PRACTICAL RESEARCH

Let me tell you about the last time I did research of a substantial nature—I bought a new motorcycle. I don't buy new cars, or even new houses, but riding motorcycles is a special passion of mine and this time I wanted to buy a new motorcycle—the latest, smoothest, best-handling, most powerful motorcycle available. My wife and I enjoy touring long distances, so I especially wanted a motorcycle adept at carrying a passenger in security and comfort as well as carrying me, alone, at faster speeds. Finding the right machine meant shopping carefully, which meant asking good questions: What kind of motorcycle was best for touring? How large should it be? What make was most reliable? How much would it cost? (How much could I afford?) Who were the best dealers? And so on.

To locate and purchase this new motorcycle I worked to answer all these questions and more; here's how I remember the process.

Over a period of several months I talked to people who knew a lot about motorcycles—some old friends, the mechanic who serviced my old motorcycle, a neighbor who owned two different touring machines.

I also subscribed to *Motorcyclist* and *Rider* magazines, finding the latter more interesting since it focuses primarily on touring motorcycles. The more I read, the more familiar I became with current terminology (ABS, fairings), brands (BMW, Harley-Davidson), models (sport tourers, roadsters), and performance data (roll-on speed, braking distance). On a corner of my desk, I piled the most useful magazines with many dog-eared pages.

I also visited my local library—not the larger one at the university—to browse through other magazines to which I did not personally subscribe: *Motorcycle Consumer News* and *Cycle World*. I made photocopies of the most relevant articles, underlined key findings, and made notes in the photocopy margins.

I also visited the local chapter of the BMW Motorcycle Club, meeting more people who seemed to be expert motorcyclists (all, of course, recommending BMWs); I also bought a copy of *BMW Motorcycle Owners News*.

I rented the video *On Any Sunday* (1972), considered by many to be the best film ever made about motorcycles, and watched it three times—this more to feed my passion than to help select a motorcycle to purchase. As far as research goes, I'd call this step filling in context and helping maintain momentum, since watching the film never played a role in my finished research.

I approached the owner of Frank's Motorcycle Shop and asked about the virtues of certain BMW models, looking especially at the R1100RT and K1100LT as possible replacements for my K100RS. I also read everything available about a new model, soon to be available, a K1200RS. I actually wrote in my journal about the dreams motorcycles inspired and the difficult choices they required.

I also checked out the local dealers who sold Honda, Kawasaki, Yamaha, and Suzuki motorcycles and listened to sales pitches explaining the features of their best touring machines. I found some too racy and others too large but still took notes about each to facilitate my comparison. (I already knew I did not want a Harley-Davidson—too loud, too slow, too much chrome, too expensive!)

I returned to Frank's and rode each of the two touring models available for test-driving—the RT and LT. Still I was undecided. I wanted to see the latest model, which had not yet arrived in the United States. At the same time, I fished for the trade-in value of my current motorcycle (more than I expected!), taking notes, making calculations, and lying awake nights pondering the possibilities.

I logged onto a Web site devoted to BMW news (http:/www.motoplex.com) and read the latest reviews of the new motorcycle as tested in Europe by *Motorrad* magazine—current information not easily available through any other source. The report was phenomenal; they named the K1200RS "Bike of the Year."

After weighing the relative merits of smoothness versus power versus money—and based on a large amount of reading and some faith in the reviews I read—I returned to Frank's Motorcycle Shop and ordered the K1200RS, leaving a deposit, continuing to dream, hoping I'd made the right decision on a motorcycle I had neither seen nor ridden. (I had. The new motorcycle, when it finally arrived, proved to be quite wonderful!)

✎ WRITING 1

What research outside of school settings have you conducted recently? Think about major changes, moves, or purchases that required you to ask serious questions. Also think about any explorations you've made on site, through interviews, in a library or on the World Wide Web. Choose one of these recent research activities and list the sources you found, the questions you asked, and any steps you took to answer them. Finally, what was the result of this research?

✎ ACTIVITIES IN THE RESEARCH PROCESS

While the search for motorcycle knowledge is practical rather than academic, it serves nevertheless to introduce most of the activities common to all research projects, argumentative or informational, in school or out.

1. **The researcher has a genuine interest in the topic.** It's difficult to fake curiosity, but it's possible to develop it. My interest in motorcycles was long-standing, but the more I investigated, the more I learned and the more I still wanted to know. Some academic assignments will allow you to pursue issues that are personally important to you; others will require that you dive into the research first and generate interest as you go.

2. **The researcher asks questions.** My first questions were general rather than specific. However, as I gained more knowledge, the questions became more sharply focused. No matter what your research assignment, you need to begin by articulating questions, finding out where the answers lead, and then asking still more questions.

3. **The researcher seeks answers from people.** I talked to both friends and strangers who knew about motorcycles. The people to whom I listened most closely were specialists with expert knowledge. All research projects profit when you ask knowledgeable people to help you answer questions or point you in directions where answers may be found. College communities abound with professors, researchers, working professionals, and nonacademic staff who can help out with research projects.

4. **The researcher visits places where information may be found.** I went to dealerships and a motorcycle club, not only to ask questions of experts but also to observe and experience firsthand. No matter how much other people told me, my knowledge increased when I made personal observations. In many forms of academic research, field research is as important as library research.

5. **The researcher examines texts.** I read motorcycle magazines and brochures and watched videotapes to become more informed. Keep in mind that texts include the vast resources of both library and the Internet. While printed texts are helpful in practical research, they are crucial in academic research.

6. **The researcher evaluates sources.** As the research progressed, I double-checked information to see if it could be confirmed by more than one source. In practical research, the researcher must evaluate sources to ensure that the final decision is satisfactory. Similarly, in academic research, you must evaluate sources to ensure that your final paper is convincing.

7. **The researcher writes.** I made notes both in the field and in the library as well as wrote in my journal—all to help the research turn out just right for me. In practical research, writing helps the researcher find, remember, and explore information. In academic research, writing is even more important, since the results must eventually be reported in writing.

8. **The researcher tests and experiments.** In my practical research, testing was simple and fairly subjective—riding different motorcycles to compare the qualities of each, in this case, eliminating those ridden as not quite right. Testing and experimentation are also regular parts of many research projects.

9. **The researcher combines and synthesizes information to arrive at new conclusions.** My accumulated and sorted-out information led to a decision to purchase one motorcycle rather than another. In academic research, your synthesis of information leads not to a purchase, but to a well-supported thesis that will convince a skeptical audience that your research findings are correct.

What this researcher did not do, of course, is the step most directly concerned with the subject matter of this handbook: report the results in a research paper. The greatest difference between practical and academic research is that the former leads to practical knowledge on which to act (the buying of a motorcycle or a sound system), the latter to theoretical knowledge written for others to read (a research report or essay). So, for academic research, writing becomes a crucial final step in the process.

10. **The researcher presents the research findings in an interesting, focused, and well-documented paper.** The remainder of this chapter explains how to select, investigate, and write about academic research topics.

✍ WRITING 2

Explain how any of the research activities described in this section were part of an investigation you once conducted, in school or out. How might you use computer-based research activities to facilitate any research project you do today—practical or academic?

✍ STRATEGIES FOR MANAGING THE RESEARCH PROCESS

A research paper is the end result of the process of finding, evaluating, and synthesizing information that is new to you from a wide variety of sources. Therefore, research essays occupy a span of weeks or months and are often the most important project you work on during a semester. It pays to study the assignment carefully, begin working on it immediately, and allow sufficient time for the many different activities involved.

Ask questions. Begin by asking questions about a subject, both of yourself and of others. Preliminary questions lead to more specific inquiries.

Read extensively. Texts of all kinds—books, journal and magazine articles, and studies—are the raw material from which you will build your research paper.

Question knowledgeable people. Start with people you know. If they can't help, ask who can. Then broaden that circle to include people with specialized knowledge.

Seek out firsthand information and experience. No matter how many answers other people offer you, seek out information yourself.

Evaluate your sources and double-check the information you find. Sources vary in their accuracy and objectivity. Try to confirm the information you gather by checking more than one reliable source.

Write at every stage. The notes you take on your reading, the field notes you write, and the research log you keep all help you gain control of your subject. Everything you have written also helps you when you start drafting.

Scope Out the Assignment

First, reflect on the course for which the paper is assigned. What is the aim of this field of study? What themes has the instructor emphasized? How would a research assignment contribute to the goals of this course? In other words, before even considering a topic, assess the instructor's probable reasons for making the assignment and try to predict what is expected from your finished paper.

Next, study the assignment directions carefully. Identify both the subject words and the direction words. Subject words specify the area of the investigation. Direction words (verbs such as *write* or *explain*) specify your purpose for writing—whether you should explain or report or argue or do something else.

Finally, examine the assignment requirements: the recommended perspective, the most appropriate style, the preferred format, the sources expected, the documentation system required, the due dates, and the paper length. Thinking about these early on may save time and help avoid false steps.

Identify a Topic

Your instructor may assign a specific topic, or the choice may be left up to you. If you need to find a topic, think about how your interests dovetail with the content of the course. What topics of the course do you enjoy most? What discussions, lectures, or labs have you found most engaging? What has happened recently in the news that both interests you and relates to the course material? Do some freewriting or clustering to find the topic that seems most promising to you, or turn on your computer and surf the Internet for ideas!

At the same time, limit or narrow your topic to one you can answer with real and specific information—information that catches attention in ways that loose generalities about vast subjects cannot. Besides, if you limit your topic, your chances of managing and answering it improve dramatically.

SUBJECT	The environment (too broad to be called a topic).
TOPIC	Harmful corporate practices hurt the American environment (narrower and better but still too broad).
TOPIC	American companies sometimes exploit people's concern for the environment by advertising "green" policies and products; the results are often false advertising, tricked consumers, and a worsened environment (narrower, still better, more focused).

Develop a Research Question

Since research essays are among the lengthiest and most complex of all college assignments, make the project interesting and purposeful by developing a research question that will be exciting for you to answer. What makes a good research question?

1. It's a question that you really want to know the answer to.
2. It requires more than a yes or no answer.

Yes or No

Do companies misrepresent the environmental advantages of their products?

Better

How do companies misrepresent the environmental advantages of their products?

3. It's sufficiently complex that you do not already know the answer:

    Some of the ways companies misrepresent "green" products are changing the names of chemical ingredients, hiding research that proves there are hazards, and creating advertising that suggests environmental purity.

4. It's a question that you have a reasonable chance of answering. For example, a quick survey of library sources on false advertising reveals a wealth of books and periodicals on the subject:

    How have U.S. companies capitalized in their advertising on American consumers' environmental consciousness?

Become an Authority

Whenever you undertake research, you join an ongoing conversation among a select community of people who are knowledgeable about the subject. As you collect information, you, too, become something of an expert. You gain an authoritative voice, becoming a stronger, more powerful writer. Plan to become enough of an expert on your topic that you can teach your classmates and instructor something they didn't know before. The best way to exercise your newly found authority is to write in your own words all the major information and ideas connected with your research project, both periodically in your journal entries and later when you draft. Finding your own language to express an idea makes that idea yours and increases both your understanding of and your commitment to the topic.

Keep an Open Mind

Don't be surprised if, once you begin doing research, your questions and answers multiply and change. You ask one question, find the answer to another, pursue that, and then stumble onto the answer to your first question. Be prepared to travel in circles at times or find dead ends where you expected thoroughfares, but also thoroughfares where you expected dead ends. For example, suppose you start out researching local recycling efforts, but you stumble upon the problem of finding buyers for recycled material. One source raises the question of manufacturing with recycled materials, while another source turns the whole question back to consumer education. All of these concerns are related, but if you attempt to study them all with equal intensity, your paper will be either very long or very superficial. Follow your strongest interests, and try to answer the question you care about most.

Similarly, what began as informational research may become argumentative as the research process turns up information to tip your original neutrality one way or another. Or a research investigation that starts out to prove a thesis may result in a more neutral, informative paper if multiple causes or complexities in what had seemed a straightforward case are uncovered.

Look for a Thesis

When an answer to your research question starts to take shape, form a working thesis, that is, a preliminary answer from which you can investigate further. If additional research leads you in a different direction, be ready to redirect your investigation and revise your working thesis.

Whether or not you already have a working thesis will determine what type of research you undertake. Informational research is conducted when you don't know the answer to the question or don't have a firm opinion about the topic. You enter this kind of investigation with an open mind, focusing on the question, not on a predetermined answer. In some cases this kind of research will lead you to an argumentative position; in others you may report the results of your research in an explanatory paper. In either case, you will eventually need to develop a working thesis, and the purpose of informational research is to help you find it. Such research might be characterized as thesis-finding.

Locate Evidence

Ask yourself, "What evidence best supports my working thesis?" Research is usually conducted to prove a point, take a stand, or make a decision. You usually enter the project with some thesis, or at least a hunch, already in mind: You suspect or favor at least a tentative answer to your question. For example, should companies be held to stricter standards of truth in commercial advertising? You might argue yes, and then conduct research to buttress your position. A research question helps focus your energies, but argumentative research is really thesis-driven since you're hunting for support to prove your point. The process of researching may lead to a revised or entirely different thesis, so remember that any thesis is subject to revision, redirection, and clarification. Nonetheless, whatever thesis you start with will point you in one direction rather than another.

As a final check, ask your instructor to respond to your intended question before you invest too much time in it. Your instructor will help direct you to projects that are consistent with the goals of the assignment or course and steer you away from questions that are too broad or too offbeat.

✐ WRITING 3

Select a topic that interests you and that is compatible with the research assignment. List ten questions that you have, or that you think others might have, about this topic. Select the question that most interests you and freewrite about it for ten more minutes. Why does it interest you? Where would you start looking for answers?

Start a Research Log

A research log can help you keep track of the scope, purpose, and possibilities of any research project. Such a log is essentially a journal in which you write to yourself about the process of doing research by asking questions and monitoring the results. Questions you might ask include the following:

What evidence challenges my working thesis?
How is my thesis changing from where it started?

Writing out answers to these questions in your log clarifies your tasks as you go along. It forces you to articulate ideas and examine supporting evidence critically. Doing this, in turn, helps you focus your research activities. Novice researchers often waste time tracking down sources that are not really useful. Answering questions in your research log as you visit the library, look for and read sources, review note cards, and write various drafts can make research more efficient.

When you keep notes in a research log, record them as if they were on separate note cards. Here, for example, are some entries from a first-year student's research log.

11/12 Checked the subject headings and found no books on ozone depletion. Ref. librarian suggested magazines because it takes so long for books to come out on new subjects. In the General Science Index I found about twenty articles—I've got them all on a printout. Need to come back tomorrow to actually start reading them.
11/17 Conference today with Lawrence about the ozone hole thesis—said I don't really have much of a thesis, rather a lot of information aiming in the same direction. Suggested I look at what I've found already and then back up to see what question it answers—that will probably point to my thesis.

When a project requires you to investigate only a few sources of information, you may keep all your bibliographic and research notes in your research log, in a separate part of your class notebook, or on a single computer file.

☞ WRITING 4

Keep a research log for the duration of your research project. Write in it daily and record everything you think of or find in relation to the project. When your project is finished, write an account of the role the log played. As an alternative, keep an online journal as a resource for recording and organizing your materials. Download any useful information from your online journal into any computer fields created for planning or drafting as well as researching.

Make Research a Process

Research writing, like all important writing, benefits from the multistage process of planning, drafting, revising, and editing. In research writing, however, managing information and incorporating sources present special problems.

After developing some sense of the range and amount of information available, write out a schedule of when you will do what. For example, plan a certain amount of time for trips to the library. If any of the books you need are checked out, allow time for the library to call those in. Arrange needed interviews well in advance, with time to reschedule in case a meeting has to be canceled. And allow enough time not only for writing but also for revising and editing. As you accomplish each task, you may want to check off that item in your plan or note the date. If you find yourself falling behind schedule or discover additional tasks to be accomplished, revise your plan.

☞ WRITING 5

On one page of your research log, design a research plan that includes both library and field investigations. In this plan, list sources you have already found as well as those you hope to find.

Research a Range of Sources

To conduct any kind of research, you need to identify appropriate sources of information, consult and evaluate these sources, and take good notes recording the information you collect. You also need to understand how each source works as evidence.

Primary sources contain original material and raw information. Secondary sources report on, describe, interpret, or analyze someone else's work. For example, if you were exploring the development of a novelist's style, the novels themselves would be primary sources and other people's reviews and critical interpretations of the novels would be secondary sources. What constitutes a primary source will differ according to the field and your research question. For example, the novel *Moby Dick* is a primary source if you are studying it as literature and making claims about what it means. It is a secondary source if you are investigating nineteenth-century whaling and referring to chapters in it that contain descriptions of harpooning.

Most research essays use both primary and secondary sources. Primary sources ground the essay in firsthand knowledge and verifiable facts; secondary sources supply the context for your discussion and provide support for your own interpretation or argument.

Many research essays are based on library research, since libraries contain so much of the collective knowledge of the academic community. However, some of the most interesting research essays are based on field research. This research includes firsthand interviews with people who have expert knowledge of your subject. You can also conduct field research simply by being in the field, such as touring a local sewage treatment plant to observe the process.

Draft a Thesis Statement

When you write your first draft, be sure to allow time for any needed further research. You, not your audience, are the expert on your research topic, so plan to explain, define, and clarify new or unusual terms and concepts. To find out what your classmates and instructor already know, ask them to read and respond to your paper at several points during its development—for example, when you have a detailed outline or thesis and when you have completed your first draft.

In research writing, as in most writing, you need to tell your reader what you're up to. An argumentative thesis statement takes a position on an issue:

> In order to reduce the annual number of violent deaths in the United States, Congress should pass a law to mandate a waiting period of ten days before all handguns can be purchased in this country.

An informational thesis statement presents information but does not advance one position over another:

> Many U.S. companies attempt to capitalize on Americans' concern for the environment by a variety of false advertising strategies.

If your research was argumentative, you probably began with a working thesis. If your research was informational, you probably began with a research question, and a working thesis evolved as you found answers.

Most research essays are long and complex enough that the thesis should be stated explicitly rather than left implicit. However, the thesis that ends up in your final paper may be different from the one you developed to help you write the paper. For example, Zoe developed this detailed statement as her thesis statement:

> To get an internship with a professional photographer in today's competitive and heavily commercialized market, you need to sell yourself effectively to busy people

by developing a small and varied portfolio, making contact early and often, and demonstrating that you understand the unglamorous work involved.

In her paper, however, her explicit thesis focused on only part of this statement:

> I've learned the steps to setting up an internship: First, establish contact through writing a letter and sending a résumé, then push your portfolio.

Zoe developed her other points—about the competitive marketplace and the unglamorous work involved in being an assistant—elsewhere in her paper. Many research essays give the thesis at the beginning, somewhere in the first or second paragraph, where it acts as a promise to the reader of what will follow. Some research papers present the thesis at the end, where it acts as a conclusion or summary. If you take the delayed-thesis approach, you still need to be sure that the topic and scope of your paper are clear to readers from the very beginning.

Revise and Adjust Your Thesis

When drafting, you must follow your strongest research interests and try to answer the question you most care about. However, sometimes what began as informational research may become argumentative as the process of drafting the paper tips your original neutrality one way or another. Or a research investigation that starts out to prove a thesis may, as you draft, become a more neutral, informative paper of various perspectives, especially when multiple causes or complications surface in what had seemed a straightforward case.

The research process must remain flexible if it is to be vital and exciting. Keep an open mind when drafting, but when revising step back from your draft and assess the direction your research paper is moving or has moved. Revise your paper to streamline all points and incorporate all source information to succeed in that direction.

Be ready to spend a great deal of time revising your draft, adding new research information, and incorporating sources smoothly into your prose. Such work takes a great deal of thought, and you'll want to revise your paper several times.

Edit Everything

Editing a research paper requires extra time. Not only should you check your own writing, but you should also pay special attention to where and how you use sources and use the correct documentation style. The editing stage is also a good time to assess your use of quotations, paraphrase, and summary to make sure you have not misquoted or used a source without crediting it.

∽ SUGGESTIONS FOR WRITING AND RESEARCH

Individual

1. Select a research topic that interests you and write an exploratory draft about it. First, write out everything you already know about the topic. Second, write out everything you want to know about the topic. Third, identify experts you can talk to. Finally, make a list of questions you need answered. Plan to put this paper through a process that includes not only planning, drafting, revising, and editing but also locating, evaluating, and using sources.

2. Keeping in mind the research topic you developed in individual assignment 1, visit the library and conduct a search of available resources. What are you looking for? Where can you find it? Show your questions to a reference librarian and ask what additional electronic databases he or she would suggest. Finally, don't forget the Internet, an ever-expanding source of information on virtually any topic.

3. After completing assignments 1 and 2 above, find a person who knows something about your topic and ask him or her for leads about doing further research: With whom would this person recommend you speak? What books or articles would he or she recommend? What's the first thing this expert would do to find out information? Finally, look for a "virtual" person, someone available through an e-mail listserv or on the Internet with whom you might chat to expand your knowledge.

Collaborative

1. Join with a classmate or classmates to write a collaborative research essay. Develop plans for dividing tasks among members of the group.

2. After completing a collaborative research essay, write a short report in which you explain the collaborative strategies your group used and evaluate their usefulness.

Chapter

✣ 16 ✣

Conducting Field Research

Talking to the author was a lot different from reading her book. She was actually there, friendly, smiling sometimes, thinking out loud, telling me inside stories and trade secrets, and when she didn't know an answer, she didn't pretend that she did. Her stories really make my paper come alive.

MARY

Research is an active and unpredictable process requiring serious investigators to find answers wherever they happen to be. Academic research is not confined to libraries. Depending on your research question, you may need to seek answers by visiting museums, attending concerts, interviewing politicians, observing classrooms, or following leads down some other trail. Investigations that take place outside the library are commonly called *field research.*

Field researchers collect information that is not yet written down, that cannot be read and examined in books. Because the information they collect has not been previously recorded or assessed, field researchers have the chance to uncover new facts and develop original interpretations. To conduct such research, you need to identify the people, places, things, or events that can give you information you need. Then you must go out in the field and either *observe* by watching carefully or *interview* by asking questions of one particular person. You should also take careful notes to record your observations or interviews and critically evaluate the information you've collected.

✣ PLANNING FIELD RESEARCH

Unlike a library, which bundles millions of bits of every kind of information in a single location, "fields" are everywhere—your room, a dormitory, cafeteria, neighborhood, theater, mall, park, playground, and so on. Field information is not bundled, cataloged, organized, indexed, or shelved for your convenience. Nor is there a person like a reference librarian to point you to the best resources. In other words, when you think about making field research a part of your research project, you start out on your own. Successful field research requires diligence, energy, and careful planning.

First, you need to find the person, place, thing, or event most helpful to you and decide whether you want to collect observations, conduct interviews, or do both. Consider your research question as you've developed it thus far and the sources most likely to give you the answers you need. Also consider what sort of information will be most effective in your final paper.

Second, you need to schedule field research in advance and allow enough time for it. People, things, and events will not always hold still or be exactly where you want them to be when it's most convenient for you. Allow time for rescheduling a visit or interview or returning for more information.

Third, you need to visit the library before conducting extensive field research. No matter who, where, or what you intend to collect information from, there's background information at the library that can help you make more insightful observations or formulate better interview questions.

Finally, you need to keep a *research log* so you can keep track of all the visits, questions, phone calls, and conversations that relate to your research project. Write in your log as soon as you receive your assignment and at each step along the way about topics, questions, methods, and answers. Even record dead-end searches to remind you not to repeat them.

✒ WRITING 1

In your research log, write about the feasibility of using field research information to help answer your research question. What kind of field research would strengthen your paper? Where would you go to collect it?

✒ INTERVIEWING

A good interview provides the researcher with timely, original, and useful information that often cannot be obtained by other means. Getting such information is part instinct, part skill, and part luck. If you find talking to strangers easy, then you have a head start on being a good interviewer. In many respects, a good interview is simply a good conversation. If you're not naturally comfortable talking to strangers, you can still learn how to ask good interview questions that will elicit the answers you need. Of course, no matter how skillful or prepared you are, whether your interview results in good answers also depends on your interview subject—his or her mood, knowledge and information, and willingness to spend time talking with you.

Your chances of obtaining good interview material increase when you've given some thought to the interview ahead of time. The following guidelines should help you conduct good interviews.

Select the right person. People differ in both the amount and kind of knowledge they have on a subject. Not everyone who knows something about your research topic will be able to give you the specific information you need. In other words, before you make an appointment with a local expert because he or she is accessible, consider whether that expert is the best person to talk to. Ask yourself (1) exactly what information you need, (2) why you need it, (3) who is likely to have it, and (4) how you might approach them to gain it. Most research projects benefit from various perspectives, so you may want to interview several people. For example, to research Lake Erie pollution, you could interview someone who lives on the lake shore, a chemist who knows about the pesticide decomposition, and the president of a paper company dumping waste into the lake. Just be sure the people you select are likely to provide the information you really need.

Do your homework. Before you talk to an expert about your topic, make sure you know something about it yourself. Be able to define or describe your interest in it, know what the general issues are, and learn what your interview subject has already said or written about it. Having become familiar with your topic, you will ask sharper questions, get to the point faster, and be more interesting for your subject to talk with.

Write out questions in advance. A good interview doesn't follow a script, but it usually starts with one. Before you begin an interview, write the questions you plan to ask and arrange them so that they build on each other—general questions first, specific ones later. If your subject digresses too much, your questions can serve as reminders about the information you need.

Ask both open and closed questions. Different kinds of questions elicit different kinds of information. Open questions place few limits on the kinds of answers given: Why did you decide to major in business? What are your plans for the future? Closed questions specify the kind of information you want and usually elicit brief responses: When did you receive your degree? From what college? Open questions usually provide general information, while closed questions allow you to zero in on specific details.

Ask follow-up questions. Listen closely to the answers you receive, and when the information is incomplete or confusing, ask follow-up questions requesting more detail or clarification. Such questions can't be scripted; you've just got to use your wits to direct your interview subject toward the information you consider most important.

Use silence. If you don't get an immediate response to a question, wait a bit before rephrasing it or asking another one. In some cases, your question may not have been clear, and you will need to rephrase it. But in many cases your subject is simply collecting his or her thoughts, not ignoring you. After a slight pause, you may hear thoughtful answers that are worth waiting for.

Read body language. Be aware of what your subject is doing when he or she answers questions. Does he look you in the eye? Does she fidget and squirm? Does he look distracted or bored? Does she smile or frown? From these visual cues you may be able to infer when your subject is speaking most frankly, when he or she doesn't want to give more information, or when he or she is tired of answering questions.

Take good notes. Most interviewers take notes, using a pad that is spiral-bound on top, which allows for quick page flipping. Don't try to write down everything, just major ideas and important statements in the subject's own language that you might want to use as quotations in your paper. Omitting small words, focusing on the language that is most distinctive and precise, and using common abbreviations (like *b/c* for *because, w/* for *with,* and *&* for *and*) can make note taking more efficient. Include notes about your subject's physical appearance, facial expressions, and clothing and about the interview setting itself. These details will be useful later to help you reconstruct the interview and represent it more vividly in your paper.

Consider a tape recorder. If you plan to use a tape recorder, ask your subject's permission in advance. The advantage of tape recording is that you have a complete record of the conversation. Sometimes on hearing the person a second time you notice important things that you missed the first time. The disadvantages are that sometimes tape recorders make subjects nervous and that transcribing a tape is time-consuming work. If you use a tape recorder, it's still a good idea to have a pen in your hand to catch points of emphasis or jot down questions that occur as the conversation continues.

Confirm important assertions. When your subject says something you judge to be especially important or controversial, read your notes back to your subject to check for accuracy and allow your subject to elaborate on the topic. Some interviewers do this as the interview progresses, others at the end of the interview.

Review notes within twenty-four hours. Notes taken during an interview are brief reminders of what your subject said, not complete quotations. You need to write out the complete information as soon after the interview as you can, certainly within twenty-four hours. Supplement the notes with other remembered details while they are still fresh, by recording the information on note cards or directly into a computer file that you can refer to as you write your paper.

Integrate interview material. When interview notes become sources in your paper, use your best material and be economical. Quote lines that show your subject's expertise and useful information that advances your paper in some significant way. Work interview sources smoothly into the context of your paper, as you would lines quoted from a printed source; paraphrase when your subject is too wordy; quote exactly when his or her words are especially sharp or colorful.

In the following example, Zoe Reynders includes short quotations from an interview with a university photography editor who assisted her in preparing a photography portfolio to help her get a job. Note that she paraphrases much that happened during the interview, quoting only short passages that convey the most important part of the Brennan interview:

> Photo editor Tom Brennan took ten minutes to sort through my images and then told me, "Most photography editors wouldn't take more than two minutes to look at a portfolio." Again, I became nervous, not because I didn't think I could handle the job, but because of the strict professionalism of these people. I question whether my photos are strong enough to make an impact, whether I have what it takes to stand apart from the crowd of would-be photographers.
>
> Brennan chose twelve images he thought would be good. They do not have a consistent theme or subject; some are in black and white, others are slides or color negatives. "What I am looking for are your photos that show some kind of feeling or atmosphere," he said. Some of the photos he chose, I would have too. Others, I wouldn't have looked at twice.

᠅ WRITING 2

Describe any experience you have had as either interviewer or interviewee. Drawing on your own experience—or on television interviews—what additional advice would you give to researchers setting out to interview a subject?

᠍ᠣ **WRITING 3**

If you are planning a research project, whom could you interview to find out useful and relevant information? Make a list of such people. Write out first drafts of possible questions to ask them.

᠍ᠣ OBSERVING

Another kind of field research calls for closely observing people, places, things, and events and then describing them accurately to show readers what you saw and experienced. Strictly speaking, the term *observation* denotes visual perception; however, it also applies to information collected on site through other senses. The following suggestions may help you conduct field observations more effectively.

Select a site. As with interviewing, observing requires that you know where to go and what to look for. You need to have your research question in mind and then identify those places where observation will yield useful information. For example, a research project on pollution in Lake Erie would be enhanced by on-the-scene observation of what the water smells, feels, and looks like. Sometimes the site you visit is the primary object of your research, as when the purpose of your paper is to profile people and activities you find there. At other times the on-site observation provides supplemental evidence but is not the focus of the paper itself.

Do your homework. To observe well, you need to know what you are looking for and what you are looking at. If you are observing a political speech, know the issues and the players; if you visit an industrial complex, know what products are manufactured there. Researching background information at the library or through other means will allow you to use your time on site more efficiently.

Plan your visit. Learn not only where the place is located on a map but also how to gain access; call ahead to ask directions. Find out where you should go when you first arrive. If relevant, ask which places are open to you, which are off limits, and which you could visit with permission. Find out about visiting hours; if you want to visit at odd hours, you may need special permission. Depending on the place, after-hours visits can provide detailed information not available to the general public.

Take good notes. At any site there's a lot going on that casual observers take for granted. As a researcher you should take nothing for granted. Keep in mind that without notes, as soon as you leave a site you forget more than half of what was there. As with interview notes, be sure to review your observation notes as soon after your site visit as possible.

You can't write everything down, though, so be selective. Keep your research question in mind and try to focus on the impressions that are most important in answering it. Some of your observations and notes will provide the background information needed to represent the scene vividly in your paper. Some will provide the specific, concrete details needed to make your paper's assertions believable. Make your notes as precise as possible, indicating the exact color, shape, size, texture, and arrangement of everything you can.

Use a notebook that has a stiff cover so you can write standing, sitting, or squatting, as a table may not be available. Double-entry notebooks are useful for site visits, because they allow you to record facts observed in one column and interpretations of those facts in the other. If visual images would be useful, you can also sketch, photograph, or videotape. You can also use a tape recorder and speak your notes to pick up the characteristic sounds of the site.

Incorporate visual images. Pertinent visual information gathered from site visits can be added to research essays to enrich, enliven, and support the written texts. Computer scanning devices and sophisticated photocopy machines are available at most university libraries or graphics departments and at commercial operations that specialize in visual reproductions. In addition, visual images can be downloaded from Web sites and printed directly as part of a written text. Consider these options carefully, and study the way visual images are used effectively in books and magazines to help you integrate images smoothly in your text. Keep in mind that sharp graphics enhance but are not substitutes for careful writing, especially in the academic world where in many disciplines, words continue to carry more weight than visual images. (While pictures may indeed be worth a thousand words, most instructors who assign research papers will also want to see the thousand words!) Be sure to document the source of any images you borrow.

Integrate descriptive writing. Site description notes taken in the field may be in shorthand or messy, so be sure to recompose these into clear visual descriptions when you integrate this information into your paper. In the following sample, taken from a collaborative research essay focusing on the institution of Ben and Jerry's Homemade Ice Cream Company, the writers describe their first impressions of entering a Ben and Jerry's ice cream shop. Note the attention to both visual detail and print information:

> When you enter the front door of Ben and Jerry's Homemade Ice Cream shop in downtown Burlington, Vermont, you find yourself standing on a clean, tiled, black-and-white checkered floor. To the left are three dark red booths with white tables. Painted on the opposite wall are eight black-and-white spotted cows standing in a lush, green field probably somewhere in Vermont. The sky above them is a bright blue with several white, puffy clouds.
>
> Upstairs, the first thing you see is a blue plastic bin with a sign that says, "We are now recycling spoons!" To the right is the ice cream counter. Behind it a colorful wooden sign reads "Today's Euphoric Flavors" and lists 29 flavors, including Cherry Garcia, Chocolate Chip Cookie Dough, Chunky Monkey, New York Super Fudge Chunk, Rainforest Crunch, and Phish Food. To the left of the flavors is a white sign with black writing that tells the prices—$1.44 for a small, $1.84 for a medium, $2.60 for a large.

✂ CONDUCTING SURVEYS

A type of field research commonly used in the social sciences is the *survey,* a structured interview in which respondents, representative of a larger group, are all asked the same questions. Their answers are then tabulated and interpreted. Researchers usually conduct surveys to discover attitudes, beliefs, or habits of the general public or segments of the population. They may try to predict how soccer moms will vote in an election, determine the popularity of a new movie with teenage audiences, compare the eating habits of students who live off campus to those of students who eat in college dining halls, or investigate an infinite number of other issues.

Respondents to surveys can be treated like experts for research purposes, because they are being asked for their own opinions or information about their own behavior. However, you must ask your questions skillfully to get useful answers. Wording that suggests a right or wrong answer does more to reveal the researcher's biases and preconceived ideas than to collect candid responses. Furthermore, the questions should be easy to understand and answer, and they should be reviewed to make sure they are relevant to the research topic or hypothesis. The format for questioning and the way the research is conducted also have an influence on the responses. For example, to get complete and honest answers about a sensitive or highly personal issue, the researcher would probably use anonymous written surveys to ensure confidentiality. Other survey techniques involve oral interviews, in person or by phone, in which the researcher records each subject's responses on a written form.

Surveys are usually short to gain the cooperation of a sufficiently large number of respondents. And to enable the researcher to compare answers, the questions are usually closed, although open-ended questions may be used to gain additional information or insights.

Surveys are treated briefly here because, in truth, the designing of good survey questions, the distributing of surveys, and the assessment of the results is a highly complex and sophisticated business. If you are interested in systematically surveying a given population, say, of a residence hall or a class, it would pay to consult a social scientist or an education professor for guidelines and help. A further complication arises if your surveys request sensitive information (personal experience with drugs, alcohol, sex, and so on) from identifiable subjects; most colleges and universities have "human subject" boards or committees that must approve any research that could compromise the privacy of students, staff, or faculty members. Ask your instructor for guidelines before launching on any sensitive surveys.

However, the simple, informal polling of people to request opinions takes place quite often in daily college life, such as every time a class takes a vote or an instructor asks the class for opinions or interpretations about texts. A good example of conducting a very simple survey took place in one of my recent writing classes when one student who was writing a self-profile wanted to find out how she was perceived by others. First, Anna made a list of ten people who knew her in different ways—her mother, father, older sister, roommate, best friend, favorite teacher, and so on. Next, she invited each to make a list of five words that best characterized her. Finally, she asked each to call her answering machine on a day when she knew she would not be home and name these five words. In this way, she was able to collect live, original outside opinion (field research) in a nonthreatening manner that she then used to weave into her profile paper, using a combination of other's opinions as well as her own self-assessments. The external points of view added an interesting (and sometimes surprising) view of herself as well as other voices to her paper.

ᠵ **WRITING 4**

Invent a simple survey (no more than five questions) to generate external information about any paper topic you are working on. To implement the survey, plan to use a one-page paper handout, a brief e-mail message, or the telephone to collect responses.

✂ SUGGESTIONS FOR FIELD RESEARCH PROJECTS

Individual

Plan a research project that focuses on a local place (park, playground, street, building, business, or institution) and make a research plan that includes going there, describing what you find, and interviewing somebody with expert knowledge about the site.

Collaborative

Form a research team with one or two other students. Plan a research project that begins with an issue of some concern to all of you. Identify a local manifestation of this issue that would profit from some field research. Conduct the research, using field techniques appropriate to your topic.

✨ 17 ✨

Conducting Library Research

I'm beginning to understand the importance of feeling safe in order to be creative and productive. Here, in the library, I feel secure, protected from real violence, and isolated from everyday distractions. There are just enough people for security's sake, but not so many that I feel crowded. And besides, I'm surrounded by all these books, all these great minds who dwell in this hallowed space! I am comfortable, safe, and beginning to get an idea for my paper.

JUDITH

Libraries are the heart—or perhaps the head—of the academic community. They provide the primary knowledge base that allows professors to teach and to conduct research in their fields. They also provide students with the sources of information that allow them to investigate any area of study.

The modern library is complex and multifaceted; you should visit your college library early and get to know it well. At first, a college library may appear intimidating, but the more you use it, the more friendly it will become. At the same time you need to learn to use the electronic library of the twenty-first century, the Internet, in a critical way. Learn its entrance and exit points, its resource rooms and reference desks, and how to navigate and browse, to answer and ask.

✨ PLANNING LIBRARY RESEARCH

To really learn about the physical library, you need to go there, walk slowly through it, read the signs that identify special rooms and departments, poke your nose into nooks and crannies, and browse through a few books or magazines. If there is an introductory video explaining the library, pause to see it. If there is a self-paced or guided tour, take it. Read informational handouts and pamphlets. Be sure to locate the following:

- The book catalog, computerized or on cards, that tells you which books your library owns and where they are located.
- The book stacks, where books, periodicals, and pamphlets are stored.
- The circulation desk, where you check out and reserve books and get information on procedures and resources.
- The periodical room, which houses current issues of magazines, journals, and newspapers.

- The reference room, which contains general reference works such as dictionaries and encyclopedias along with guides and indexes to more specific sources of information.

To take full advantage of library resources, keep the following suggestions in mind:

Visit the library early and often. As soon as you receive a research assignment, visit the library to find out what resources are available for your project, and plan to return often. Even if your initial research indicates a wealth of material on your topic, you may not be able to find everything the first time you look. A book you need may be checked out, or your library may not subscribe to a periodical containing important information, and you may need to order it on interlibrary loan, a process that takes some time.

Prepare in advance to take notes. If you take careful notes from the sources you find, you will save yourself time and write a better paper. Bring index cards to the library—three-by-five-inch cards for bibliographical information and four-by-six-inch cards for notes—from your first visit on. Also keep a research log. Writing in it as you work on your research project will help you find ideas to research, plan your course of action, pose and solve problems related to your topic, and keep track of where you have been so far.

Check general sources before specific ones. During your first or second visit to the library, check general sources—dictionaries, encyclopedias, atlases, and yearbooks—for information about your topic. An hour or two spent with these general sources will give you a quick overview of the scope and range of your topic and will lead you to more specific information.

Ask for help. What this really means is *talk to librarians.* At first you might show them your assignment and describe your topic and your research plans; later you might ask them for help in finding a particular source or ask whether they know of any sources that you have not checked yet. Keep in mind, however, that reference librarians are busy people; don't ask questions that you haven't tried to answer for yourself. The following suggestions may help:

1. Before you ask for help, try to answer your questions yourself.
2. Bring with you a copy of the research assignment to show the librarian.
3. Be ready to explain the assignment in your own words: purpose, format, length, number of sources, and due date.
4. Identify any special requirements about sources: Should information come from government documents? Rare books? Films?
5. Describe the particular topic you are researching and the tentative question you have framed to address the topic.
6. Describe any work you have done so far: books read, periodicals looked at, log entries written, people interviewed, and so on.

WRITING 1

Visit your college library and locate the areas and materials discussed in this section. Then look for a comfortable place, sit down, and write about which of these might be of most use to you as you begin a research project. Then conduct a similar tour of the World Wide Web via your computer and construct a map of what such a space looks like.

ॐ FINDING SOURCES OF INFORMATION

Most of the information you need will be contained in reference books, other books, and periodicals (journals, magazines, and newspapers). Reference books are fairly easy to locate: There are relatively few of them and they are usually placed in one room or section of the library. However, even a moderately sized college library owns hundreds of thousands of books and periodicals. To simplify the researcher's task of finding the relevant ones, bibliographies and indexes have been developed. These resources either indicate which books or articles have been published on a given topic or present comprehensive lists of sources in an easy-to-use format. Once you know which book or periodical might be useful, you still need to find it. The library's catalogs tell you whether it owns the source. To find and use all these resources efficiently, follow this five-step process:

1. Consult general reference works to gain background information and basic facts.
2. Consult bibliographies and indexes to learn which books, periodicals, and articles are relevant.
3. Consult computerized databases.
4. Consult your library's catalogs to see whether it owns the books and periodicals you want.
5. Consult other sources as needed.

Searching through indexes, databases, and catalogs is much easier if you have identified the key words for your research topic. A key word is an important word describing your topic—either a word for the topic itself, a word for the general subject, or a word describing constituent parts of the topic.

Key words are sometimes nothing more than authors' names or titles of books. For example, key words for a research paper investigating the pottery of Native Americans in the western United States could include *Indian, art,* and *California.* The key words you select can mean the difference between success and failure. For example, a search including the word *Sioux* might turn up nothing, but a search including the word *Dakota* (the preferred term for this group) might result in a number of sources. To find good key words for your topic, consult the Library of Congress Subject Headings.

Reference Works

General reference works provide background information and basic facts about a topic. The summaries, overviews, and definitions in these sources can help you decide whether to pursue a topic further and where to turn next for information. The information in these sources is necessarily general and will not be sufficient by itself as the basis for most research projects; you will need to consult specialized sources as well. General reference works do not make strong sources to cite in research papers.

Specialized reference works contain detailed and technical information in a particular field or discipline. They often contain articles by well-known authorities and sometimes have bibliographies and cross-references that can lead to other sources. Two useful guides to finding specialized reference books are *Guide to Reference Books,* edited by Eugene P.

Sheehy (10th ed., 1986) and *Walford's Guide to Reference Material* (5th ed., 1989). Also see your library's online guide to reference books.

While many reference works are published as books, increasingly they are available on CD-ROM. The following lists suggest common and useful references, although there are many more in each category.

Almanacs and Yearbooks

Almanacs and yearbooks provide up-to-date information on politics, agriculture, economics, and population along with statistical facts of all kinds.

Facts on File: News Digest (1941–present). A summary and index to current events reported in newspapers worldwide. (Also CD-ROM, 1980–present.)
Statesman's Year-Book (1863–present). Annual statistics about government, agriculture, population, religion, and so on for countries throughout the world.
World Almanac and Book of Facts (1868–present). Review of important events of the past year as well as data on a wide variety of topics, including sports, government, science, business, and education.

Atlases

Atlases such as the *Hammond Atlas,* the *National Geographic Atlas of the World,* and the *New York Times Atlas of the World* can help you identify places anywhere in the world and provide information on population, climate, and industry.

Biographical Dictionaries

Biographical dictionaries contain information on people who have made some mark on history in many different fields; biographical indexes tell you how to locate additional sources.

Contemporary Authors (1967–present). Contains short biographies of authors who have published during the year.
Current Biography (1940–present). Contains articles and photographs of people in the news.
Who's Who in America (1899–present). The standard biographical reference for living Americans.

Dictionaries

Dictionaries contain definitions and histories of words along with information on their correct usage.

Encyclopedias

Encyclopedias provide elementary information, explanations, and definitions of virtually every topic, concept, country, institution, historical person or movement, and cultural artifact imaginable. One-volume works such as the *Random House Encyclopedia* and *The New Columbia Encyclopedia* give brief overviews. Larger works such as *Collier's*

Encyclopedia (24 volumes) and the *New Encyclopaedia Britannica* (32 volumes) contain more detailed information.

Specialized Reference Works

Although these works provide information that is more detailed and technical than that found in general reference works, you should still use them primarily for exploratory research and background information. Each discipline has many reference works; here is a small sampling.

LANGUAGES AND LITERATURE

Cassell's Encyclopedia of World Literature
Handbook to Literature
McGraw-Hill Encyclopedia of World Drama
Oxford Companion to American Literature

HUMANITIES

Cambridge Ancient History
Dictionary of the Bible
Encyclopedia of American History
Encyclopedia of Philosophy
Encyclopedia of Religion
Encyclopedia of World History
New Grove Dictionary of Music and Musicians
Oxford Companion to Art

SOCIAL SCIENCES

Dictionary of Education
Encyclopedia of Anthropology
Encyclopedia of Crime and Justice
Encyclopedia of Psychology
Encyclopedia of Social Work
International Encyclopedia of the Social Sciences
Political Handbook and Atlas of the World

SCIENCES

Encyclopedia of Biological Sciences
Encyclopedia of Chemistry
Encyclopedia of Computer Science and Technology
Encyclopedia of Physics
Larousse Encyclopedia of Animal Life
McGraw-Hill Dictionary of Science and Technology

BUSINESS

Encyclopedia of Banking and Finance
Encyclopedia of Economics
McGraw-Hill Dictionary of Modern Economics

༉ **WRITING 2**

Look up information on your research topic, using at least three of the reference sources described in this section.

༉ BIBLIOGRAPHIES AND INDEXES

Bibliographies and indexes are tools: They help you locate books and periodicals that contain the information you need. Periodicals consist of magazines, journals, and newspapers, which are published at set periods throughout the year. They focus on particular areas of interest, and their information is more current than that found in books. Because so many periodical issues are published each year and because every issue can contain dozens of articles on various topics, using a periodical index or database is essential to finding the article you need.

Most bibliographies and indexes are available in electronic form, either through an online service (which your library's computers access through a telephone line and a modem) or on a CD-ROM disk.

Bibliographies

Bibliographies list books alphabetically by title, by author, or by subject. Many books include a bibliography of the works consulted by the author in researching the book; always consult the bibliography of a book you have found helpful. Other bibliographies are published separately as reference tools. Some of the most useful are listed here:

Bibliographic Index: A Cumulative Bibliography of Bibliographies. New York: Wilson (1938–present). This index lists the page numbers of bibliographies in books over a wide variety of subjects. Such bibliographies provide you with lists of related sources already compiled by another author on a subject similar to your own.

Books in Print. New York: Bowker (1948–present). The latest edition of this yearly index lists by author, subject, and title all books currently in print. It is also available online and on CD-ROM.

MLA Bibliography of Books and Articles in the Modern Languages and Literature (1921–present). This is also available online and on CD-ROM.

Paperbound Books in Print. This semiannual index lists paperback books currently in print by author, subject, and title. It is also available online and on CD-ROM.

Indexes

Indexes are guides to the material published within works, sometimes within books but more often within periodicals. Each index covers a particular group of periodicals. Make sure that the index you select contains the journals, magazines, and newspapers that you want to use as sources.

Indexes list works alphabetically by author or by subject. To conduct an effective subject search, use the key words you have identified for your topic. Check under every subject heading that might be relevant. Many periodicals use the subject headings in the Library of Congress Subject Headings, but others use their own lists.

Most indexes are available in both printed and computerized forms. Many are also available on microfiche or microfilm—media that simulate the printed page but must be read on special machines. Indexes in book form are usually the most comprehensive; those presented on microfilm, microfiche, and computer usually cover only the past ten or twenty years.

Computerized indexes allow you to focus your search strategy most effectively. By combining key words in certain ways, you can have the computer generate a list of works that closely match your topic. For example, if your research topic is the art of the Dakota Indians, you might search for all works with the key word *Dakota* in their subject description. This search would result in hundreds of works, not only on art but on politics, economics, history, and many other topics. Something similar would happen if you searched for *art.* But if you search for *Dakota and art,* the computer will list only those works with both words in their subject descriptions, a much more useful list for your research topic. You can also combine key words using *or.* Searching for *Dakota and art or fiction* would result in a list of works that have the word *Dakota* in their subject descriptions, some of which also have *art* and some of which also have *fiction.*

General Periodical Indexes

These indexes list articles published in a variety of periodicals of interest to the general public.

InfoTrac. This monthly index, available only on CD-ROM, contains three separate indexes. The Academic Index covers nearly one thousand commonly used scholarly publications. The General Periodical Index covers over one thousand general-interest publications. The Newspaper Index covers large-circulation newspapers. Many entries include summaries.

New York Times Index (1851–present). This bimonthly index lists every article that appears in the *New York Times.* Short summaries are provided for many articles. It is also available online.

Readers' Guide to Periodical Literature (1900–present). This semimonthly index lists articles in more than two hundred magazines of general interest, such as *Newsweek, Popular Science,* and *Rolling Stone.* Online and CD-ROM versions are also available (1983–present).

Specialized Periodical Indexes

These indexes list articles in periodicals in specific disciplines or fields of interest. They are usually much more helpful than general indexes for college-level research. Here are some of the most common:

America: History and Life
Applied Science and Technology Index
Art Index

Biological and Agricultural Index
Business Periodicals Index
Dissertation Abstracts International
Education Index
Essay and General Literature Index
General Science Index
Humanities Index
Index to Legal Periodicals
Music Index
Psychological Index
Social Science Index

✎ COMPUTERIZED RESOURCES

Databases are large collections of electronically stored information that function like indexes. Often they provide summaries in addition to bibliographic information on the sources they list; occasionally they contain copies of the sources themselves. Databases can be either online or on CD-ROM. Many library catalogs today are computerized as well, in the form of online databases.

DIALOG Database

A major online system commonly found in college libraries is DIALOG, which keeps track of more than a million sources of information. DIALOG offers many specialized databases, 987 of which are currently listed in the manual *DIALOG Blue Sheets.* Some of the most commonly used databases within DIALOG are Arts and Humanities Search (1980–present), ERIC (Educational Resources Information Center, 1965–present), PsychINFO (1967–present), Scisearch (1974–present), and Social Scisearch (1972–present). You must decide which specialized database you need before you begin your search.

DIALOG and other search services are available online. To use an online database, you usually need the assistance of a reference librarian, who will ask you to fill out a form listing the key words you have identified for your project. The library is charged a fee for each search, calculated according to the time spent and the number of entries retrieved. Some libraries have the person requesting the search pay the fee; others limit the time allotted for each search. Be sure to ask what your library's policy is.

Some databases, including some of the specialized databases within DIALOG, are also available on CD-ROM. You can usually search through these databases without the aid of a librarian.

Internet Links

Don't forget, as well, that Internet-to-library and library-to-library links (for example, Ohio-Link) make exploring the vast resources of libraries such as Harvard's or Michigan's almost as quick and easy as finding a book in your campus library's electronic catalog.

The Library Catalog

The library catalog lists every book a library owns; many libraries also catalog their periodicals. At one time all catalogs were card catalogs, with the information printed on small cards and stored in drawers. More recently, libraries began computerizing their catalogs, which are now known as online catalogs. Often online catalogs can be accessed through telephone lines and modems from locations outside the library. These catalogs list books by author, title, and subject; provide basic information about its physical format and content; and tell you where in the library to find it and whether or not it is checked out.

During the course of your research project, you will probably use the library catalog in two different ways. Sometimes you will already know the title of a work you want to find. The catalog can confirm that your library owns the work and can tell you where to find it. At other times you will use the catalog as you would an index or a database, searching for works that are relevant to your topic; this process is called *browsing*. Online catalogs are particularly good for browsing and usually have several special features designed to facilitate it.

Even if you discover that your library doesn't own the source you want, don't despair. Many libraries can obtain a work owned by another library through an interlibrary loan, although this usually takes a few days. Ask your librarian.

Consulting Online Catalogs

Online catalog systems vary slightly from library to library, though all systems follow the same general principles. Locating works through author and title is much like the corresponding procedure with a card catalog, with one important exception: Most online catalogs allow you to search with partial information. For example, if you know that the title of a novel begins with the words *Love Medicine* but you can't remember the rest of it, you can ask the catalog computer to search for the title *Love Medicine*. It will present you with a list of all works that begin with those words (see Figure 17.1).

Most online catalogs also allow you to perform key word searches, much like the searches conducted on computerized databases. The advantage of this kind of search is that the computer can search all three categories (author, title, and subject) at once. To perform a key word search, use the words you've identified as describing your topic, linked by *and* or *or* as appropriate. For example, if you're trying to research fictional accounts of Dakota Indians, you can search for *Dakota Indians* and *fiction*. The computer will present you with a list of works that fit that description (see Figure 17.2). As with all computer searches, making your key word search request as specific as possible will result in the most useful list.

```
Search Request: T=LOVE MEDICINE
Search Results: 4 Entries Found                        Title Index
-----------------------------------------------------------T257
1  LOVE MEDICINE: A NOVEL.  ERDRICH LOUISE <1984>  (BH)

   LOVE MEDICINE AND MIRACLES
2     SIEGEL BERNIE S <1986>  (BH)
3     SIEGEL BERNIE S <1988>  (DA)

   LOVE MEDICINE AND MIRACLES LESSONS LEARNED ABOUT SELFHEALING
   FROM A SURGEON'S EXPERIENCE WITH EXCEPTIONAL PATIENTS
4     SIEGEL BERNIE S <1986>  (BH)
-----------------------------------------------------------
COMMANDS:      Type line # to see individual record
               O  Other options
               H  Help

NEXT COMMAND:
```

Figure 17.1   Results of a title search in an online catalog

Once you have found your book, you can ask the online catalog to provide complete information on it. For the most part, this is the same information you would see in the card catalog. However, many online catalogs also show circulation information, letting you know whether the book is checked out.

Using Call Numbers

Once you have determined through the catalog that your library owns a book you want to consult, use the book's call number to locate it in the stacks. Most academic libraries use the Library of Congress system, whose call numbers begin with letters. Some libraries still use the older Dewey Decimal system, whose call numbers consist entirely of numbers. In either case, the first letters or numbers in a call number indicate the general

```
Search Request: T=LOVE MEDICINE
BOOK - Record 1 of 5 Entries Found                    Long View
--------------------Screen 1 of 1--------------------T259
Author:        Erdrich, Louise.
Title:         Love medicine : a novel
Edition:       1st ed.
Published:     New York : Holt, Rinehart, and Winston, c1984.
Description:   viii, 275 p. ; 22cm.
Subjects(LC):  Indians of North America--North Dakota--Fiction.
-----------------------------------------------------------
   LOCATION:            CALL NUMBER          STATUS
1  Halley Stacks        PS3555.R42 L6 1984   Not checked out


COMMANDS:      P  Previous screen
               O  Other options
               H  Help

NEXT COMMAND:
```

Figure 17.2   Full information on a book in an online catalog

subject area. Because libraries shelve all books for a general subject area together, this portion of the call number tells you where in the library to find the book you want.

Be sure to copy a book's call number exactly as it appears in the catalog. Most libraries have open stacks, allowing you to retrieve the book yourself; one wrong number or letter could lead you to the wrong part of the library. If your library has closed stacks, you will need to give the call number to a librarian, who will retrieve the book for you; the wrong call number will get you the wrong book.

> ### ᧐ WRITING 4
>
> Use the library's catalog to see what holdings the library has on your topic. Retrieve one of these books from the stacks, and check to see if it contains a bibliography that could lead you to other books. Record these findings in your research log.

Other Sources of Information

Many libraries own materials other than books and periodicals. Often these do not circulate. If you think one of the sources listed here might contain information relevant to your research, ask a librarian about your library's holdings.

Government Documents

The U.S. government publishes numerous reports, pamphlets, catalogs, and newsletters on most issues of national concern. Reference books that can lead you to these sources include the *Monthly Catalogue of United States Government Publications* and the *United States Government Publications Index,* both available on CD-ROM and online.

Nonprint Media

Records, audiocassettes, videotapes, slides, photographs, and other media are generally catalogued separately from book and periodical collections.

Pamphlets

Pamphlets and brochures published by government agencies and private organizations are generally stored in a library's vertical file. The *Vertical File Index: A Subject and Title Index to Selected Pamphlet Material* (1932/35–present) lists many of the available titles.

Special Collections

Rare books, manuscripts, and items of local interest are commonly found in a special room or section of the library.

↝ WRITING 5

Identify one relevant source of information in your library's holdings other than a book or periodical. Locate it and take notes on the usefulness of the source and the process you used to obtain it.

↝ SUGGESTIONS FOR LIBRARY RESEARCH PROJECTS

Individual and Collaborative

Your school experience of a dozen or more years already tells you that library research can find information about a virtually unlimited number of topics. See what types of library information you can locate to supplement the field research you conducted either individually or as a collaborative project for your work in Chapter 16.

Chapter

✦ 18 ✦

Conducting Internet Research

The Internet provides a wide variety of information unavailable either in the library or in the field. Within the past decade, businesses and individuals have gained widespread access to the Internet, especially to that portion known as the World Wide Web (WWW). Although academic and scientific use of the Internet continues, commercial applications drive most Web development—which makes it a messy, crowded, but ever-so-useful place to conduct research, academic or otherwise. Many Web sites are secondary sources, written by people describing, analyzing, or interpreting the work of others. Less easy to find are primary sources, the original words and accounts of people describing their own experiences or findings.

Much of the difficulty and delight of the Internet stems from the fact that no one is in charge. No single agency or company is responsible for organizing or policing the Internet. No one knows exactly what is on it, nor is there a central card catalog or index showing what's available, from whom, or where it's located. The unscreened nature of the Internet makes it essential that you supplement Internet information with other sources. If you are unfamiliar with Web searching for noncommercial information, the following suggestions will help.

✦ SELECT A SEARCH ENGINE

To begin an Internet search from off campus, log on via your local service provider. From on campus, use the campus network to arrive at the same place. If your computer logs first to a university home page, find the Internet link or type in the electronic address (URL) of a major search engine in the home page locator box. A search engine will help you locate information. Any of the following are useful for academic searches:

- **Alta Vista** <http://www.altavista.digital.com>: A powerful and comprehensive site; allows restricted searches, full texts often available.
- **Excite** <http://www.excite.com>: Good for first and general searches; ranks sites by frequency of key words; suggests related sites.
- **Hotbot** <http://www.hotbot.com>: Powerful; good for customized searches by date, media, specific domains. Includes newsgroups, classified ads, Yellow Pages, e-mail directories.

- **Infoseek** <http://infoseek.go.com>: Identifies sites by frequency of key words. Allows searching by title, URL, directory, and newsgroup databases.
- **Lycos** <http://www.lycos.com>: Allows advanced searches; directory covers 90 percent of the Web; includes newsgroups, Reuters news service.
- **Net Search** <http://netscape.com>: Located on the Netscape task bar; powerful, easy to use, and connected to other Web search engines.
- **Northern Light** <http://www.northernlight.com>: Sorts search results into customized folders. Special collections of abstracts (free) and periodical articles (small charge to download).
- **Yahoo!** <http://www.yahoo.com>: The most comprehensive subject directory; good for browsing; allows restricted searches; includes full-text downloads.

In addition, check out search engines that search the search engines: All in One <http://www.albany.net/allinone/>, Dogpile <http:www.dogpile.com>, MetaCrawler <http://metacrawler.com>, and 37.com <http://37.com>. However, these collective search engines sometimes provide too many sites with too few restrictions, making it difficult to limit your search to relevant information.

Limit the Search

First searches often provide too much information—sometimes several thousand (or million!) sites—making it hard to locate those you can use. For example, a broad search term such as *Vietnam* may locate everything from travel information to news articles to geography and language—thousands of sites, too many to explore.

If your subject is the Vietnam War, most search engines will retrieve a list of sites whose titles, Web addresses, or text contains only one of the two search words—thus, some will be about Vietnam but not the war, others about war but not the Vietnam War. Again, too many sites. To search for two or more terms at the same time, most search sites offer logical search tools called *Boolean operators.* You can also group terms to limit searches.

- Use quotation marks to limit a specific phrase, title, or name. Put the words in double quotation marks: *"Vietnam War"* (subject), *"The Best and the Brightest"* (title), *"David Halberstam"* (author). You will find sites that include both words.
- Use *and* between words to limit the search to sources that include both terms (not already combined in a phrase). Typing *crime and punishment* will return any documents that include both words *crime* and *punishment,* while using quotation marks will get you *Crime and Punishment,* the novel by Dostoevsky. (Some sites, such as Excite and Alta Vista, use a plus sign (+) instead of *and*—try both to be sure.)
- Use *or* between words to retrieve documents that include any, rather than most, of the search words. Type *puma or "mountain lion" or cougar or panther*—all different names for the same animal.
- Use *not* after a term to exclude a word that must *not* appear in the documents. Example: You want dolphin, the fish, so you type *dolphin not mammal or NFL.*

- Use an asterisk (*) to substitute for letters or word endings that might vary: *univers** could stand for *university* or *universities.*
- Use parentheses to group and combine search expressions: (*treaty or armistice or police action*) *and "Korean War."* (The Korean War was technically called a police action, and no side won.)

There is no limit to the level of nesting which you can use in a query: (*treaty or armistice*) *and Korean* (*war or "police action"*) *and "Joseph McCarthy."* The more you limit up front, the more specific your information will be.

Search Strategies

Type words that describe or name your topic, click the search button ("Search now," "Go"), and wait for results to tell you how many possible sites exist. You'll get a list of sites to examine, one by one. Try searching several different ways, using related terms in combination or alone. Narrow your search by using other, more specific terms, such as *"Vietnam and war and infantry,"* or narrow further to *"Tet Offensive"* or *"Ho Chi Minh Trail."* These searches may return shorter, more useful lists.

Try several different search engines and directories. No search engine or directory catalogs the entire Web, and no two engines search exactly the same way. Depending on your topic, you may find one search engine much more useful than the others. The law of the Web is trial and error. If you are still not getting what you expect, click on the help or advanced search option included on every site, read it, and try again.

༈ WRITING 1

Search for a topic of special interest to you, using three different search engines. Make notes to share with classmates about the results of each search, and explain why one engine performed better than another in this particular case.

Surf Strategies

Research profits from curiosity and exploration. No place on earth takes you faster to more information than the Internet. To explore it, log on and type one of the following URLs into the command line of your browser, hit the enter key, and see what you find.

Amazon.com <http://www.amazon.com>: Lists books in print along with out-of-print titles. Useful to locate or buy books; includes all relevant publishing data and reviews.

Argus Clearinghouse <http://www.clearinghouse.net>: Selects, describes, and evaluates Internet resources; a good site to begin any research project.

Biographical Dictionary <http://www.s9.com/biography>: Searchable by name, birth and death years, profession, works, and other terms.

E.span <http://www.espan.com/doot/doot.html>: Job hunting and the Occupational Outlook Handbook.

Encyclopaedia Britannica <http://www.britannica.com>: Information on any subject under the sun.

Frequently Asked Questions (FAQ) About the Web <http://www.boutell.com/openfaq/browsers/>.

Information Please <http://www.infoplease.com>: A searchable online almanac with topics from architecture to biography to historical statistics to weather.

Learn the Net Inc. <http://www.learnthenet.com/english/index.html>: Web tours and training.

Peterson's Education Center <http://www.petersons.com:8080/>: Information on colleges.

Research-It <http://www.iTools.com/research-it/research-it.html>: A reference toolkit of online dictionaries, and a thesaurus.

Shakespeare Web <http://www.shakespeare.com>: Links to sites for searching the works of Shakespeare.

Statistical Abstract of the United States <http://www.census.gov/stat_abstract/>: From the Census Bureau.

ABC News <http://www.abcnews.com/>.

MSNBC (NBC News) <http://www.msnbc.com>.

ESPN <http://espn.sportszone.com>: Cable sports news.

◌ WRITING 2

Explore the World Wide Web by logging on to three of the sites listed above. In your journal, note your observations about these sites. Compare them on the amount of information offered, the visual presentation (how easy it was to find information), and links between pages of the site and to other sites.

Check E-mail and Newsgroups

Once you have an e-mail address, you can correspond electronically with any of millions of people around the world who can, in turn, write to you. Of course, you need your correspondents' e-mail addresses. Try using a Web search site such as Search.com <http://www.search.com> that lists multiple e-mail search routines. Enter a name and check the search results against other information you already have. For example, if you were looking for Toby Fulwiler's address and found one with the domain *uvm.edu,* you could be fairly sure you had found the author of this book, who works at the University of Vermont (UVM), an educational institution.

E-mail can make a good interview tool. If you have questions for the author of an article you're citing in your research, you may be able to e-mail your questions and get quick answers. If you write to someone in search of information, be sure to identify yourself and describe your research project. In addition to your specific questions, don't forget to ask general questions such as "Can you think of other important sources (or questions or issues) I should be aware of?"

Another kind of continuing discussion group on the Internet is the *newsgroup* or *Usenet group*. A newsgroup consists of a collection of postings on a single topic. As with mailing lists, there are thousands of newsgroups. Most Internet service providers include newsgroup access as part of basic service.

To access newsgroups, you must tell your browser program the name of your Internet provider's news server, which your provider, campus librarian, or computer department can give you. To see a list of all newsgroups, go to your browser's news reader program. Many academically oriented newsgroups contain abbreviations such as *lit* for literature, *sci* for science, or *soc* for sociology. You can pick the groups you wish to read by subscribing to each one, then see the messages in that group by double-clicking the group name.

Rather than browse through lists of newsgroups, you can instruct some Web sites to search through newsgroups' current and archived postings. Two sites that specialize in newsgroups are Reference.COM <http://www.reference.com> and Deja News <http://www.dejanews.com>.

✣ WRITING 3

Conduct an e-mail interview with an expert in your research field. Introduce yourself, explain your purpose, and ask three good questions. See Chapter 16 for tips on asking interview questions; then modify those for an online interview.

Troubleshooting

Following are frequently asked questions about Internet research.

1. **What if I can't find a Web site?**

If you type in a URL but cannot locate the site, or you get an error message saying "Object not found," check your typing very carefully and try again. If you still get an error message, try a simplified version of the URL to take you to the site's home page, and then click on a link for more information. Even if the specific file you were seeking has been moved, chances are you can find related or updated information you can use.

Also try searching several different sites. No search engine or directory catalogs the entire Web, and no two engines search exactly the same way. So any comprehensive search effort requires several different search tools. Depending on your topic, you may find one search engine much more useful than the others. In any case, check the online help file that each search engine offers to find out the exact conventions of that engine.

2. **How do I cite a Web source?**

It can be difficult to retrace your steps to a valuable Internet resource. Use a notebook to keep track of which search engines and which search terms you use so that you can reproduce a search easily. When you find a useful Web page, print a copy for your records. If your browser doesn't automatically do so, write the URL of the page on your printout along with the date and time you accessed the page. These will help to document your paper and "freeze" the contents of a site that changes between visits.

3.   **How can I tell from an Internet address whether the source is reliable?**

The reliability of information found on the Internet varies considerably. For starters, notice the type of organization that sponsors the site—information available by reading the domain (dot) suffix. As of this writing, the following domain suffixes are in operation:

.aero	aviation groups
.biz	businesses
.com	commercial/business
.coop	credit unions and rural cooperatives (co-ops)
.edu	educational
.gov	government institutions
.info	information, open to public
.museums	accredited museums worldwide
.mil	military
.name	second-level names
.net	news and other networks
.org	nonprofit agency
.pro	professionals
~	personal home page

Each abbreviation suggests the potential bias in sites: *.com* and *.biz* sites are usually selling something; *.coop* and *.pro* may be selling something, but may have a stronger interest in promoting the public welfare; *.mil* and *.org* are nonprofit, but each has an agenda to promote and defend; *.edu, .gov, .museum,* and *.net* should be neutral and unbiased, but still need careful checking; and *.info, .name,* or ~ could be anyone with any idea. For more on evaluating Internet sources, see Chapter 20.

Chapter

✌ **19** ✌

Working with Sources

I really don't mind doing the research. The library, once you get to know it, is a
very friendly place. But when I go to write the paper I keep forgetting all the rules
about what to quote, and what not to quote, and how to introduce quotations and
not make them too long or too short. It seems like that part could be less confusing.

JASON

Locating potential sources for a research project is one thing; deciding which ones to include, where to use them, and how to incorporate them is something else. Some writers begin making use of their sources in their very earliest explanatory drafts, perhaps by trying out a pithy quotation to see how it brings a paragraph into focus. Others prefer to wait until they have all or most of their note cards in neat stacks in front of them before making any decisions about what to include. No matter how you begin writing with sources, there comes a time when you need to incorporate them finally, smoothly, effectively, and correctly into your paper.

✌ **ORGANIZING SOURCE INFORMATION**

Once your research is under way, you need to decide which sources to use and how to use them. You can't make this decision how much work you spent finding and analyzing each source; you have to decide how useful the source is in answering your research question. In other words, you need to control your sources rather than let them control you.

Papers written in an effort "to get everything in" are source-driven and all too often read like patch jobs of quotations loosely strung together. Your goal should be to remain the director of the research production, your ideas on center stage and your sources the supporting cast.

You may find that a potential source on which you decided not to take notes has become crucial, while notes that initially seemed central are now irrelevant. Don't be discouraged. Real research about real questions is vital and dynamic, which means it's always changing. Just as you can't expect your first working thesis to be your final thesis, you can't expect to know in advance which sources are going to prove most fruitful. And, of course, you can collect more information once you've begun drafting. At each step in the process you see your research question and answer more clearly, so the research you conduct as you draft may be the most useful of all.

Selecting Sources

The best way to ensure that you and your thesis remain in control is to make an outline first and then organize your notes according to it. (If you compose your outline on a word processor, it will be easier to make changes later.) If you do it the other way around—organizing your notes in a logical sequence and then writing an outline based on the sequence—you'll be tempted to find a place for every note and to gloss over areas where you haven't done enough research. By outlining first, you let the logical flow of your ideas create a blueprint for your paper. (Of course, your outline may change as your ideas continue to develop.) If you can't outline before you write, then be sure to begin writing before you arrange your note cards.

Once you've outlined or begun drafting and have a good sense of the shape of your paper, take time to organize your notes. Set aside bibliographic and note cards for sources that you don't think you're going to use. Put the rest of the bibliographic cards in alphabetical order by the authors' last names and arrange your note cards so that they correspond to your outline. Integrate field research notes as best you can. Finally, you may want to go back to your outline and annotate it to indicate which source goes where. This will also show you whether there are any ideas that need more research.

Keep in mind that referring more than two or three times to a single source—unless it is itself the focus of your paper—undercuts your credibility and suggests overreliance on a single point of view. If you find you need to refer often to one source, make sure that you have sufficient references to other sources as well.

Integrating Information

Once you know which sources you want to use in your paper, you still have to decide how to use them. The notes you made during your research are in many forms. For some sources, you will have copied down direct quotations; for others, you will have paraphrased or summarized important information. For some field sources you may have made extensive notes on background information, such as your interview subject's appearance. For some library sources you may have photocopied whole pages, highlighting useful passages. Simply because you've quoted or paraphrased a particular source in your notes, however, doesn't mean you have to use a quotation or paraphrase from this source in your paper. Once again, you must remain in control. Make decisions about how to use sources based on your goals, not on the format of your research notes.

Whenever you quote, paraphrase, or summarize, you must acknowledge your source through documentation. Different disciplines have different conventions for documentation. The examples in this chapter use the documentation style of the Modern Language Association (MLA), the style preferred in the languages and literature. (See the Writer's References section for details about the MLA system as well as the American Psychological Association [APA] style used in the social sciences.)

Synthesizing Sources

Once you have outlined or begun drafting and have a good sense of the shape of your paper, organize your notes so that they support the case your paper makes—if you've been working from intuition rather than outline, now may be a good time to organize that intuition. Put bibliographic cards in alphabetical order by the author's last name, since this is the order in which they need to appear on the "Works Cited" or "References" page at the end of your paper. Finally, go back to your outline and annotate it to indicate which source goes where. By doing this, you can see if there are any ideas that need more research.

As you prepare to draft, you need to assess all the information you have found and decide which sources are useful. Read your notes critically to evaluate each source, and synthesize the material into a new, coherent whole.

Synthesizing material involves looking for connections among different pieces of information and formulating ideas about what these connections mean. The connections may be similar statements made by several sources, or contradictions between two sources. Try to reach some conclusions on your own that extend beyond the information in front of you; then use those conclusions to form the goals for your paper.

Figure out how you will use the source information. Base these decisions on your goals for the paper and not on the quantity of your research notes. Your synthesis, not your sources, makes your paper creative and unique.

Documenting Sources

Whether you quote, paraphrase, or summarize, you must acknowledge your source through documentation. Different disciplines have different conventions for documentation. The examples in this chapter use the MLA documentation style preferred in the languages and literature field. To document a paper according to MLA conventions, see Writer's Reference 1. To document a paper according to APA conventions, see Writer's Reference 2.

Using Internet Sources

The Internet makes available a nearly unlimited number of resources for research purposes; however, the quality of those sources varies greatly, reflecting as they do commercial or, sometimes, crackpot biases that are screened out of cataloged library sources. As a researcher, then, you need to evaluate Internet sources with an especially critical eye (see Chapter 20). Be aware that many informed readers—college instructors among them—are skeptical about the reliability or credibility of sources found exclusively on the Internet. If you find Internet information you believe to be useful, you can enhance its credibility by providing in the text of your paper a brief context for the source, perhaps explaining how you located the source, the keywords used, the number of references, and your limiting criteria. For example, to write a paper about graffiti artists, a student named Sarah found interesting information at a Web site called *Art Crimes*. However, in addition to simply identifying the source at the end of her paraphrase (with full information on the Works Cited page), she explained how she found and used the Web information.

In searching the Web for information about people who call themselves graffiti artists, I found the most relevant information categorized under the heading of "Society/Subcultures/Hip-Hop/Graffiti/United_States/California." The most informative of the 57 California sites was *Art Crimes,* the Web page of a Los Angeles art gallery, which listed, in alphabetical order, the home pages of nearly one thousand graffiti artists with names such as Amok, Billboard Liberation Front, Crayone, and Daddy Cool (*Art Crimes*). I have drawn many of my conclusions about the so-called art of graffiti from a close examination of these home pages.

By providing detailed contextual information, Sarah reassured her readers that she had used the Web carefully and found information there impossible to locate any other way. How much information you supply to support a source is, ultimately, a judgment call. At the same time, Internet sources that reproduce cataloged print sources require no special treatment in the text of a research paper except an additional notation on the reference page that identifies the URL at which the source was found, indicating it was located on the Internet instead of in a library or in another print source.

✒ WRITING 1

Describe any experience you have had using the Internet for personal or academic research. If you used the research to write an academic paper, how did your readers respond to your use of Internet sources?

✒ QUOTING SOURCES DIRECTLY

To quote a source directly, you reproduce the writer's or speaker's own words, exactly as they were in the original source. Direct quotation provides strong evidence and can add both life and authenticity to your paper. However, too much quotation can make it seem as though you have little to say for yourself. Long quotations also slow readers down and often have the unintended effect of inviting them to skip over the quoted material. Unless the source quoted is itself the topic of the paper (as in a literary interpretation), limit brief quotations to no more than two per page and long quotations to no more than one every three or four pages.

Direct quotations should be reserved for cases in which you cannot express the ideas better yourself. Using only strong, memorable quotations will make your writing stronger and more memorable as well. Use them when the original words are especially *precise, clear, powerful,* or *vivid.*

- *Precise.* Use direct quotations when the words are important in themselves or when they've been used to make fine but important distinctions.
- *Clear.* Use quotations when they are the clearest statement available.
- *Powerful.* Let people speak for themselves when their words are especially strong. Powerful words are memorable; they stay in the reader's mind long after the page is turned.

- *Vivid*. Use direct quotation when the language is lively and colorful, when it reveals something of the author's or speaker's character and individuality.

Direct quotation provides strong evidence and can add both life and authenticity to your paper.

To quote, you must use an author's or speaker's exact words. Slight changes in wording are permitted in certain cases (see the next section), but these changes must be clearly marked. Although you can't change what a source says, you do have control over how much of it you use. Too much quotation can imply that you have little to say for yourself. Use only as long a quotation as you need to make your point. Remember that quotations should be used to support your points, not to say them for you.

Be sure that when you shorten a quotation, you have not changed its meaning. If you omit words within quotations, you must indicate that you have done so by using ellipsis points. Also any changes or additions must be indicated with brackets. (In MLA style, but not APA, the omission of words is such a change and requires therefore a bracketed ellipsis.) The following example passage is quoted from Stewart Brand's *The Media Lab,* page 258:

ORIGINAL PASSAGE

The human communication environment has acquired biological complexity and planetary scale, but there are no scientists or activists monitoring it, theorizing about its health, or mounting campaigns to protect its resilience. Perhaps it's too new, too large to view as a whole, or too containing—we swim in a sea of information, in poet Gary Snyder's phrase. All the more reason to worry. New things have nastier surprises, big things are hard to change, and containing things is impossible.

INACCURATE QUOTATION

In *The Media Lab,* Stewart Brand describes the control that is exerted by watchdog agencies over modern telecommunications: "The human communication environment has [. . .] activists monitoring it, theorizing about its health [. . .]" (258).

By omitting important words, the writer changed the meaning of the original source.

ACCURATE QUOTATION

In *The Media Lab,* Stewart Brand notes that we have done little to monitor the growth of telecommunications. Modern communication technology may seem overwhelmingly new, big, and encompassing, but these are reasons for more vigilance, not less: "New things have nastier surprises, big things are hard to change, and containing things is impossible" (258).

Integrating Quotations

Direct quotations will be most effective when you integrate them smoothly into the flow of your paper. You can do this by providing an explanatory "tag" or by giving one or more sentences of explanation. Readers should be able to follow your meaning easily and to see the relevance of the quotation immediately.

Brief quotations should be embedded in the main body of your paper and enclosed in quotation marks. According to MLA style guidelines, a brief quotation consists of four or fewer typed lines.

> Photo editor Tom Brennan took ten minutes to sort through my images and then told me, "Most photography editors wouldn't take more than two minutes to look at a portfolio."

Long quotations, those of five lines or more, should be set off in block format. Begin a new line, indent ten spaces (for MLA), and do not use quotation marks.

> Kate Kelly focuses on Americans' peculiarly negative chauvinism, in this case, the chauvinism of New York residents:
>
> > New Yorkers are a provincial lot. They wear their city's accomplishments like blue ribbons. To anyone who will listen they boast of leading the world in everything from Mafia murders to porno movie houses. They can also boast that their city produces more garbage than any other city in the world. (89)

Introducing Quotations

Introduce all quoted material so that readers know who is speaking, what the quotation refers to, and where it is from. If the author or speaker is well-known, it is especially useful to mention his or her name in an introductory signal phrase.

> **Henry David Thoreau asserts in** *Walden,* **"The mass of men lead lives of quiet desperation" (5).**

There are certain *signal phrases* to tell the reader that the words or ideas that follow come from another source. A signal phrase is a verb that indicates the tone or intention of the author. To avoid monotony, vary the placement and words of the signal phrases you use. Note the differences, both slight and significant, in the following signal phrases:

acknowledges	denies	points out
admits	emphasizes	refutes
agrees	endorses	reports
argues	finds	reveals
asserts	grants	says
believes	illustrates	shows
claims	implies	states
comments	insists	suggests
concedes	maintains	thinks
concludes	notes	writes
declares	observes	

If your paper focuses on written works, you can introduce a quotation with the title rather than the author's name, as long as the reference is clear.

> *Walden* **sets forth one individual's antidote against the "lives of quiet desperation"
> led by the working class in mid-nineteenth-century America (Thoreau 5).**

If neither the author nor the title of a written source is well-known (or the speaker in a field source), introduce the quotation with a brief explanation to give your readers some context.

> Mary Catherine Bateson, daughter of anthropologist Margaret Mead, has become, in her own right, a student of modern civilization. In *Composing a Life* she writes, "The twentieth century has been called the century of the refugee because of the vast numbers of people uprooted by war and politics from their homes" (8).

Explaining and Clarifying Quotations

Sometimes you will need to explain a quotation in order to clarify why it's relevant and what it means in the context of your discussion.

> In *A Sand County Almanac,* Aldo Leopold invites modern urban readers to confront what they lose by living in the city: "There are two spiritual dangers in not owning a farm. One is the danger of supposing that breakfast comes from the grocery, and the other that heat comes from the furnace" (6). Leopold sees city-dwellers as self-centered children, blissfully but dangerously unaware of how their basic needs are met.

You may also need to clarify what a word or reference means. Do this by using square brackets, as explained in the following section.

Adjusting Grammar When Using Quotations

A passage containing a quotation must follow all the rules of grammatical sentence structure: Tenses should be consistent, verbs and subjects should agree, and so on. If the form of the quotation doesn't quite fit the grammar of your own sentences, you can quote less of the original source, change your sentences, or make a slight alteration in the quotation. Use this last option sparingly, and always indicate any changes with brackets.

UNCLEAR

In *A Sand County Almanac,* Aldo Leopold follows various animals, including a skunk and a rabbit, through fresh snow. He wonders, "What got him out of bed?" (5).

CLEAR

In *A Sand County Almanac,* Aldo Leopold follows various animals, including a skunk and a rabbit, through fresh snow. He wonders, "What got [the skunk] out of bed?" (5).

GRAMMATICALLY INCOMPATIBLE

If Thoreau believed, as he wrote in *Walden* in the 1850s, "The mass of men lead lives of quiet desperation" (5), then what would he say of the masses today?

The verb *lead* in Thoreau's original quotation is present tense, but the sentence might call for the past tense form, *led*.

GRAMMATICALLY COMPATIBLE

If Thoreau believed, as he wrote in *Walden* in the 1850s, that the masses led "lives of quiet desperation" (5), then what would he say of the masses today?

GRAMMATICALLY COMPATIBLE

In the nineteenth century, Thoreau stated, "The mass of men lead lives of quiet desperation" (*Walden* 5). What would he say of the masses today?

GRAMMATICALLY COMPATIBLE

If Thoreau thought that in his day, the "mass of men [led] lives of quiet desperation" (*Walden* 5), what would he say of the masses today?

✍ WRITING 2

Read through your research materials, highlighting any quotations you might want to incorporate into your paper. Use your research log to explore why you think these words should be quoted directly. Also note where in your essay a quotation would add clarity, color, or life; then see if you can find one to serve that purpose.

✍ PARAPHRASING SOURCES

Although it is generally wiser to write as many research notes as possible in your own words, you may have written down or photocopied many quotations instead of taking the time to put an author's or speaker's ideas into your own words. To *paraphrase,* you restate a source's ideas in your own words. The point of paraphrasing is to make the ideas clearer (both to your readers and to yourself) and to express the ideas in the way that best suits your purpose. In paraphrasing, attempt to preserve the intent of the original statement and to fit the paraphrased statement smoothly into the immediate context of your essay.

When to Paraphrase

Paraphrases generally re-create the original source's order, structure, and emphasis and include most of its details.

- *Clarity.* Use paraphrase to make complex ideas clear to your readers.
- *Details.* Use paraphrase only when you need to present details that an author or speaker has described at greater length.
- *Emphasis.* Use paraphrase when including an author's or speaker's point suits the emphasis you want to make in your paper.

The best way to make an accurate paraphrase is to stay close to the order and structure of the original passage, to reproduce its emphasis and details. However, don't use the same sentence patterns or vocabulary or you risk inadvertently plagiarizing the source.

If the original source has used a well-established or technical term for a concept, you do not need to find a synonym for it. If you believe that the original source's exact words are the best possible expressions of some points, you may use brief direct quotations within your paraphrase, as long as you indicate these with quotation marks. Be careful not to introduce your own comments or reflections in the middle of a paraphrase, unless you make it very clear that these are your thoughts, not the original author's or speaker's.

The following quotation is taken from Naomi Wolf, *The Beauty Myth,* page 9:

ORIGINAL PASSAGE

The affluent, educated, liberated women of the First World, who can enjoy freedom unavailable to any woman ever before, do not feel as free as they want to. And they can no longer restrict to the subconscious their sense that this lack of freedom has something to do with—with apparently frivolous issues, things that really should not matter. Many are ashamed to admit that such trivial concerns—to do with physical appearance, bodies, faces, hair, clothes—matter so much.

INACCURATE PARAPHRASE

In *The Beauty Myth,* Naomi Wolf argues that First World women, who still have less freedom than they would like to have, restrict to their subconscious those matters having to do with physical appearance—that these things are not really important to them (9).

ACCURATE PARAPHRASE

In *The Beauty Myth,* Naomi Wolf asserts that First World women, despite their affluence, education, and liberation, still do not feel very free. Moreover, many of these women are aware that this lack of freedom is influenced by superficial things having primarily to do with their physical appearance—things which should not matter so much (9).

ᴥᴧ **WRITING 3**

Read through your note cards for any passages you quoted directly from an original source. Find notes that now seem wordy, unclear, or longer than necessary. Paraphrase notes that you expect to use in your paper. Exchange your paraphrases and the originals with a classmate, and assess each other's work.

ᴥᴧ **SUMMARIZING SOURCES**

To summarize, you distill a source's words down to the main ideas and state these in your own words. A summary includes only the essentials of the original source, not the supporting details, and is consequently shorter than the original. As you draft, summarize often so that your paper doesn't turn into a string of undigested quotations. The following guidelines may help:

- *Main points.* Use summary when your readers need to know the main point the original source makes but not the supporting details.
- *Overviews.* Sometimes you may want to devise a few sentences that will effectively support your discussion without going on and on. Use summary to provide an overview or an interesting aside without digressing too far from your paper's focus.
- *Condensation.* You may have taken extensive notes on a particular article or observation only to discover in the course of drafting that you do not need all that detail. Use summary to condense lengthy or rambling notes into a few effective sentences.

Keep in mind that summaries are generalizations and that too many generalizations can make your writing vague and tedious. You should occasionally supplement summaries with brief direct quotations or evocative details collected through observation to keep readers in touch with the original source.

Summaries vary in length, and the length of the original source is not necessarily related to the length of the summary you write. Depending on the focus of your paper, you may need to summarize an entire novel in a sentence or two, or you may need to summarize a brief journal article in two or three paragraphs. Remember that the more material you attempt to summarize in a short space, the more you will necessarily generalize and abstract it. Reduce a text as far as you can while still providing all the information your readers need to know. Be careful, though, not to distort the original's meaning.

The following quotation is taken from Jane Goodall's *Through a Window,* page 12:

ORIGINAL PASSAGE

For a long time I never liked to look a chimpanzee straight in the eye—I assumed that, as is the case with most primates, this would be interpreted as a threat or at least as a breach of good manners. Not so. As long as one looks with gentleness, without arrogance, a chimpanzee will understand and may even return the look.

INACCURATE SUMMARY

Goodall learned from her experiences with chimpanzees that they react positively to direct looks from humans (12).

ACCURATE SUMMARY

Goodall reports that when humans look directly but gently into chimpanzees' eyes, the chimps are not threatened and may even return the look (12).

ᘒ WRITING 4

Review any sources on which you have taken particularly extensive notes. Would it be possible to condense these notes into a briefer summary of the entire work? Would it serve your purpose to do so? Why or why not?

✴ UNDERSTANDING AND AVOIDING PLAGIARISM

Acknowledging your sources through one of the accepted systematic styles of documentation is a service to your sources, your readers, and future scholars. Knowledge in the academic community is cumulative, with one writer's work building on another's. After reading your paper, readers may want to know more about a source you cited, perhaps in order to use it in papers of their own. Correct documentation helps them find the source quickly and easily.

Failure to document your sources is called *plagiarism*. Plagiarism is taking someone's ideas or information and passing them off as your own. The practice of citing sources for "borrowed" ideas or words is both customary and expected in academic writing.

Much plagiarism is not intentional; many writers are simply unaware of the standard guidelines for indicating that they have borrowed words or ideas from someone else. Nevertheless, it is the writer's responsibility to learn these guidelines and follow them.

Using a Documentation Style

Remember that each area of academic study has developed its own conventions for documentation, a standardized set of guidelines that continue to evolve as the discipline evolves. This book includes detailed discussion of the MLA system (see Writer's Reference 1) and the APA system (see Writer's Reference 2). Other humanities areas, such as history and art history, use a system of endnotes or footnotes following *The Chicago Manual of Style* while the natural sciences use the style recommended by the Council of Science Editors (CSE) or a related style. You should use the documentation of the discipline for which you are writing; if you are in doubt, ask your instructor.

Basically, you must attribute any idea or wording you use in your writing to the source through which you encountered that idea or those words, if the material is not original to yourself. This means that you must give credit to other people's (1) words, sentences, documents, speeches, lectures, or expressions as well as their (2) ideas, concepts, plans, arguments, policies, jokes, or illustrations—no matter where you found, heard, read, or saw them. It is also important to note that even "nonpublished" ideas or words should be attributed to their sources whenever such documentation is feasible. For example, if you use opinions from a sidewalk poll, conversation, Internet commentaries, or a World Wide Web (WWW) page in your writing, you must cite that original source. At the same time, since memory is faulty, and since we've all borrowed countless terms and ideas throughout our lives, you may not remember where you heard an idea or expression and so cannot reliably document it. That's fine. The point is to give credit in good faith, when you know the source of a borrowing.

Avoiding Plagiarism

You need to document other people's ideas and language when those ideas and words, to the best of your knowledge, originate with those people. If you don't, you are guilty of plagiarism—passing off someone else's writing as your own. Plagiarism is grounds for dismissal from reputable academic institutions. So, to avoid plagiarism, you need to know what needs to be documented and what does not—which means you need to learn the common knowledge rule.

The Common Knowledge Rule

You do not need to document common historical knowledge or geographical information that an educated person can be expected to know and factual information that appears in multiple sources such as the dates of historical events (the fall of Rome in 410 A.D.), the names and locations of states and cities, the general laws of science (gravity, motion), and statements of well-known theories (feminism, liberalism).

You don't need to document ideas that are common current knowledge and in widespread use in your culture (hip hop, poly sci, pregnant chads).

You do not need to document ideas commonly known to the field or discipline in which you are writing and that can be found in several sources. For example, don't document the term *libido* (associated with Sigmund Freud) in a psychology paper, *means of production* (associated with Karl Marx) in a political science paper, or *postmodern* in an English, history, art, or philosophy paper.

However, if an author offers opinions or interpretations about any type of common knowledge, these should be credited using the proper documentation style.

How to Avoid Plagiarism

Intentional plagiarism is simply cheating, which all honest students avoid.

Unintentional plagiarism, a far more common problem, occurs when a writer paraphrases or summarizes another author's idea, but does not give credit for the author's language.

ORIGINAL

Notwithstanding the widely different opinions about Machiavelli's work and his personality, there is at least one point in which we find a complete unanimity. All authors emphasize that Machiavelli is a child of his age, that he is a typical witness to the Renaissance. (Ernst Cassirer, *The Myth of the State*)

PLAGIARIZED PARAPHRASE

Despite the widely different opinions about Machiavelli's work and personality, everyone agrees that he was a representative witness to the Renaissance (Cassirer 43).

Even though Cassirer is credited with the idea, the writer does not credit him with the wording.

ACCEPTABLE PARAPHRASE

Although views on the work and personality of Machiavelli vary, everyone agrees that he was "a typical witness to the Renaissance" (Cassirer 43).

The best way to avoid plagiarism is to make careful notes when you are consulting sources in the first place: In your computer file or on your note card, put all borrowed language in quotation marks along with author, title, publisher, date, and page number. When you need to use the source, all the information is there, eliminating the need for questionable shortcuts.

The following guidelines will help you avoid inadvertent plagiarism:

- Place all quoted passages in quotation marks and provide source information, even if it is only one phrase.
- Identify the source from which you have paraphrased or summarized ideas, just as you do when you quote directly.
- Give credit for any creative ideas you borrow from an original source. For example, if you use an author's anecdote to illustrate a point, acknowledge it.
- Replace unimportant language with your own, and use different sentence structures when you paraphrase or summarize.
- Acknowledge the source if you borrow any organizational structure or headings from an author. Don't use the same subtopics, for example.
- Put any words or phrases you borrow in quotation marks, especially an author's unique way of saying something.

↵ WRITING 5

Read the following quotation from Mike Rose's *Lives on the Boundary*, page 192. Then explain why each of the sentences that follow is an example of plagiarism.

> The discourse of academics is marked by terms and expressions that represent an elaborate set of shared concepts and orientations: alienation, authoritarian personality, the social construction of self, determinism, hegemony, equilibrium, intentionality, recursion, reinforcement, and so on. This language weaves through so many lectures and textbooks, it is integral to so many learned discussions, that it is easy to forget what a foreign language it can be.

1. The discourse of academics is marked by expressions that represent shared concepts.
2. Academic discourse is characterized by a particular set of coded words and ideas that are found throughout the college community.
3. Sometimes the talk of professors is as difficult for outsiders to understand as a foreign language is to a native speaker.

Chapter

✃ 20 ✃

Evaluating Research Sources

Good sources inform your papers and make them believable. For a source to be "good," it needs to answer yes to two questions: (1) Is the source itself credible? (2) Is it useful in my paper? This chapter provides guidelines for evaluating the credibility and usefulness of sources found in the library, on the Internet, and in the field.

✃ EVALUATING LIBRARY SOURCES

The sources you find in a college library are generally credible because experts have already screened them. The books, periodicals, documents, special collections, and electronic sources have been recommended for library acquisition by scholars, researchers, and librarians with special expertise in the subject areas the library catalogs. Consequently, library resources have been prejudged credible, at some level, before you locate them. However, just because *some* authorities judged a source to be credible *at one time* does not necessarily mean it still is, or that it's the best available, or that it's not contested, or that it's especially useful to the paper you are writing. Two of the main reasons for distrusting a source found in the library have to do with *time* (When was it judged true?) and *perspective* (Who said it was true and for what reason?).

Identifying Dated Sources

Most library documents include their date of publication inside the cover of the document itself, and in most cases this will be a fact that you can rely on. In some cases, such as articles first published in one place, now reprinted in an anthology, you may have to dig harder for the original date, but it's usually there (check the permissions page).

One of the main reasons any source may become unreliable—and incredible—is the passage of time. For example, any geographical, political, or statistical information true for 1950 or even 1999 will be more or less changed by the time you examine it—in many cases, radically so. (See atlas or encyclopedia entries for Africa or Asia from 1950!) Yet at one time this source was judged to be accurate.

Check the critical reception of books when published by reading reviews in *The Book Review Digest* (also online); often you can tell if the critical argument over the book twenty years ago is still relevant or has been bypassed by other events and publications.

At the same time, dated information has all sorts of uses. In spite of being "dated," works such as the Bible, the I Ching, the novels of Virginia Woolf, and the beliefs of Malcolm X are invaluable for many reasons. In studying change over time, old statistical information is crucial. Knowing the source date lets you decide whether to use it.

Identifying Perspective

Who created the source and with what purpose or agenda? Why has someone or some organization written, constructed, compiled, recorded, or otherwise created this source in the first place? This second critical question is difficult to answer by reviewing the source itself. While most library texts include the dates they were published, few accurately advertise their purpose or the author's point of view—and when they do, this information cannot always be believed.

To evaluate the usefulness of a text, ask questions about (1) the assumptions it makes, (2) the evidence it presents, and (3) the reasoning that holds it together. Finding answers to these critical questions reveals an author's bias.

- What is this writer's purpose—scholarly analysis, political advocacy, entertainment, or something else?
- Can you classify the author's point of view (liberal, conservative, radical) and differentiate it from other points of view?
- What does the writer assume about the subject or about the audience? (What does unexplained jargon tell you?)
- How persuasive is the evidence? Which statements are facts, which inferences drawn from facts, which matters of opinion? (See Chapter 12.)
- Are there relevant points you are aware of that the writer *doesn't* mention? What does this tell you?
- How compelling is the logic? Are there places where it doesn't make sense? How often?

Your answers to these questions should reveal the degree to which you accept the author's conclusions.

Cross-Referencing Sources

While at first it may seem daunting to answer all these questions, have patience and give the research process the time it needs. On a relatively new subject, you won't know many answers; however, the more you learn, the more you learn! As you read further, you begin to compare one source to another and to notice differences, especially if you read carefully and take notes to keep track of each source's timeliness and perspective. The more differences you note, the more answers to the preceding questions you will find, and the better you can judge whether a source might be useful.

ᴽ WRITING 1

Select one of the sources you are planning to use in your paper about which you know very little, and write out answers to the preceding questions. What does the close scrutiny tell you about the source? Should other sources on your list be examined as carefully?

ᴽ EVALUATING ELECTRONIC SOURCES

You need to apply the same critical scrutiny to Internet sources as to library sources, only more so! With no editor, librarian, or review board to screen for accuracy, reliability, or integrity, anyone with a computer and a modem can publish personal opinions, commercial pitches, bogus claims, bomb-making instructions, or smut on the World Wide Web. While the Internet is a marvelous source of research information, it's also a trap for unwary researchers. So, in addition to timeliness and perspective, what do you need to look out for? First, look at the domain name at the end of the electronic address (URL) to identify the type of organization sponsoring the site. Page 208 tells you how to analyze this information.

Second, ask as many critical questions as you would of a library source. An easy way to do this is to ask reporter's questions (*who, what, where, when, why,* and *how?*) and see what the answers tell you.

WHO IS THE SITE'S AUTHOR?

- Look for an individual's name (check the beginning or end of the site).
- Look for expert credentials: scholar, scientist, physician, college degrees, experience.
- Look for the author's connection to an organization or agency: university, government, Sierra Club, National Rifle Association?
- If there is no individual name, look for a sponsoring organization. What does it stand for?
- Look for links to the author/agency home page.
- Look for a way to contact the author or agency by e-mail, phone, or mail to ask further questions.
- If you cannot tell who created the site or contact its sponsors, credibility is low. Don't rely on this site's information.

WHAT IDEAS OR INFORMATION DOES THE SITE PRESENT?

- Look for concepts and terminology you know.
- Summarize the claim or central idea in your own language. How does it match your research needs?
- Look for facts versus inferences versus opinions versus speculations—be wary of opinion and speculation.
- Look for balanced versus biased points of view—what tips you off? Which would you trust more?
- Look for missing information. Why is it not there?
- Look for advertising. Is it openly identified and separated from factual material?

- Look for a "hit count" to suggest the popularity of the site—a sign that others have found it useful.

HOW IS THE INFORMATION PRESENTED?

- Look at the care with which the site is constructed, an indication of the education level of the author. If it contains spelling and grammar errors or is loaded with unexplained jargon, do you trust it? Will your readers?
- Look at the clarity of the graphics and/or sound features. Do they contribute to the content of the site?
- Look for links to other sites that suggest a connected, comprehensive knowledge base.

WHERE DOES THE INFORMATION COME FROM?

- Identify the source of the site: *.edu, .gov, .com,* and so on (see p. 208).
- Identify the source of site facts. Do you trust it?
- Look for prior appearance as a print source. Are you familiar with it? Is it reputable?
- Look for evidence that the information has been refereed. If so, where?

WHEN WAS THE SITE CREATED?

- Look for a creation date; a date more than a year old suggests a site not regularly updated.
- Does absence of a creation date affect the reliability of the information on the site?
- Is the site complete or still under construction?

WHY IS THE INFORMATION PRESENTED?

- Look for clues to agenda of the site: Is it to inform? Persuade? Entertain? Sell? Are you buying?
- Does getting information from the site cost money? (You should not have to pay for reference material for a college paper.)

✺ WRITING 2

When you locate an anonymous Web source—no owner claiming credit—try contacting Network Solutions, Inc., the company responsible for administering most domain names: <http://rs.internic.net>. In your journal, speculate on possible reasons for the anonymity. Be wary of a site that makes you search for such information rather than providing it up front.

✺ EVALUATING FIELD SOURCES

If people and places were as carefully documented, reviewed, cataloged, and permanent as library sources, or as widely available as Internet sources, the first part of this chapter would cover everything you need to know. Unfortunately, the reliability and credibility of field sources is problematic because it is often more difficult for readers to track down field sources than textual sources. An interview is a one-time event, so a subject available one day may not be the next. A location providing information one day may change or become off limits the next. To critically examine field sources, you need to freeze them and make them hold still. Here's what to do.

Interviews

To freeze an interview, use a tape recorder and transcribe the whole session. Once an interview is taped, apply to it the critical questions you would to a written source. If you cannot tape-record, then take careful notes, review main points with your subject before the interview ends, and apply these same critical questions.

Site Observations

To freeze a site, make photographic or video records of what it looked like and what you found, in addition to taking copious notes. Include details about time, location, size, shape, color, number, and so forth. If you cannot make photo records, then sketch, draw, or diagram what you find. Pictures and careful verbal descriptions add credibility to papers by providing specific details that would be difficult to invent had the writer not been present. Even if you don't use them directly in your paper, visual notes will jog your memory of other important events.

Personal Bias

Evaluating personal observations is complicated since you are both the creator and evaluator of the material. In and of itself, the material has no meaning or value until you, the recorder, assign it. This material has not been filtered through the lens of another writer. You are the interpreter of what you witness, and when you introduce field evidence, it's your own bias that will show up in the way you use language; you will lead your reader one way or another depending on the words you select to convey what you saw. Was the lake water *cloudy, murky,* or *filthy?* Was the electric car *slow, relaxed, hesitant,* or *a dog?* You also shape interview material by the questions you ask, the manner in which you conduct the interview, and the language of your notes. In other words, in field research, the manner in which you collect, record, and present information is likely to introduce the bias that is most difficult to control—your own.

✍ WRITING 3

Locate a passage in your research paper in which you have cited a field source. Closely examine the language in which you present this information. Do you detect personal biases or value judgments? What other word choices are possible? What change in emphasis would happen if you switched some of your language?

Chapter

ᔕ 21 ᔕ

Research Essays:
A Sampler

This chapter provides samples of three research papers, two following Modern Language Association (MLA) format and documentation conventions and one following American Psychological Association (APA) conventions. Each of the student papers has been edited slightly for publication but remains largely as originally written. Each contains annotations explaining the format and documentation conventions. (For a full discussion of MLA style, see Writer's Reference 1; for a full discussion of APA style, see Writer's Reference 2.) Note that proportions shown in margins of papers are not actual but are adjusted to fit space limitations of this book. Follow actual dimensions indicated in margins and your instructor's directions.

ᔕ LITERARY RESEARCH ESSAY: MLA STYLE

In literary research essays, students are expected to read a work of literature, interpret it, and support their interpretation through research. The research in such papers commonly uses two kinds of sources: primary sources (the literary works themselves, from which students cite pages), and secondary sources (the opinions of literary experts found in books or periodicals).

The paper that begins on page 230 was written by Andrew Turner in a first-year English class. Students were asked to focus on a topic of personal interest about the author Henry David Thoreau and to support their own reading of Thoreau's work with outside sources, including library, Internet, and interview.

TITLE CENTERED, ONE-THIRD DOWN PAGE

The Two Freedoms of Henry David Thoreau

by

NAME

Andrew Turner

INSTRUCTOR   Professor Fulwiler

COURSE   English 2

DATE   3 October 2001

1"

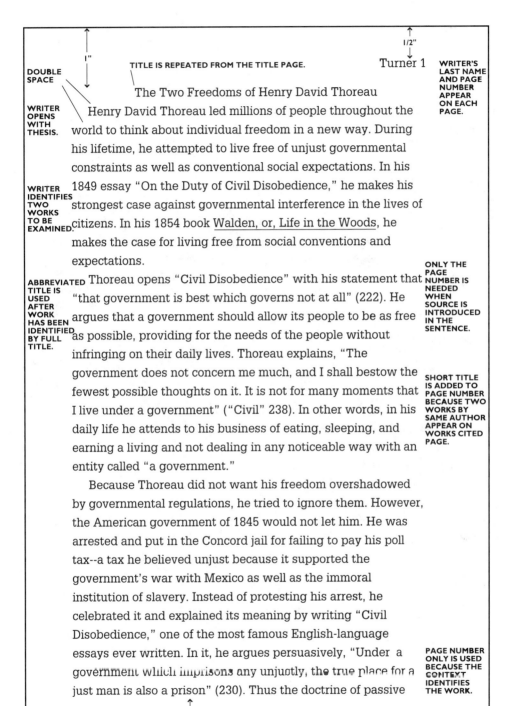

DOUBLE SPACE

TITLE IS REPEATED FROM THE TITLE PAGE.

1/2"

Turner 1

The Two Freedoms of Henry David Thoreau

WRITER OPENS WITH THESIS.

Henry David Thoreau led millions of people throughout the world to think about individual freedom in a new way. During his lifetime, he attempted to live free of unjust governmental constraints as well as conventional social expectations. In his

WRITER IDENTIFIES TWO WORKS TO BE EXAMINED.

1849 essay "On the Duty of Civil Disobedience," he makes his strongest case against governmental interference in the lives of citizens. In his 1854 book Walden, or, Life in the Woods, he makes the case for living free from social conventions and expectations.

ABBREVIATED TITLE IS USED AFTER WORK HAS BEEN IDENTIFIED BY FULL TITLE.

Thoreau opens "Civil Disobedience" with his statement that "that government is best which governs not at all" (222). He argues that a government should allow its people to be as free as possible, providing for the needs of the people without infringing on their daily lives. Thoreau explains, "The government does not concern me much, and I shall bestow the fewest possible thoughts on it. It is not for many moments that I live under a government" ("Civil" 238). In other words, in his daily life he attends to his business of eating, sleeping, and earning a living and not dealing in any noticeable way with an entity called "a government."

ONLY THE PAGE NUMBER IS NEEDED WHEN SOURCE IS INTRODUCED IN THE SENTENCE.

SHORT TITLE IS ADDED TO PAGE NUMBER BECAUSE TWO WORKS BY SAME AUTHOR APPEAR ON WORKS CITED PAGE.

Because Thoreau did not want his freedom overshadowed by governmental regulations, he tried to ignore them. However, the American government of 1845 would not let him. He was arrested and put in the Concord jail for failing to pay his poll tax--a tax he believed unjust because it supported the government's war with Mexico as well as the immoral institution of slavery. Instead of protesting his arrest, he celebrated it and explained its meaning by writing "Civil Disobedience," one of the most famous English-language essays ever written. In it, he argues persuasively, "Under a government which imprisons any unjustly, the true place for a just man is also a prison" (230). Thus the doctrine of passive

PAGE NUMBER ONLY IS USED BECAUSE THE CONTEXT IDENTIFIES THE WORK.

Turner 2

resistance was formed, a doctrine that advocated protest against the government by nonviolent means:

INDENTED 10 SPACES

QUOTATION OF MORE THAN 4 LINES IS PRESENTED IN BLOCK FORMAT.

> How does it become a man to behave toward this American government today? I answer that he cannot without disgrace be associated with it. I cannot for an instant recognize that political organization as my government which is the slave's government also. (224)

PAGE NUMBER IS OUTSIDE PERIOD FOR INDENTED PASSAGES.

SIGNAL PHRASE INTRODUCES THE NAME OF THE SECONDARY SOURCE AUTHOR.

According to Charles R. Anderson, Thoreau's other writings, such as "Slavery in Massachusetts" and "A Plea for Captain John Brown," show his disdain of the "northerners for their cowardice on conniving with such an institution" (28). He wanted all free American citizens, north and south, to revolt and liberate the slaves.

PARTIAL QUOTATION IS WORKED INTO SENTENCE IN A GRAMMATICALLY CORRECT WAY.

In addition to inspiring his countrymen, Thoreau's view of the sanctity of individual freedom affected the lives of later generations who shared his beliefs (King). "Civil Disobedience" had the greatest impact because of its "worldwide influence on Mahatma Gandhi, the British Labour Party in its early years, the underground in Nazi-occupied Europe, and Negro leaders in the modern south" (Anderson 30). For nearly one hundred fifty years, Thoreau's formulation of passive resistance has been a part of the human struggle for freedom.

WRITER SWITCHES TO DISCUSSION OF A SECOND WORK AFTER DISCUS-SION OF FIRST WORK IS COMPLETED.

Thoreau also wanted to be free from the everyday pressure to conform to society's expectations. He believed in doing and possessing only the essential things in life. To demonstrate his case, in 1845 he moved to the outskirts of Concord, Massachusetts, and lived by himself for two years on the shore of Walden Pond (Spiller et al. 396-97). Thoreau wrote Walden to explain the value of living simply, apart from the unnecessary complexity of society: "Simplicity, simplicity, simplicity! I say, let your affairs be as two or three, and not a hundred or a thousand" (66). At Walden, he lived as much as possible by this

IDENTIFI-CATION FOR WORK WITH MORE THAN THREE AUTHORS.

ABBREVIATED POPULAR TITLE IS LISTED AFTER WORK'S FIRST REFERENCE.

Turner 3

statement, building his own house and furniture, growing his
own food, bartering for simple necessities, attending to his own
business rather than seeking employment from others (Walden

PAGE NUMBERS FOR PARAPHRASE ARE INCLUDED. 16-17).

Living at Walden Pond gave Thoreau the chance to
formulate many of his ideas about living the simple, economical
life. At Walden, he lived simply in order to "front only the
essential facts of life" (66) and to center his thoughts on living
instead of unnecessary details of mere livelihood. He developed
survival skills that freed him from the constraints of city
dwellers whose lives depended upon a web of material things
and services provided by others. He preferred to "take rank
hold on life and spend my day more as animals do" (117).

PAGE NUMBERS ALONE ARE SUFFICIENT WHEN CONTEXT MAKES THE SOURCE CLEAR.

While living at Walden Pond, Thoreau was free to occupy his
time in any way that pleased him, which for him meant writing,
tending his bean patch, and chasing loons. He wasn't troubled
by a boss hounding him with deadlines or a wife and children
who needed support. In other words, "he wasn't expected to be
anywhere at any time for anybody except himself (Franklin)."
His neighbors accused him of being selfish and did not
understand that he sought most of all "to live deliberately" (66),
as he felt all people should learn to do.

Then as now, most people had more responsibilities than
Thoreau had and could not just pack up their belongings and
go live in the woods--if they could find free woods to live in.
Today, people are intrigued to read about Thoreau's
experiences and inspired by his thoughts, but few people can
actually live or do as he suggests in Walden. In fact, most
people, if faced with the prospect of spending two years
removed from society, would probably think of it as a
punishment or banishment, rather than as Thoreau thought of
it, as the good life (Poyer).

Practical or not, Thoreau's writings about freedom from

Turner 4

government and society have inspired countless people to reassess how they live their lives. Though unable to live as he advocated, readers everywhere remain inspired by his ideal, that one must live as freely as possible.

WRITER'S CONCLUSION REPEATS THESIS ASSERTION.

Turner 5

HEADING CENTERED

Works Cited

Anderson, Charles Roberts, ed. <u>Thoreau's Vision: The Major</u>   **DOUBLE SPACE**
    <u>Essays</u>. Englewood Cliffs: Prentice, 1973.

Franklin, George. Professor of American Literature, Northfield
    College. Personal interview. 5 April 2000.

"Ghandi." <u>Britannica Online</u>. Vers. 97.1. Mar. 1997.
    Encyclopedia Britannica. 2 Mar. 1998 <http://www.eb.com:180>.

King, Jr., Martin Luther. "Letter from Birmingham Jail." 28 Aug. 1963.
    <u>Civil Disobedience in Focus</u>. Ed. Hugo Adam Bedau.
    New York: Routledge, 1991. 68-84.

Poger, Ralph. <u>A Postmodern Point of View</u>. 2 April 2000
    <http://www.acu.edu/~rpoger>.

Spiller, Robert E., et al. <u>Literary History of the United States:</u>
    <u>History</u>. 3rd ed. New York: Macmillan, 1963.

Thoreau, Henry David. "On the Duty of Civil Disobedience."
    <u>Walden and Civil Disobedience</u>. New York: NAL, 1960.

---. "Walden, or, Life in the Woods." <u>Walden and Civil</u>
    <u>Disobedience</u>. New York: Harcourt, 1987. 6-187.

WORK WITH MORE THAN THREE AUTHORS IS CITED WITH FIRST AUTHOR'S NAME AND "ET AL."

FORMAT FOR WORKS WITHIN AN ANTHOLOGY

INDENTED 5 SPACES

✤ PERSONAL RESEARCH ESSAY: MLA STYLE

The following research essay was the result of an assignment asking students to explore any topic of strong personal interest about which they had substantial and real questions that could be answered by some combination of field, Internet, and library research. Students were asked to cast the paper in the form of an *I-search essay,* requiring the final draft to report as much on the process of the search as the answer or answers discovered. Amanda Kenyon Waite investigated what it would be like to work as a resident assistant (RA) in her college dormitory.

For a topic of personal interest that includes visiting places and interviewing people, writing often takes a first-person point of view in the writer's personal voice. Such a paper commonly omits a title page and includes on the first page, in the top left-hand corner, the author's name, instructor's name, course, and date. The title is centered, followed by a double space, and the complete text, including quotations, is double-spaced as well. Page numbers are included at the top right of each page. At the same time, Amanda followed the academic conventions of citing her sources according to current MLA documentation style, complete with brief in-text references to a Works Cited page at the end of her essay. (See Writer's Reference 1.) (For an example of a more formal MLA research paper, see page 230.)

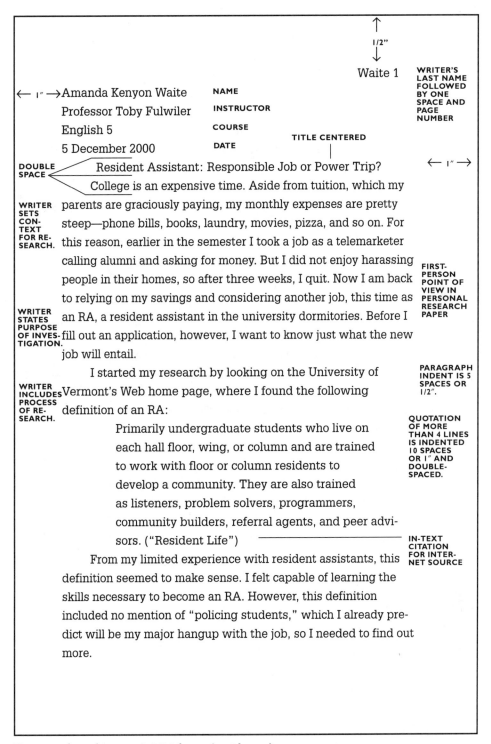

↑
1/2"
↓

Waite 1

WRITER'S
LAST NAME
FOLLOWED
BY ONE
SPACE AND
PAGE
NUMBER

← 1" → Amanda Kenyon Waite   NAME

Professor Toby Fulwiler   INSTRUCTOR

English 5   COURSE

5 December 2000   DATE

TITLE CENTERED

DOUBLE SPACE   Resident Assistant: Responsible Job or Power Trip?   ← 1" →

College is an expensive time. Aside from tuition, which my

WRITER SETS CONTEXT FOR RESEARCH. parents are graciously paying, my monthly expenses are pretty

steep—phone bills, books, laundry, movies, pizza, and so on. For

this reason, earlier in the semester I took a job as a telemarketer

calling alumni and asking for money. But I did not enjoy harassing   FIRST-PERSON POINT OF VIEW IN PERSONAL RESEARCH PAPER

people in their homes, so after three weeks, I quit. Now I am back

to relying on my savings and considering another job, this time as

WRITER STATES PURPOSE OF INVESTIGATION. an RA, a resident assistant in the university dormitories. Before I

fill out an application, however, I want to know just what the new

job will entail.

I started my research by looking on the University of   PARAGRAPH INDENT IS 5 SPACES OR 1/2".

WRITER INCLUDES PROCESS OF RESEARCH. Vermont's Web home page, where I found the following

definition of an RA:

QUOTATION OF MORE THAN 4 LINES IS INDENTED 10 SPACES OR 1" AND DOUBLE-SPACED.

> Primarily undergraduate students who live on
> each hall floor, wing, or column and are trained
> to work with floor or column residents to
> develop a community. They are also trained
> as listeners, problem solvers, programmers,
> community builders, referral agents, and peer advi-
> sors. ("Resident Life")

IN-TEXT CITATION FOR INTERNET SOURCE

From my limited experience with resident assistants, this

definition seemed to make sense. I felt capable of learning the

skills necessary to become an RA. However, this definition

included no mention of "policing students," which I already pre-

dict will be my major hangup with the job, so I needed to find out

more.

First page of a student essay in MLA format (no title page)

Waite 2

I made an appointment to talk to Sarah Daniels, a first-year RA who lives nearby. When I asked why she became an RA, she explained:

> Most of us do it for the money, but for me there was more to it. I was sick of all the hate crimes in the residence halls. People writing graffiti all over the place. I felt that by not doing anything it might as well have been me who was writing it. I saw becoming an RA as a way to educate people so that kind of stuff would stop.

After hearing Sarah's reasons, I felt a little guilty for thinking about this job just for the money. However, since the pay for being an RA is free room and board, it actually will be my parents, not I, who save money. (Of course, if I do this, I'm sure I can talk them into paying my phone bills!) Still, I am all for stopping students from messing up the walls, so this chance makes the work meaningful. Sarah went on to describe the RA training as intense: "[It] never really stops. All RAs have to take a ten-week course, too, and discuss a theme each week. Last week we talked about sexuality. This week I think it's alcohol."

I knew that students can get written up for anything from lighting candles and incense to possessing a toaster or stumbling around drunk, but when I asked Sarah about busting students, she said, "I really haven't had too much experience with that."

Later, when I talked with friends in the dining hall, I found several who believed that RAs are just students on power trips, using their position to cause problems for people they don't like. I understood this bias, since RAs are in a position of authority over people the same age or even older, but I'm sure I would not use my position unfairly. I really do not enjoy getting other people into trouble, so if I were an RA, I don't think that students would feel this way about me.

Waite 3

I also learned that some students live much closer to RAs than they would like to. In the Living and Learning Center, for example, rooms are set up in suite style, with six rooms surrounding a common dining room, so residents sometimes live in the same complex as their RAs. One student in this situation who wished to remain anonymous said, "It feels like we need to sneak around, like she's our mom almost." Another, who also wants her name withheld, said, "They're so hypocritical. They bust you for being underage and drinking, but you can find the same 18-, 19-, 20-year-old RAs drinking off campus at their friends' houses." Karen Corey, however, went on record to say, "It's good to live with [an RA] because she keeps us informed about events and policies."

ANONYMOUS SOURCE IS NOT CITED ON WORKS CITED PAGE.

Still, I felt a little discouraged talking to these students—discouraged mostly because I tended to agree with them. I do not want to be the person who has to spoil everyone else's fun, and I do not want to spoil my own fun either. I do not want to be a hypocrite who busts a student for something that I might do myself.

A few days later, I attended a meeting of the Future Educators program at which I had the chance to talk with Ben King, who had been an RA for two years. While we were doing the project of the night, basket weaving, I asked him how much time he spent each week on RA duties. He answered in detail:

> It takes a fair amount of time from your schedule. Aside from the required class, I would say the job takes five to ten extra hours a week, attending weekly meetings, preparing bulletin boards, programming, counseling, being on duty—which means making rounds in your whole building, which everyone is required to do four times a month. An average round takes half an hour to an hour and a half, depending on how rowdy the night is.

Waite 4

Ben's answer relieved some of my anxiety, since I know I could find the extra time to take on this job. I also have experience organizing activities and sending out information, so I know I can do that too. But hesitations about busting students were still on my mind, and Ben's comments didn't help:

> Enforcing policy—a nice way of saying busting—is a very small part of the job, but unfortunately, it's the thing RAs are most known for. I don't like busting students—it's definitely my least favorite part of the job. It is really confrontational, and I find that nerve-racking.

My search seemed to be leading me back and forth. Whenever I thought about my own leadership ability compared to that needed by an RA, I felt encouraged. For instance, one site on the World Wide Web, "The Study of Leadership Behaviors of Resident Assistants," defined a leader as "someone who challenges the process, inspires shared vision, enables others to act, models the way, and encourages the heart" (DeBenedictis et al., pars. 6-7). I feel that I have all these positive skills—to challenge, inspire, enable, model, and encourage.

But when I looked beneath the positive aspects of the job, the enforcement elements made me anxious. Another Web site posted a letter from RA Kate Lupton to the Daily Stanford college newspaper that raised concerns similar to mine about the role of the RAs on college campuses:

> The greatest tragedy of all would be a change in the RA role from community builder to police [officer]. At Stanford, RAs have the opportunity to develop student trust; if the situation continues on its present track, forcing us to assume an enforcer role, we will no longer have that chance.

WRITER'S DOUBTS CONTINUE AS CONSTANT THEME.

WRITER ACKNOWLEDGES AMBIGUITY DURING HER INVESTIGATION.

SOURCE IS INTRODUCED WITH SIGNAL PHRASE; NO FURTHER CITATION IS NEEDED.

Waite 5

I finally visited the university library to see what else I could find on the enforcement duties of RAs. A subject search turned up two books by M. Lee Upcraft that looked especially useful, Residence Hall Assistants in College and Learning to Be a Resident Assistant. According to Upcraft, before you sign on to become an RA, you need to ask yourself, "Can [I] turn in someone who is my friend?" (Learning 7). After reading the question, I immediately felt uncomfortable. According to Upcraft, if you couldn't turn in your friends, you shouldn't be an RA. This is exactly what I was afraid of; I know I do not want to put myself in this situation.

LIBRARY SOURCES CITED.

SHORTENED TITLE IS INCLUDED WHEN CITING MORE THAN ONE BOOK.

Recently, a friend of mine was written up for drinking in a suite with seven other UVM students. The Incident Report described the event this way: "As we entered the suite we all noticed beer cans and bottles and people drinking. We asked them to collect all of the beer, and Chad watched them pour it down the sink" (Hart). No matter how hard I thought about it, I couldn't imagine myself doing this.

INTERNAL UNIVERSITY DOCUMENT IS USED AS A SOURCE.

After this month-long search, I'm still undecided. Almost everything about being an RA sounds like something I could do and be good at. I have leadership, organizational, and communication skills. But I'm not sure I possess the ability to walk into students' rooms and force them to pour their good time down the drain. So, I have to answer one more question myself before I decide whether to apply: Is the risk of turning in my friends worth saving my parents $5,000 next year?

WRITER CONCLUDES SEARCH PROCESS BETTER INFORMED BUT STILL UNDECIDED.

Waite 6

Works Cited   HEADING CENTERED

← 1″ →   Corey, Karen. Personal interview. 12 Oct. 1998.

AUTHORS LISTED ALPHABETICALLY, LAST NAMES FIRST.   Daniels, Sarah. Personal interview. 10 Oct. 1998.

DeBenedictis, Michelle, et al. "The Study of Leadership Behav-

SECOND AND SUBSEQUENT LINES INDENT 5 SPACES OR 1/2″.   iors of Resident Assistants." West Chester U. Spring

1997. 13 pars. 15 Oct. 1988 <http://albie. wcupa.edu/

ttreadwell/971gp4.html>.

INTERNET SOURCES CONCLUDE WITH URL IN ANGLE BRACKETS.

Hart, Darrell. "Incident Report: U. of Vermont Resident Life

Services." 1 Oct. 1998.

King, Benjamin. Personal interview. 14 Oct. 1998.

Lupton, Kate. "Letter: Resident assistants fear role will

include policing students' drinking." 5 Nov. 1996. Stan-

ford U. 15 Oct. 1998 <http//:www.dailystanford.org/

Daily96-97/11-5/OP>.

"Resident Life," Aug. 1998. U. of Vermont. 10 Oct. 1998

<http://www.reslife.uvm.edu/resstaff/overview.htm>.

← 1″ →

AUTHOR'S NAME IN SECOND REFERENCE IS NOT REPEATED.   Upcraft, M. Lee. Learning to Be a Resident Assistant. San

Francisco: Jossey, 1982.

---. Residence Hall Assistants in College. San Francisco:

Jossey, 1982.

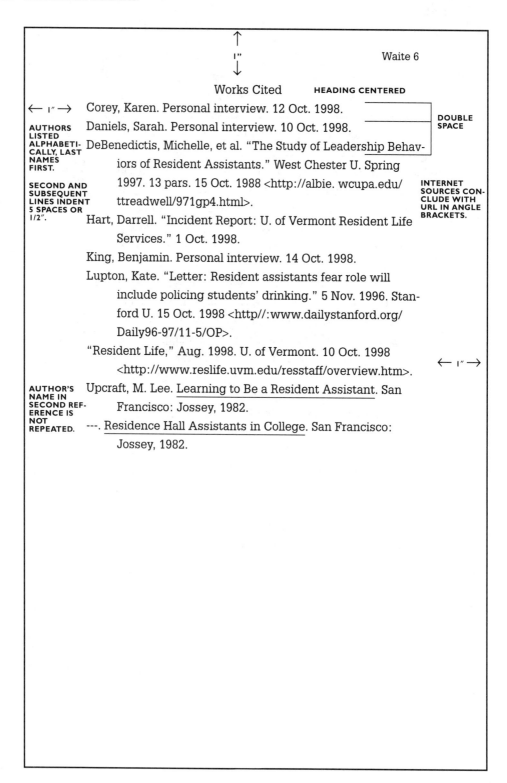

✂ ARGUMENTATIVE RESEARCH PAPER: MLA STYLE

For an example of an argumentative research essay with MLA documentation, see Chapter 12, "On the Trail: Can the Hikers Share with the Bikers?"

✂ INFORMATIONAL RESEARCH PAPER: APA STYLE

The research essay that follows, "Green Is Only Skin Deep: False Environmental Marketing," by Elizabeth Bone, was written in response to an assignment to identify and explain one problem in contemporary American culture. Bone's essay is documented according to the conventions of the American Psychological Association (APA). This sample includes title page, abstract, and outline page; check with your instructor to find out if these are or are not required for course papers.

ABBREVIATED TITLE

Green Is 1

PAGE NUMBERING
BEGINS ON TITLE
PAGE.

Green Is Only Skin Deep: TITLE

False Environmental Marketing

Elizabeth Bone   AUTHOR

Professor John Clark   INSTRUCTOR

TITLE PAGE IS
CENTERED TOP TO
BOTTOM, RIGHT TO
LEFT, AND DOUBLE-
SPACED.

Social Issues 2200   COURSE

December 6, 2000   DATE

**ABSTRACT SHOULD BE PRINTED ON A SEPARATE PAGE
FOLLOWING THE TITLE PAGE.**

Green Is 2

**HEADING
CENTERED**

Abstract

**NO
PARAGRAPH
INDENT**

**DOUBLE
SPACE**

**THE
ABSTRACT
SUMMARIZES
THE
MAIN
POINT OF
THE PAPER.**

Most Americans consider themselves environmentalists and favor supporting environmentally friendly or "green" companies. However, companies use a number of false advertising practices to mislead the public about their green practices and products by (1) exaggerating claims, (2) masking false practices behind technical terminology, (3) missponsoring green events, (4) not admitting responsibility for real problems, (5) advertising green by association, and (6) solving one problem while creating others. Consumers must be skeptical of all commercial ads and take the time to find out the truth behind advertising.

OUTLINE SHOULD FOLLOW THE TITLE PAGE AND
CONFORM TO TRADITIONAL OUTLINE FORMAT.

1/2"

Green Is 3

Outline — HEADING CENTERED

DOUBLE
SPACE

I.   Environmental consciousness is strong in Americans.

   A.   Gallup poll finds 75% are environmentalists.

   B.   False advertising betrays consumers.

II.   Definitions are exaggerated by the media and
government.

ROMAN
NUMERALS
INDICATE
MAJOR
DIVISIONS.

LETTERS
INDICATE
SUBDIVISIONS
AND
SUBORDINATE
POINTS.

   A.   Biodegradable plastic is false advertising.

   B.   Federal Trade Commission regulates definitions.

III.   Terminology is highly technical.

   A.   CFCs threaten our ozone layer.

   B.   Chrysler advertising misleads us about chemicals.

IV.   Some companies are green by sponsorship yet not green.

OUTLINE
USES
SENTENCE
FORMAT.

   A.   Ford supports the Smithsonian Institute Ocean
   Planet.

   B.   Ford is guilty of pollution in Michigan.

V.   "It's not my problem."

   A.   CFCs are not caused by natural gas.

   B.   Natural gas causes other pollution.

VI.   Many companies are only green by association.

   A.   Advertising has nothing to do with product.

   B.   Chevrolet logo implies relationship.

VII.   Some companies are single-minded in their
environmentalism.

   A.   Chevron employees do good in Mississippi.

   B.   Chevron pollutes Santa Monica Bay.

VIII.   Environmental image does not match reality.

   A.   Earth First! educates consumers.

   B.   Federal Trade Commission regulates.

   C.   Consumers beware!

TITLE IS REPEATED FROM TITLE PAGE.

Green Is Only Skin Deep:

False Environmental Marketing

A recent Gallup poll reported that 75% of Americans

AUTHOR'S NAME, DATE, AND PAGE NUMBERS ARE IN PARENTHESES.

DOUBLE SPACING

consider themselves to be environmentalists (Smith & Quelch, 1993). In the same study, nearly half of the respondents said they would be more likely to purchase a product if they perceived it to be environmentally friendly or "green." According to Smith and Quelch (1993), since green sells, many companies have begun to promote themselves as marketing products that are either environmentally friendly or manufactured from recycled material. Unfortunately, many of these companies care more about appearance than reality.

INFORMATIONAL THESIS IS AT END OF FIRST PARAGRAPH.

The most common way for a company to market itself as pro environment is to stretch the definitions of terms such as "biodegradable" so that consumers believe one thing but the product delivers something else. For example, so-called biodegradable plastic, made with cornstarch, was introduced to ease consumers' fears that plastic lasts forever in the environment. However, the cornstarch plastic broke down only in specific controlled laboratory conditions, not outdoors and not in compost bins. The Federal Trade Commission has updated its regulations to prevent such misrepresentations, so that now Glad and Hefty trash bags are no longer advertised as biodegradable (Carlson, Grove, & Kangun, 1993).

FIRST EXAMPLE OF FALSE ADVERTISING INTRODUCED.

The use of technical terms can also mislead average consumers. For example, carbon fluoride compounds, called CFCs, are known to be hazardous to the protective layer of ozone that surrounds the earth, so that their widespread use in air conditioners is considered an environmental hazard (Decker & Stammer, 1989). Chrysler Corporation advertises that it uses CFC-free refrigerant in its automobile air conditioners to appeal to environmentally concerned consumers ("Ozone layer," 1994).

SECOND EXAMPLE IS GIVEN.

PAGE NUMBER IS NOT REQUIRED FOR A PARAPHRASE, BUT IS RECOMMENDED.

Green Is 5

However, Weisskopf (1992) points out that the chemical compounds that replace CFCs in their air conditioners pose other environmental hazards that are not mentioned.

Another deceptive greening tactic is the sponsoring of highly publicized environmental events such as animal shows, concerts, cleanup programs, and educational exhibits. For example, Ocean Planet was a well-publicized exhibit put together by the Smithsonian Institution to educate people about ocean conservation. Ford Motor Company helped sponsor the event, which it then used in its car advertisements: "At Ford, we feel strongly that understanding, preserving, and properly managing natural resources like our oceans should be an essential commitment of individuals and corporate citizens alike" ("Smithsonian Institution's Ocean Planet," 1995, p. 14).

While sponsoring the exhibit may be a worthwhile public service, such sponsorship has nothing to do with how the manufacture and operation of Ford automobiles affect the environment. In fact, Ford was ranked as among the worst polluters in the state of Michigan in 1995 (Parker, 1995).

Some companies court the public by mentioning environmental problems and pointing out that they do not contribute to those problems. For example, the natural gas industry describes natural gas as an alternative to the use of ozone-depleting CFC's ("Don't you wish," 1994). However, according to Fogel (1985), the manufacture of natural gas creates a host of other environmental problems from land reclamation to the carbon-dioxide pollution, a major cause of global warming. By mentioning problems they don't cause, while ignoring ones they do, companies present a favorable environmental image that is at best a half truth, at worst an outright lie.

Green Is 6

Other companies use a more subtle approach to misleading green advertising. Rather than make statements about environmental compatibility, these companies depict the product in unspoiled natural settings or use green quotations that have nothing to do with the product itself. For example, one Chevrolet advertisement shows a lake shrouded in mist and quotes an environmentalist: "From this day onward, I will restore the earth where I am and listen to what it is telling me" ("From this day," 1994). Below the quotation is the Chevy logo with the words "Genuine Chevrolet." Despite this touching appeal to its love of nature, Chevrolet has a history of dumping toxic waste into the Great Lakes (Allen, 1991). Has this company seriously been listening to what the earth has been telling it?

QUOTATION OF FEWER THAN 40 WORDS IS INTEGRATED INTO THE TEXT.

The most common manner in which companies attempt to prove they have a strong environmental commitment is to give a single example of a policy or action that is considered environmentally sound. Chevron has had an environmental advertising campaign since the mid-1970s. More recent ads feature Chevron employees doing environmental good deeds (Smith & Quelch, 1993). For example, one ad features "a saltwater wetland in Mississippi at the edge of a pine forest . . . the kind of place nature might have made" and goes on to explain that this wetland was built by Chevron employees ("The shorebirds who found," 1990, p. 000). However, during the time this advertisement was running in magazines such as *Audubon*, LaGanga (1993) points out that Chevron was dumping millions of gallons of nasty chemicals (carcinogens and heavy metals) into California's Santa Monica Bay, posing a health risk to swimmers. The building of the wetland in one part of the country does not absolve the company of polluting water somewhere else.

ELLIPSIS POINTS INDICATE MISSING WORDS IN QUOTATION.

It should be clear that the environmental image a company

projects does not necessarily match the realities of the company's practice. The products produced by companies such as Chrysler, Ford, General Motors, and Chevron are among the major causes of air and water pollution: automobiles and gasoline. No amount of advertising can conceal the ultimately negative effect these products have on the environment (Kennedy & Grumbly, 1988). According to Shirley Lefevre, president of the New York Truth in Advertising League:

DOUBLE SPACING

> It probably doesn't help to single out one automobile manufacturer or oil company as significantly worse than the others. Despite small efforts here and there, all of these giant corporations, as well as other large manufacturers of metal and plastic material goods, put profit before environment and cause more harm than good to the environment. (personal communication, May 1995)

QUOTATION OF 40 WORDS OR MORE IS INDENTED 5 SPACES

COLON IS USED TO INTRODUCE A LONG QUOTATION.

INTERVIEW CONDUCTED BY AUTHOR AND OTHER PPERSONAL COMMUNICATIONS ARE CITED IN TEXT ONLY, NOT ON THE REFERENCE PAGE.

Consumers who are genuinely interested in buying environmentally safe products and supporting environmentally responsible companies need to look beyond the images projected by commercial advertising in magazines, on billboards, and on television. Organizations such as Earth First! attempt to educate consumers to the realities by writing about false advertising and exposing the hypocrisy of such ads ("Do people allow," 1994), while the Ecology Channel is committed to sharing "impartial, unbiased, multiperspective environmental information" with consumers on the Internet (Ecology, 1996). Meanwhile the Federal Trade Commission is in the process of continually upgrading truth-in-advertising regulations (Carlson et al., 1993). Americans who are truly environmentally conscious must remain skeptical of simplistic and misleading commercial advertisements while continuing to educate themselves about the genuine needs of the environment.

SECOND CITATION OF MORE THAN THREE AUTHORS IS SHORTENED TO FIRST AUTHOR'S NAME AND "ET AL."

THESIS IS REPEATED IN MORE DETAIL AT END.

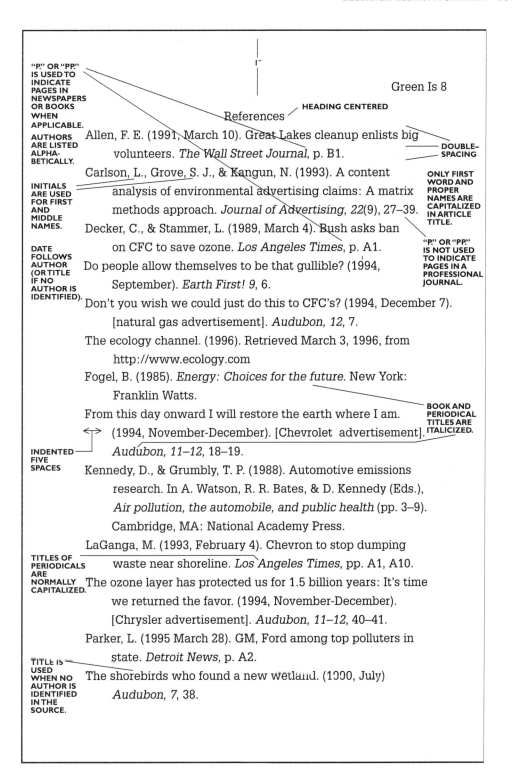

Green Is 8

References

Allen, F. E. (1991, March 10). Great Lakes cleanup enlists big volunteers. *The Wall Street Journal*, p. B1.

Carlson, L., Grove, S. J., & Kangun, N. (1993). A content analysis of environmental advertising claims: A matrix methods approach. *Journal of Advertising, 22*(9), 27–39.

Decker, C., & Stammer, L. (1989, March 4). Bush asks ban on CFC to save ozone. *Los Angeles Times*, p. A1.

Do people allow themselves to be that gullible? (1994, September). *Earth First! 9*, 6.

Don't you wish we could just do this to CFC's? (1994, December 7). [natural gas advertisement]. *Audubon, 12*, 7.

The ecology channel. (1996). Retrieved March 3, 1996, from http://www.ecology.com

Fogel, B. (1985). *Energy: Choices for the future*. New York: Franklin Watts.

From this day onward I will restore the earth where I am. (1994, November-December). [Chevrolet advertisement]. *Audubon, 11–12*, 18–19.

Kennedy, D., & Grumbly, T. P. (1988). Automotive emissions research. In A. Watson, R. R. Bates, & D. Kennedy (Eds.), *Air pollution, the automobile, and public health* (pp. 3–9). Cambridge, MA: National Academy Press.

LaGanga, M. (1993, February 4). Chevron to stop dumping waste near shoreline. *Los Angeles Times*, pp. A1, A10.

The ozone layer has protected us for 1.5 billion years: It's time we returned the favor. (1994, November-December). [Chrysler advertisement]. *Audubon, 11–12*, 40–41.

Parker, L. (1995 March 28). GM, Ford among top polluters in state. *Detroit News*, p. A2.

The shorebirds who found a new wetland. (1990, July) *Audubon, 7*, 38.

Annotations (left and right margins):

"P." OR "PP." IS USED TO INDICATE PAGES IN NEWSPAPERS OR BOOKS WHEN APPLICABLE.

AUTHORS ARE LISTED ALPHABETICALLY.

INITIALS ARE USED FOR FIRST AND MIDDLE NAMES.

DATE FOLLOWS AUTHOR (OR TITLE IF NO AUTHOR IS IDENTIFIED).

INDENTED FIVE SPACES

TITLES OF PERIODICALS ARE NORMALLY CAPITALIZED.

TITLE IS USED WHEN NO AUTHOR IS IDENTIFIED IN THE SOURCE.

HEADING CENTERED

DOUBLE-SPACING

ONLY FIRST WORD AND PROPER NAMES ARE CAPITALIZED IN ARTICLE TITLE.

"P." OR "PP." IS NOT USED TO INDICATE PAGES IN A PROFESSIONAL JOURNAL.

BOOK AND PERIODICAL TITLES ARE ITALICIZED.

Smith, N. C., & Quelch, J. A. (1993. *Ethics in marketing*.
Boston: Richard D. Irwin.

Smithsonian Institution's Ocean Planet: A Special report. (1995,
March). *Outdoor Life, 3*, 13–22.

Weisskopf, M. (1992, February 23). Study finds CFC alternatives
more damaging than believed. *The Washington Post*,
p. A3.

PART FOUR

REVISING AND EDITING

Chapter

ֆ 22 ֆ

Strategies for Revision

Writing. I'm more involved in it. But not as attached. I used to really cling to my writing and
didn't want it to change. Now I can see the usefulness of change. I just really like my third
draft, but I have to let it go. I can still enjoy my third draft and make an even better fourth.

KAREN

A first draft is a writer's first attempt to give shape to an idea, argument, or experience. Occasionally, this initial draft is just right and the writing is done. More often, however, the first draft shows a broad outline or general direction that needs further thinking and further revision. An unfocused first draft, in other words, is not a mistake but rather a start toward a next, more focused draft.

No matter how much prior thought writers give to complex composing tasks, once they begin writing, the draft begins to shift, change, and develop in unexpected ways. Each act of writing produces new questions and insights that must be dealt with and incorporated into the emerging piece of writing; it is during this process that active and aggressive revision strategies can help. Inexperienced writers often view revising as an alien activity that neither makes sense nor comes easy. However, most experienced writers view revising as the essence of writing, the primary way of developing thoughts to be shared with others.

ֆ UNDERSTANDING REVISION

The terms *revising, editing,* and *proofreading* are sometimes used to mean the same thing, but there is good reason to understand each as a separate process, each in its own way contributing to good finished writing. *Revising* is reseeing, rereading, rethinking, and reconstructing your thoughts on paper until they match those in your mind. It's conceptual work, generally taking place beyond the sentence, at the level of the paragraph and higher.

In contrast, *editing* is changing language more than ideas. You edit to make precise what you want to say, testing each word or phrase to see that it is accurate, appropriate, necessary. Editing is stylistic and mechanical work, generally taking place at the level of the sentence or word.

Proofreading is checking a manuscript for accuracy and correctness. It is the last phase of the editing process, completed after conceptual and stylistic concerns have been addressed. When you proofread, you review spelling, punctuation, capitalization, and usage to make sure no careless mistakes have occurred that might confuse or distract readers.

There are two good reasons to revise before you edit. First, in revising you may cut out whole sections of a draft because they no longer suit your final purpose. If you have already edited those now-deleted sections, all that careful work goes for naught. Second, once you have invested time in carefully editing sentences, you become reluctant to cut them, even though these sections may no longer suit your purpose. Of course, writers are always circling back through the stages, editing when it makes more sense to revise, inventing when they mean to edit. Nonetheless, you will save time if you revise before editing, and edit before proofreading.

> ### ᠵ WRITING 1
>
> Describe any experience you've had with revising papers: Was it for a school assignment or some writing on your own? Why did you revise? How many drafts did you do? Were you pleased? Was your audience?

ᠵ PLANNING TO REVISE

You cannot revise if you haven't first written, so write early and leave time to revise later. Good college papers are seldom written in one draft the night before they are due. When you plan in advance to revise, the following tools and techniques will serve you well:

Keep a revision notebook. When you begin any substantial writing project, keep a notebook or journal in which to capture all ideas related to your paper, including invention, drafting, research, and revision ideas. Over the span of several days or weeks, your revision may profit from your returning to earlier information, ideas, or insights. (See Chapter 3 for more about keeping a journal.)

Impose due dates. Write the due date for a final draft on your calendar, then add earlier, self-imposed due dates for first, second, or third drafts. Your self-imposed or false due dates will guarantee you the time you need to revise well.

Write and rewrite with a computer. Computers make revising easier and more effective. Any kind of word-processing program allows you to change your text infinitely before ever calling it finished. The computer allows you to change words and sentences as well as move blocks of text from one part of your paper to another with ease—all essential acts of revising.

Read hard copy. When revising with a computer, print out hard copy of your drafts and see how they read on paper. Hard copy lets you scan several pages at a time and quickly flip pages in search of certain patterns or information.

Save draft copies. Make backup files of old drafts on floppy disks or on your hard drive; if you become unhappy with your revisions, you can always return to the earlier copy.

> ᘜ **WRITING 2**
>
> Describe your approach to writing a paper from the time it's assigned to the time you hand it in. Do you do any of the prerevision work described above? Which of these general strategies makes sense in view of your current writing habits?

ᘜ ASKING REVISION QUESTIONS

To begin revising, return to the basic questions of purpose, audience, and voice: Why am I writing? To whom? In what voice?

Questions of Purpose

It is often easier to see your purpose—or lack thereof—most clearly after you have written a draft or two. Ask the following questions:

1. Why am I writing this paper? (Review the assignment.)
2. Do all parts of the paper advance this purpose? (Outline by paragraph and make sure they do.)
3. What is my rhetorical strategy: to narrate, explain, interpret, argue, reflect, or something else? (Review appropriate chapters to fine-tune strategy.)
4. Have I stated the paper's theme or thesis clearly? (If not, do so, or have a good reason for not doing so.)

Questions of Audience

Make sure your paper is aimed accurately at your readers by asking the following questions:

1. What does my audience know about this subject? (Avoid repeating elementary information.)
2. What does my audience need to know to understand the point of my paper? (Provide full context and background for information your audience is not likely to know.)
3. What questions or objections do I anticipate my audience raising? (Try to answer them before they are asked.)

Questions of Voice

Make sure your paper satisfies you. Revise it as necessary so you say what you intend in the voice you intend by asking the following questions:

1. Do I believe everything I've written in this paper? (Eliminate nonsense and filler.)
2. What doubts do I have about my paper? (Address these; don't avoid them.)

> ‿ゆ **WRITING 3**
>
> Can you think of other questions to ask about purpose, audience, and voice?

3. Which passages sound like me speaking and which don't? (Enjoy those that do; fix those that don't.)

‿ゆ REVISING STRATEGIES

For many writers, revising seems to be an instinctive or even unconscious process—they just do it. However, even experienced writers might profit by pausing to think deliberately about what they do when they revise.

This section lists more than a dozen time-tested revision strategies that may be useful to you. While they won't all work for you all the time, some will be useful at one time or another. Notice that these suggestions start with larger concerns and progress toward smaller ones.

Four revising strategies—*limiting, adding, switching,* and *transforming*—are complicated enough that I'll treat them in detail in the next chapter. Limiting is focusing on a narrow portion of a paper or concept and eliminating extraneous material; adding is incorporating new details and dialogue to make writing more vivid and powerful. Switching and transforming are more innovative strategies for revision: by changing the tense, point of view, form, or genre of a piece of writing, writers can gain insights into their writing and present their ideas in a new light. (For more information on focused revision, see Chapter 23.)

Establish Distance

Let your draft sit for a while, overnight if possible, then reread it to see if it still makes sense. A later reading provides useful distance from your first words, allowing you to see if there are places that need clarification, explanation, or development that you did not see when drafting. You can gain distance also by reading your draft aloud—hearing instead of seeing it—and by sharing it with others and listening to their reactions. No matter how you gain it, with distance you revise better.

Reconsider Everything

Reread the whole text from the beginning: Every time you change something of substance, reread again to see the effect of these changes on other parts of the text. If a classmate or instructor has made comments on some parts of the paper and not on others, do not assume that those are the only places where revision is needed.

Believe and Doubt

First, reread your draft as if you wanted to believe everything you wrote (imagine you are a supportive friend), putting checks in the margins next to passages that create the most belief—the assertions, the dialogue, the details, the evidence. Next, reread your draft as if you were suspicious and skeptical of all assertions (imagine your most critical teacher), putting question marks next to questionable passages. Be pleased with the checkmarks, and answer the question marks.

Test Your Theme or Thesis

Most college papers are written to demonstrate a theme or thesis (to outlaw handguns; to legalize marijuana). However, revision generates other ideas, raises new questions, and sometimes reshapes your thesis (license handguns; legalize hemp as a cash crop). Make sure to modify all parts of your paper to keep up with your changing thesis.

Evaluate Your Evidence

To make any theme or thesis convincing, you need to support it with evidence. Do your facts, examples, and illustrations address the following questions?

1. Does the evidence support my thesis or advance my theme? (In states that license handgun ownership, crime rates have decreased.)
2. What objections can be raised about this evidence? (The decrease in crime rates has other causes.)
3. What additional evidence will answer these objections? (In states that do not license handguns, crime has not decreased.)

Make a Paragraph Outline

The most common unit of thought in a paper is the paragraph, a group of sentences set off from other groups of sentences because they focus on the same main idea. Make a paragraph outline to create a map of your whole paper and see whether the organization is effective: Number each paragraph and write a phrase describing its topic or focus. Does the subject of each paragraph lead logically to the next? If not, reorganize.

Rewrite Introductions and Conclusions

Once started, papers grow and evolve in unpredictable ways: An opening that seemed appropriate in an earlier draft may no longer fit. The closing that once ended the paper nicely may now fail to do so. Examine both introduction and conclusion to be sure they actually introduce and conclude. Sometimes it is more helpful to write fresh ones than to tinker with old ones. (For more on openings and conclusions, see Chapter 24.)

Listen for Your Voice

In informal and semiformal papers, your language should sound like a real human being speaking. Read your paper out loud and see if the human being speaking sounds like you. If it doesn't, revise so that it does. In more formal papers, the language should sound less like you in conversation and more like you giving a presentation—fewer opinions, more objectivity, no contractions.

Let Go

View change as good, not bad. Many writers become overly attached to the first words they generate, proud to have found them in the first place, now reluctant to abandon them. Learn to let your words, sentences, and even paragraphs go. Trust that new and better ones will come.

Start Over

Sometimes revising means starting over completely. Review your first draft, then turn it face down and start fresh. Starting over generates your best writing, as you automatically delete dead-end ideas, making room for new and better ones to emerge. (Many writers have discovered this fact accidentally, by deleting a file on a computer and thus being forced to reconstruct, almost always writing a better draft in the process.)

---

༠ WRITING 4

Look over the suggestions for revision in this section. Which of them have you used in the past? Which seem most useful to you now? Which seem most far-fetched?

---

༠ A REVISION CHECKLIST

1. Reread the assignment and state it in your own words. Does your paper address it?
2. Restate your larger purpose. Has your paper fulfilled it?
3. Consider your audience. Have you told your readers what they want and need to know?
4. Read the text out loud and listen to your voice. Does the paper sound like something you would say?
5. Restate the paper's thesis or theme in a single sentence. Is it stated in the paper? Does the paper support it?

6. List the specific evidence that supports this thesis or theme. Is it sufficient? Is it arranged effectively?
7. Outline the paper paragraph by paragraph. Is the development of ideas clear and logical?
8. Return to your introduction. Does it accurately introduce the revised paper?
9. Return to your conclusion. Does it reflect your most recent thoughts on the subject?
10. Return to your title and write five alternative titles. Which is the best one?

↭ SUGGESTIONS FOR WRITING AND RESEARCH

Individual

Select any paper that you previously wrote in one draft but that you believe would profit from revision. Revise the paper by following some of the revision strategies and suggestions in this chapter.

Collaborative

Have each member of the class go to the library and research the revision habits of a favorite or famous writer. If you cannot find such information, interview a professor, teacher, or person in your community who is known to write and publish. Find out about the revision process he or she most often uses. Write a report in which you explain the concept of revision as understood and used by the writer you have researched, and publish all the reports in a class anthology.

Chapter

ᔐ *23* ᔐ

Focused Revision

I never realized before that in revising you can do drafts from totally different perspectives and keep experimenting with your ideas. When you write the final draft, it could be totally different from how you expected it to come out in the beginning.

GARY

I want my audience to feel like they're actually attending the game, that they're sitting just behind the bench, overhearing Coach telling us how to defend against the in-bounds pass, and I can do this if I just close my eyes while I write and remember being there—I can put you at the game.

KAREN

Have you ever found yourself running out of ideas, energy, or creativity on what seemed to be a perfectly good topic for a paper? Have you ever been told to rewrite, revise, review, redo, rethink a paper, but didn't know exactly what those suggestions meant? Have you ever written a paper you thought was carefully focused and well researched but also was dull and lifeless?

Odds are you're not alone. When anyone writes a first draft—especially on an assigned topic to which he or she has given little prior thought—it's easy to summarize rather than analyze, to produce generalities and ignore specifics, to settle for clichés rather than invent fresh images, to cover too much territory in too little time. In fact, most first drafts contain more than their share of summary, generalization, superficiality, and cliché, since most first-draft writers are feeling their way and still discovering their topic. In other words, it's seldom a problem if your first draft is off the track, wanders a bit, and needs refocusing. However, it *is* a problem if, for your second draft, you don't know what to do about it.

The best way to shape a wandering piece of writing is to return to it, reread it, slow it down, take it apart, and build it back up again, this time attending more carefully to purpose, audience, and voice. Celebrate first-draft writing for what it is—a warm-up, a scouting trip—but plan next to get on with your journey in a more deliberate and organized fashion. Sometimes you already know—or your readers tell you—exactly where to go.

Other times, you're not sure and need some strategies to get you moving again. This chapter offers four specific strategies for restarting, reconceiving, and refocusing a stuck paper.

✧ LIMITING

Broad topics lead to superficial writing. It's difficult to recount a four-week camping trip, to explain the meaning of *Hamlet,* or to solve the problems of poverty, crime, or violence in a few double-spaced pages. You'll almost always do better to cover less ground in more pages. Instead, can you *limit* your focus to one pivotal day on the trip? Can you explain and interpret one crucial scene? Can you research and portray one real social problem in your own backyard?

Limit Time, Place, and Action

When a first draft attempts to describe and explain actions that took place over many days, weeks, or months, try limiting the second draft to actions that took place on one day, on one afternoon, or in one hour. Limiting the amount of time you write about automatically limits the action (what happened) and place (where it happened) as well. For example, in the first draft of a paper investigating the homeless in downtown Burlington, Vermont, Dan began with a broad sweep:

> In this land of opportunity, freedom takes on different meaning for different people. Some people are born to wealth, others obtain it by the sweat of their brows, while average Americans always manage to get by. But others, not so fortunate or talented, never have enough food or shelter to make even the ends of their daily lives meet.

While there is nothing inherently wrong with this start, neither is there anything new, interesting, or exciting. The generalizations about wealth and poverty tell us only what we already know; there are no new facts, information, or images to catch our attention and hold it for the pages still to come.

Before writing his second draft, Dan visited the downtown area, met some homeless people, and observed firsthand the habits of a single homeless man named Brian; then he limited his focus and described what he witnessed on one morning:

> Dressed in soiled blue jeans and a ragged red flannel shirt, Brian digs curiously through an evergreen bush beside a house on Loomis Street. His yellow mesh baseball cap bears no emblem or logo to mark him a member of any team. He wears it low, concealing any expression his eyes might disclose. After a short struggle, he emerges from the bush, a Budweiser can in hand, a grin across his face. Pouring out the remaining liquid, he tosses the can into his shopping cart among other aluminum, glass, and plastic containers. He pauses, slides a Marlboro out of the crumpled pack in his breast pocket, lights it, and resumes his expedition.

While only one small act happens in this revised first paragraph—the retrieving of a single beer can—that act anticipates Dan's story of how unemployed homeless people earn money. By starting with a single detailed—and therefore convincing—scene, Dan writes more about less; in the process, he teaches his readers specific things about people he originally labeled "not so fortunate or talented." By describing instead of evaluating or interpreting this scene, he invites readers to make their own inferences about what it means. In other words, writing one specific, accurate, nonjudgmental scene asks readers to interpret and therefore engage more deeply in the text.

Limit Scope

In the process of Dan's researching and writing, the scope of his paper became progressively more restricted: In draft one, he focused on the homeless in America; in draft two, he focused on the homeless in downtown Burlington; in draft three, he focused on those homeless people who collect cans for income. From his initial limitation in time came a consequent limitation in scope, and a distinct gain in specificity, detail, and reader interest.

One technique for limiting the scope of any type of paper is to identify the topic of any one page, paragraph, or sentence in which something important or interesting is introduced. Begin your next paper with that specific topic, focusing close now and limiting the whole draft to only that topic. For example, in a paper arguing against the clear-cutting of forests, focus on one page describing the cutting of Western red cedar in one specific place; limit the next whole draft to that single subject. In a paper examining the exploitation of women in television advertising, focus on one paragraph describing a single soda ad; limit the next whole draft to that single subject. In a paper examining your high school soccer career, focus on one sentence describing the locker room after the loss of an important game. When you limit scope, you gain depth.

> ⌇ **WRITING 1**
>
> Devote a portion of your journal or class notebook exclusively to exploring the revision possibilities of one paper. For your first entry, reread the paper you intend to revise, and limit either the time or scope that you intend to cover in the second draft.

⌇ ADDING

A sure way to increase reader interest in a paper, and your own interest as well, is to *add* new and specific material to that overly general first draft. Whether you are arguing about the effects of mountain bikes on the wilderness, explaining the situation of the homeless people downtown, or interpreting the poems of Gwendolyn Brooks, it is your job to become the most informed expert on this subject in your writing class. It's your job to read the necessary articles, visit the appropriate places, interview the relevant people who will make you the authority to write the paper. On first drafts, neither your instructor nor classmates expect you to be this authority; on subsequent drafts, their expectations increase.

Add Expert Voices

The surest way to locate new information to add to next drafts is to read widely and listen carefully. Get to the library and locate sources that supplement and substantiate your own knowledge. Enlist the support of experts by citing them in your paper, identifying who they are and why they should be listened to. Also get out into the field and talk to people who are the local experts on your subject. Quote these experts, too, and include their voices in your next draft.

Although textual quotations are helpful and expected in academic papers, they are seldom so locally specific or lively as interview quotations from local people. In many instances where little may have been published on local issues, the only way to get up-to-date local information is from talking to people. Quoting people directly not only adds new and credible information to your paper, it invariably adds a sense of life as well. For example, as Dan continued his story of Burlington's homeless people, he interviewed a number of people, such as police officer Pat Hardy, who had firsthand knowledge of the homeless can collectors:

> "They provide a real service to the community," he explains. "You'd see a lot more cans and bottles littering the streets if they weren't out here working hard each day. I've never had a problem with any of them. They are a real value."

While Dan himself could have made the same observation, it has greater authority and life coming from a cop on the beat.

In another instance, a team of first-year students collaborated to write a profile of the local Ronald McDonald House, a nonprofit organization providing free room and board for the families of hospital patients. In their first draft, they researched the local newspaper for introductory information on the origins of this institution. It was useful information, but without much life:

> The McDonald's corporation actually provided less than 5 percent of the total cost of starting the Ronald McDonald House. The other 95 percent of the money came from local businesses and special-interest groups.

For their second draft, the group interviewed the director of the Ronald McDonald House and used her as an additional and more current source of information. In fact, they devoted the entire second draft to material collected through interviews with the director and staff at the house. In the following sample, the director substantiates the information from the initial newspaper story but adds more specific, local, and lively details.

> "Our biggest problem is that people think we're supported by the McDonald's corporation. We have to get people to understand that anything we get from McDonald's is just from the particular franchise's generosity—and may be no more than is donated by other local merchants. Martins, Hood, and Ben and Jerry's provide much of the food. McDonald's is not obligated to give us anything. The only reason we use their name is because of its child appeal."

The final profile of the Ronald McDonald House included information ranging from newspaper and newsletter stories to site descriptions and interviews with staff, volunteers, and family.

Add Details

If you quickly review this chapter's samples of revision by *limiting* and *adding,* you will notice the increase in specific detail. Focusing close, interviewing people, and researching texts all produce specific information, which adds both energy and evidence to whatever paper you are writing. In the can-collecting paper, the visual details make Brian come alive—*soiled blue jeans, red flannel shirt, yellow mesh baseball cap, Budweiser can, Marlboro.* In the Ronald McDonald revisions, the newspaper statistics add authority (*5 percent of the total cost*) while the interview information adds specificity (*Martins, Hood, and Ben and Jerry's*), both of which help explain the funding of this nonprofit organization.

> ### ↷ WRITING 2
>
> Identify texts, places, or people that contain information relevant to your paper topic and go collect it. If you are writing a paper strictly from memory, close your eyes and visit this place in your imagination: Describe the details and re-create the dialogue you find there.

↷ SWITCHING

Another strategy for focusing a second or third draft is to deliberately alter your customary way of viewing and thinking about this topic. One sure way to change how you see a problem, experience, story, issue, or idea is to *switch* the perspective from which you view it (the point of view), the language in which you portray it (the verb tense), or the audience for whom you are writing.

Switch Point of View

Switching the point of view from which a story, essay, or report is written means changing the perspective from which it is told. For example, in recounting personal experience, the most natural point of view is the first person (*I, we*) as we relate what happened to us. Here Karen writes in the first person in reporting her experience participating in the Eastern Massachusetts women's basketball tournament.

> We lost badly to Walpole in what turned out to be our final game. I sat on the bench most of the time.

However, Karen opened the final draft of her personal experience basketball narrative with a switch in point of view, writing as if she were the play-by-play announcer broadcasting the game at the moment, in this case moving to third person *and* adopting a new persona as well:

> Well folks, it looks as if Belmont has given up; the coach is preparing to send in his subs. It has been a rough game for Belmont. They stayed in it during the first

quarter, but Walpole has run away with it since then. Down by twenty with only six minutes left, Belmont's first sub is now approaching the table.

In her final draft, Karen opened from the announcer's point of view for one page, then switched for the remainder of the paper to her own first-person perspective, separating the two by white space. Karen's switch to announcer is credible (she *sounds* like an announcer); if she chose to narrate the same story from the perspective of the bouncing basketball, it might seem silly. (See Chapter 10 for Karen's complete essay.)

In research writing, as opposed to personal narrative, the customary point of view for reporting research results is third person (*he, she, it*) to emphasize the information and de-emphasize the writer. For example, the profile of the Ronald McDonald House begins, as you might expect, with no reference to the writers of the report:

> The Ronald McDonald House provides a home away from home for out-of-town families of hospital patients who need to visit patients for extended periods of time but cannot afford to stay in hotels or motels.

However, in one of their drafts, the writers switched to first person and wrote a more personal and impressionistic account explaining their feelings about reporting on this situation. While the impressionistic draft did not play a large part in the final profile, some of it remained purposefully in their final draft as they reported where they had had difficulties.

> In this documentary, we had a few problems with getting certain interviews and information. Since the house is a refuge for parents in distress, we limited the kinds of questions we asked. We didn't want to pry.

Switch Tense

Switching verb tense means switching the time frame in which a story or experience occurs. While the present tense is a natural tense for explaining information (see the Ronald McDonald example above), the most natural tense for recounting personal experience is the past tense, as we retell occurrences that happened sometime before the present moment—the same tense Karen adopted in the first draft of her basketball essay. However, her final draft is written entirely in the present tense, beginning with the announcer and continuing through to the end of her own narrative.

> It's over now, and I've stopped crying, and I'm very happy. In the end I have to thank—not my coach, not my team—but Walpole for beating us so badly that I got to play.

The advantage of switching to the present tense is that it lets you reexperience an event, and doing that, in turn, allows you to reexamine, reconsider, and reinterpret it—all essential activities for successful revision. At the same time, readers participate in the drama of the moment, waiting along with you to find out what will happen next. The disadvantage is that the present tense is associated with fiction—it's difficult or impossible to write while you're doing something else, like playing basketball. It's also difficult to reflect on experience if you're pretending it's occurring as you write.

Switch Sides

Another way to gain a new revision perspective is to switch sides in arguing a position: Write your first draft supporting the "pro" side, then write a second draft supporting the "con" side. For example, Issa, a dedicated mountain bike enthusiast, planned to write in favor of opening up more wilderness trails for use by mountain bikers. However, before writing his final draft, he researched the arguments against his position and wrote from that point of view:

> The hikers and other passive trail users argue against allowing mountain bikes onto narrow trails traditionally traveled only by foot and horse. They point out that the wide, deeply treaded tires of the mountain bikes cause erosion and that the high speeds of the bikers startle and upset both hikers and horses. According to hiker Donald Meserlain, the bikes "ruin the tranquillity of the woodlands and drive out hikers, bird watchers, and strollers" (Hanley 4).

For the writer, the main advantage of switching sides for a draft may be a better understanding of the opposition's point of view, making for a more effective argument against it in the final draft.

For his final draft, Issa argues his original position in favor of mountain bikes, but he does so with more understanding, empathy, and effectiveness because he spent a draft with the opposition. His final draft makes it clear where he stands on the issue:

> Educated mountain biking, like hiking and horseback riding, respects the environment and promotes peace and conservation, not noise and destruction. Making this case has begun to pay off, and the battle over who walks and who rides the trails should now shift in favor of peaceful coexistence. Buoyed by studies showing that bicycle tires cause no more erosion or trail damage than the boots of hikers, and far less than horses' hooves, mountain bike advocates are starting to find receptive ears among environmental organizations (Schwartz 78).

The tone of the mountain bike essay is now less strident and more thoughtful—an approach apparently brought about by his spending time seriously considering the objections of the opposition. (See Chapter 12 for the complete essay.)

Another switch that pays good dividends for the writer is changing the audience to whom the paper is being written. In college writing situations, the final audience always includes the instructor, so such a change may simply be a temporary but useful fiction. Had a draft of the mountain bike essay been aimed at the different constituencies mentioned in the essay—the Sierra Club, mountain bicycle manufacturers, property owners, or local newspapers—the writer might have gained a useful perspective in attempting to switch language and arguments to best address this more limited readership. Likewise,

---

ᢅ WRITING 3

Write in your journal about a past experience, using the present tense and/or third-person point of view. Then reread the passage and describe its effect on you as both writer and reader.

drafts of various papers written to young children, empathetic classmates, skeptical professors, or sarcastic friends may also provide useful variations in writer perspective.

✣ TRANSFORMING

To *transform* a text is to change its form by casting it into a new form or genre. In early drafts, writers often attend closely to the content of their stories, arguments, or reports but pay little attention to the form in which these are presented, accepting the genre as "school paper" only. However, recasting ideas and information into different and more public genres presents them in a different light. The possibilities for representing information in different genres are endless, since anything can become anything else. Consequently, keep in mind that some transformations are useful primarily to help you achieve a fresh perspective during the revision process, while others are more appropriate for presenting the information to readers.

In the world outside of college, it is actually common for research information to be reported in different genres to different audiences. For example, in a business or corporate setting, the same research information may be conveyed as a report to a manager, a letter to a vice president, a pamphlet for the stockholders, and a news release for public media—and show up still later in a feature article in a trade publication or newspaper. As in the working world, so in college: Information researched and collected for any paper can be presented in a variety of forms and formats.

Transform Personal Experience from Essay to Journal

The journal form encourages informal and conversational language, creates a sense of chronological suspense, is an ideal form for personal reflection, substitutes dates for more complex transitions, and proves especially useful for conveying experience over a long period of time. For example, after several essay-like narratives written from a past-tense perspective, Jeff used the journal format to tell the story of his month-long camping trip with the organization Outward Bound. Following is an excerpt, edited for brevity, in which he describes his reactions to camping alone for one week, using a mix of past and present tenses:

Day 14   I find myself thinking a lot about food. When I haven't eaten in the morning, I tend to lose my body heat faster than when I don't. . . . At this point, in solo, good firewood is surprisingly tough to come by. . . .

Day 15   Before I write about my fifth day of solo, I just want to say that it was damn cold last night. I have a –20 degree bag, and I froze. It was the coldest night so far, about –25. . . .

Day 17   I haven't seen a single person for an entire week. I have never done this before, and I really don't want to do it again—not having anyone to talk to. Instead of talking, I write to myself. . . . If I didn't have this journal, I think I would have gone crazy.

Transform to Letters

An issue might be illuminated in a lively and interesting way by being cast as a series or exchange of letters. Each letter allows a different character or point of view to be expressed. For example, Issa's argumentative paper on mountain bike use in wilderness areas could be represented as a series of letters to the editor of a local paper arguing different sides of the controversy: from the perspective of a hiker, a horseback rider, a mountain bike rider, a forest ranger, a landowner, among others.

Transform to a Documentary

Radio, film, and television documentaries are common vehicles for hearing news and information. Virtually any research paper could be made livelier by being cast as a documentary film or investigative feature story. Full research and documentation would be required, as for formal academic papers; however, writers would use the style of the popular press rather than the MLA or APA. In fact, the final form of the profile of the Ronald McDonald House was written as a script for *Sixty Minutes* and opened with a Mike Wallace–type of reporter speaking into a microphone:

> Smith: Hello, this is John Smith reporting for <u>Sixty Minutes.</u> Our topic this week is the Ronald McDonald House. Here I am, in front of the house in Burlington, Vermont, but before I go inside, let me fill you in on the history of this and many other houses like it.

The final paper included sections with the fictional Smith interviewing actual staff members as well as some sections presented neutrally from the camera's point of view:

> Toward the back of the house, three cars and one camper are parked in an oval-shaped gravel driveway. Up three steps onto a small porch are four black plastic chairs and a small coffee table containing a black ashtray filled with cigarette butts.

Transform to a Book with Chapters

Teams of student writers can collaborate on writing short books with "chapters" exploring issues of common interest. Such a form could include a table of contents, preface, foreword, afterword, introduction, and so on. For example, Dan's report on the life of a can collector could become one chapter in a collaborative "book" investigating how the homeless live:

1. Housing for the Homeless
2. Dinner at the Salvation Army
3. Shopping at Goodwill
4. Brian: Case Study of a Can Collector
5. Winter Prospects

Transform to a Magazine Article

If you are investigating consumer products, such as mountain bikes and CD's, consider writing the final draft as a report for *Consumer Reports*. If you are investigating an issue such as homelessness, write it as an article for *Time* or *Newsweek*. Likewise, a campus story on the Greek system could be aimed at the campus newspaper or the profile of a classmate in the style of *People* magazine or *The New Yorker*. Before writing the final draft, be sure to study the form and conventions of the periodical for which you are writing.

Transform to a Talk-Show Debate

An especially good genre for interpretive or argumentative papers would be a debate, conversation, or panel discussion. For example, students recently wrote a paper as a debate on the advantages versus disadvantages of clear-cutting timber: On one side were the environmentalists and tourist industry, on the other side the paper companies and landowners; each side had valid points in its favor. The debate format was real, as it echoed very closely a similar debate in Congress.

Transform to Any Medium of Expression

The possibilities for reshaping college papers are endless: song, play, poem, editorial, science fiction story, laboratory report, bulletin, brochure, television or radio commercial, public address, political speech, telephone conversation, e-mail exchange, World Wide Web page, poster, "Talk of the Town" for *The New Yorker,* sound bite, environmental impact statement, conference paper, video game, philosophical debate, or bar stool argument.

✌ WRITING 4

Propose a transformation for a paper you are writing or have recently written. List the advantages and disadvantages of this transformation. Recast your paper (or a part of it) in the new genre and describe the effect.

✌ EXPERIMENTATION VERSUS CONVENTION

Standard *academic conventions* are accepted ways of doing certain things, such as using an objective voice in research reports and placing the thesis first in position papers. These conventions have evolved over time for good reasons. When carefully done, they transmit ideas and information clearly and predictably, thereby avoiding confusion and misunderstanding. Although in many cases these conventions work well, successful writers sometimes invent unorthodox strategies and experiment with new or alternative forms to better express new or complicated ideas. In order to decide whether a conventional or experimental form is preferable in any part of your paper, try both to see which more appropriately presents your ideas in their best light. Sometimes an act as simple as changing time, tense, point of view, or genre can effectively change the impact of a piece of writing.

The strategies described in this chapter are useful revising tools because they force writers to resee the events in different languages and from different perspectives. Writing in new forms is also intriguing, exciting, and fun—which is often what writers need after working long and hard to put together a first draft.

When and under what circumstances should you limit, add, switch, or transform? While there are no rules, you might try using these strategies whenever you feel stuck or in need of new energy or insight. But be sure to weigh gains and losses whenever you use new focusing techniques.

Disregarding academic conventions in early drafts should seldom be a problem; however, disregarding them in final drafts could be if it violates the assignment, so check with your instructor. Be sure that in gaining reader attention in this way, you do not lose credibility or cause confusion.

⅍ SUGGESTIONS FOR WRITING AND RESEARCH

Individual

1. Write the first draft of a personal experience paper as a broad overview of the whole experience. Write the second draft by limiting the story to one day or less of this experience. Write the third draft using one of the other techniques described in this chapter: adding, switching, or transforming. Write the final draft in any way that pleases you.
2. Write the first draft of a research-based paper as an overview of the whole issue with which you intend to deal. In the second draft, limit the scope to something you now cover in one page, paragraph, or sentence. In the third draft, adopt one of the focused revision strategies described in this chapter: adding, switching, or transforming. For your final draft, revise in any way that pleases you.

Collaborative

1. For a class research project, interview college instructors in different departments concerning their thoughts about transforming academic papers into other genres. Write up the results in any form that seems useful.
2. As a class, compile a catalog in which you list and describe as many alternative forms for college papers as you can.

Chapter
✣ 24 ✣
Creative Nonfiction

Creative nonfiction is a term that describes much of the good writing that appears in magazines such as *The New Yorker, Harper's,* and *GQ,* as well as many works on nonfiction best-seller lists. Writers of creative nonfiction commonly borrow stylistic and formal techniques from the fast-paced visual narratives of film and television as well as from the innovative language of poetry, fiction, and drama—these influences encouraging a multifaceted, multidimensional prose style to keep pace with the multifaceted and multidimensional world in which we live. In short, many current nonfiction prose writers find the traditions of continuity, order, consistency, and unity associated with conventional prose insufficient to convey the chaotic truths of the postmodern world. This chapter examines some of the writing strategies associated with alternative or experimental prose and suggests appropriate venues within the academic curriculum in which such prose strategies could be useful. (Many of the ideas presented here were first articulated by Winston Weathers in his groundbreaking book *An Alternate Style: Options in Composition* [Hayden, 1980].)

✣ LISTS

Lists can break up and augment prose texts in useful, credible, and surprising ways. Lists of names, words, numbers add variety, speed, depth, and humor to texts. And lists are everywhere we look, as Joan Didion illustrates in making the case that Las Vegas weddings are big business in this excerpt from "Marrying Absurd":

> There are nineteen such wedding chapels in Las Vegas, intensely competitive, each offering better, faster, and, by implication, more sincere services than the next: Our Photos Best Anywhere, Your Wedding on a Phonograph Record, Candlelight with Your Ceremony, Honeymoon Accommodations, Free Transportation from Your Motel to Courthouse to Chapel and Return to Motel, Religious or Civil Ceremonies, Dressing Rooms, Flowers, Rings, Announcements, Witnesses Available, and Ample Parking.

Didion's list of competitive wedding services convinces us she has observed carefully; she is not making this stuff up. Without her saying it, we see some level of absurdity in the way this town promotes marriage.

Lists need not be clever so much as purposeful. That is, you include a list of names, items, quotations, and so on to show readers you know what you're talking about: You have done your homework, read widely or observed carefully, taken good notes, and made sense from what you've found. Lists deepen a text by providing illustrations or examples. And they add credibility by saying, in effect, "Look at all this evidence that supports my point."

On the printed page, sometimes lists are presented simply as lists, not embedded in prose paragraphs. Such is the case in *Blue Highways: A Journey into America,* when author William Least Heat-Moon overhears people describing the desert as full of "nothing."

> **Driving through miles of nothing, I decided to test the hypothesis and stopped somewhere in western Crockett County on the top of a broad mesa, just off Texas 29. . . . I made a list of nothing in particular:**
>
>   1. mockingbird
>   2. mourning dove
>   3. enigma bird (heard not saw)
>   4. gray flies
>   5. blue bumblebee
>   6. two circling buzzards (not yet boys)
>   7. orange ants
>   8. black ants
>   9. orange-black ants (what's been going on?)
> 10. three species of spiders
> 11. opossum skull
> 12. jackrabbit (chewed on cactus)
> 13. deer (left scat)
> 14. coyote (left tracks)

Heat-Moon's list continues through thirty items, ending this way:

> 28. earth
> 29. sky
> 30. wind (always)

Itemized lists such as this change the visual shape of prose and call extra attention to the items listed; in this case, Heat-Moon is being mildly humorous by using a list to "prove" there is always something, even in nothing.

When Craig, a student in my advanced writing class, examined sexist stereotypes in children's toys, he made the following list of dolls and accessories on a single shelf at a local Woolworth's store:

> To my left is a shelf of Barbies: Animal Lovin' Barbie, Wet 'n Wild Barbie, Barbie Feelin' Pretty Fashions, Barbie Special Expressions (Woolworth Special Limited Edition), Super Star Barbie Movietime Prop Shop, Step 'n Style Boutique, My First Barbie (Prettiest Princess Ever), Action Accents Barbie Sewing Machine, Barbie Cool Times Fashion, Barbie and the All-Stars, Style Magic Barbie, a Barbie Ferrari, and tucked away in a corner, slightly damaged and dusty, Super Star Ken.

This list simply documents by name the products on the toy shelf, actually adding a dimension of authenticity and believability to the writer's case that, yes, the Barbie image and influence on children is considerable.

Creating an extended list is a bold, even audacious move, breaking up prose sentences, surprising readers and therefore picking up their interest, engagement, involvement. The purposeful use of lists may make readers pause to note the change in the form of words on the page; at the same time, lists allow readers to pick up speed—reading lists is fast.

However, lists that are quick to read may not be quick to write; an effective list that appears to be written by free association may, in fact, have been laboriously constructed as the writer ransacked his memory or her thesaurus for words, then arranged and rearranged them to create the right sound or sense effect.

↭ SNAPSHOTS

Writing prose snapshots is analogous to constructing and arranging a photo album composed of many separate visual images. Photo albums, when carefully assembled from informative snapshots, tell stories with clear beginnings, middles, and endings, but with lots of white space between one picture and the next, with few transitions explaining how the photographer got from one scene to the next. In other words, while photo albums tell stories, they do so piecemeal, causing the viewer to fill in or imagine what happened between shots. Or think of snapshots as individual slides in a slide show or pictures in an exhibition—each the work of the same maker, each a different view, each by some logic connected, the whole a story.

Prose snapshots function the same way as visual snapshots, each connected to the other by white space and leaps of reader logic and faith, the whole making self-explanatory story structure. Or you might imagine written snapshots as a series of complete and independent paragraphs, each a whole thought, without obvious connections or careful transitions to the paragraph before or after.

Sometimes individual snapshots are numbered to suggest deliberate connectedness; other times each is titled, to suggest an ability to stand alone, as are chapters within books; sometimes they appear on a page as block paragraphs, the reading alone revealing the lack of transitions, the necessity for active reader interpretation. As such, they are satisfying for fast readers, each containing a small story unto itself, the whole a larger story, in part of the reader's making.

Margaret Atwood wrote snapshots to emphasize the dangers of men's bodies in the following passage from her essay "Alien Territory" (1983):

> The history of war is a history of killed bodies. That's what war is: bodies killing other bodies, bodies being killed.
>
> Some of the killed bodies are those of women and children, as a side effect you might say. Fallout, shrapnel, napalm, rape and skewering, anti-personnel devices. But most of the killed bodies are men. So are most of those doing the killing.
>
> Why do men want to kill the bodies of other men? Women don't want to kill the bodies of other women. By and large. As far as we know.
>
> Here are some traditional reasons: Loot. Territory. Lust for power. Hormones. Adrenaline high. Rage. God. Flag. Honor. Righteous anger. Revenge. Oppression. Slavery. Starvation. Defense of one's life. Love; or, a desire to protect the men and women. From what? From the bodies of other men.
>
> What men are most afraid of is not lions, not snakes, not the dark, not women. What men are most afraid of is the body of another man.
>
> Men's bodies are the most dangerous things on earth.

Note how the white space between one snapshot and another gives readers breathing space, time out, time to digest one thought before supping at the next. The white space between snapshots actually exercises readers' imaginations, as they participate in constructing some logic that makes the text make sense—the readers themselves to supply the connectives, construct the best meaning, which, nevertheless, will be very close to what skillful authors intend.

Snapshots allow busy writers to compose in chunks, in five- and ten-minute blocks between appointments, schedules, classes, or coffee breaks. And, as we've seen, five or ten or twenty chunks—reconsidered, rearranged, revised—can tell whole stories. For examples of students composing in snapshots, see an excerpt from Paige Kaltsas's essay later in this chapter and Rebecca's complete essay at the end of Chapter 9.

While it's fun to write fast, random, and loose snapshots, the real secret to a successful snapshot essay is putting them together in the right order—some right order—some pattern that, by the end, conveys your theme as surely as if you had written straight narration or exposition. Writing snapshots on a computer is especially fun, since you can order and reorder indefinitely until you arrive at a satisfying organization. Composing snapshot essays on three-by-five-inch cards also works. In either case, assemble and arrange as you would pictures in a photo album, playfully and seriously: Begin at the beginning, alternate themes, begin in the middle, alternate time, begin with flashbacks, alternate voices, consider frames, alternate fonts, reinforce rhythms, experiment with openings and closings, change type, and try different titles.

✌ PLAYFUL SENTENCES

No matter what your form or style, sentences are your main units of composition, explaining the world in terms of subjects, actions, and objects, suggesting that the world operates causally: Some force (a subject) does something (acts) that causes something else to happen (an object). English prose is built around complete and predictable sentences such as those in which this paragraph and most of this book are written. Sometimes, however, writers use sentences in less predictable, more playful ways, which we will explore here.

Fragment sentences suggest fragmented stories. Stories different from the stories told by conventional subject-verb-object sentences. Fragmented information. Fragment sentences, of course, can be used judiciously in conventional writing—even academic writing, so long as the purpose is crystal clear and your fragment is not mistaken for fragmentary grammatical knowledge. However, alternate-style writers use fragments audaciously and sometimes with abandon to create the special effects they want. A flash of movement. A bit of a story. A frozen scene. Fragments force quick reading, ask for impressionistic understanding, suggest parts rather than wholes. Like snapshots, fragments invite strong reader participation to stitch together, to move toward clear meaning.

Fragment sentences suggest, too, that things are moving fast. Very fast. Hold on! Remember the snapshot passage from Margaret Atwood's "Alien Territory"? Note that she used fragments to emphasize the sharp dangers of men's bodies:

> **Some of the killed bodies are those of women and children, as a side effect you might say. Fallout, shrapnel, napalm, rape and skewering, anti-personnel devices. But most of the killed bodies are men. So are most of those doing the killing.**
>
> **Why do men want to kill the bodies of other men? Women don't want to kill the bodies of other women. By and large. As far as we know.**

Atwood's fragments make the reader notice sharply the brutal and jarring truths she is writing about. In this example, the lack of conventional connections between words mirrors the disconnectedness she sees in her subject: men, violence, and war. Notice, too, that some of her fragments illustrate another use of lists.

Write fragments in such a way that your reader knows they are not mistakes. Not ignorance. Not sloppiness or printer error or carelessness. Purposeful fragments can be powerful. Deliberate. Intentional. Careful. Functional. And usually brief. (See the student samples of collage writing later in this chapter.)

Labyrinthine sentences tell stories different from either conventional or fragment sentences. In fact, a labyrinthine sentence is quite the opposite of the fragment sentence because it seems never to end; it won't quit, but instead it goes on and on and on, using all sorts of punctuational and grammatical tricks to create compound sentences (you know, two or more independent clauses joined by a comma and a conjunction such as *and* or *but*) and complex sentences (you know these, too: one independent clause with one or more dependent clauses) and is written to suggest, perhaps, that things are running together and are hard to separate—also to suggest the "stream of consciousness" of the human mind, in which thoughts and impressions and feelings and images are run together without an easy separation into full sentences or paragraphs complete with topic sentences—the power (and sometimes confusion) of which you know if you have read James Joyce or Virginia Woolf or William Faulkner or Toni Morrison.

Or James Agee, who in the following passage imaginatively enters the thoughts of the people he is profiling in *Let Us Now Praise Famous Men* (1941), the poor Alabama tenant farmers:

> **But I am young; and I am young and strong and in good health; and I am young and pretty to look at; and I am too young to worry; and so am I for my mother is kind to me; and we run in the bright air like animals, and our bare feet like plants in the wholesome earth: the natural world is around us like a lake and a wide smile and we are growing: one by one we are becoming stronger, and one by one in the terrible emptiness and the leisure we shall burn and tremble and shake with lust, and one by one we shall loosen ourselves from this place, and shall be married, and it will be different from what we see, for we will be happy and love each other, and keep the house clean, and a good garden, and buy a cultivator, and use a high grade of fertilizer, and we will know how to do things right; it will be very different:) (?:)**

Agee's long connected sentence creates the run-together, wishful, worried, desperate internal dream of his subjects in a way a conventional paragraph could not. Notice, too, that punctuation and grammar are mostly conventional and correct—even though at the end, they are used in unexpected ways to suggest something of the confusion and uncertainty these people live with daily.

You may also write run-on or fused sentences—where punctuation does not function in expected ways the missing period before this sentence is an example of that. However, such writing more often suggests error than experiment, so be careful. I use both fragments and labyrinthine sentences to create certain effects, since each conveys its information in an unmistakable way; but I never, deliberately, write with run-on sentences, and when I encounter them as a reader, they make me suspicious.

ᶜ⁰ **REPETITION/REFRAIN**

Writers repeat words, phrases, or sentences for emphasis. They repeat words to remind us to think hard about the word or phrase repeated. They repeat words to ask us to attend and not take for granted. They repeat words to suggest continuity of idea and theme. They repeat words to hold paragraphs and essays together. And, sometimes, they repeat words to create rhythms that are simply pleasing to the ear.

The following paragraph opens Ian Frazier's book-length study *The Great Plains*:

> Away to the Great Plains of America, to that immense Western shortgrass prairie now mostly plowed under! Away to the still empty land beyond newsstands and malls and velvet restaurant ropes! Away to the headwaters of the Missouri, now quelled by many impoundment dams, and to the headwaters of the Platte, and to the almost invisible headwaters of the slurped up Arkansas! Away to the land where TV used to set its most popular dramas, but not anymore! Away to the land beyond the hundredth meridian of longitude, where sometimes it rains and sometimes it doesn't, where agriculture stops and does a double take! Away to the skies of the sparrow hawks sitting on telephone wires, thinking of mice and flaring their tail feathers suddenly, like a card trick! Away to the airshaft of the continent, where weather fronts from two hemispheres meet and the wind blows almost all the time! Away to the fields of wheat and milo and Sudan grass and flax and alfalfa and nothing! Away to parts of Montana and North Dakota and South Dakota and Wyoming and Nebraska and Kansas and Colorado and New Mexico and Oklahoma and Texas! Away to the high plains rolling in waves to the rising final chord of the Rocky Mountains!

Frazier's singing chant invites us, in one sweeping passage, to think about the Great Plains as geography, biology, history, and culture. Along the way he uses fragments and lists and a plentitude of exclamation marks to invite readers to consider this arid and often overlooked part of America.

While *refrain* is a term more often associated with music, poetry, and sermons, it is a form of repetition quite powerful in prose as well. A refrain is a phrase repeated throughout a text to remind readers (or listeners) of an important theme. For example, the phrase "I have a dream" is a refrain in Martin Luther King's famous speech by the same name. In "Report from the Bahamas," June Jordan reflects upon her experience as a black woman being waited on by black maids and waiters while staying at the Sheraton British Colonial Hotel. Throughout the essay Jordan returns to a phrase to remind herself, as well as the reader, of her troubled and complicated situation as a black feminist writer. The refrain is repeated six times at different points throughout the essay. Here are the first four, which occur within a span of six pages:

> This is my consciousness of race as I unpack my bathing suit in the Sheraton British Colonial.

> This is my consciousness of class as I try to decide how much money I can spend on Bahamian gifts for my family back in Brooklyn.

> This is my consciousness of race, class, and gender identity as I notice the fixed relations between these other Black women and myself.

> This is my consciousness of race, class, and gender identity as I collect wet towels, sunglasses, wristwatch, and head towards a shower.

Repetition and refrains, along with lists, snapshots, fragments, and labyrinthine sentences, are all stylistic devices that add an emotional dimension to the otherwise factual material of nonfictional prose—without announcing, labeling, or dictating what those emotions need be. The word play of alternate-style composing allows nonfictional prose to convey themes more often conveyed only through more obviously poetic forms.

↷ DOUBLE VOICE

Good nonfiction writing usually (I'd like to say *always* but don't dare) expresses something of the writer's voice. But all writers are capable of speaking with more than one voice (how many more?)—or maybe with a single voice that has wide range, varied registers, multiple tones, and different pitches. No matter how you view it, writers project more than one voice from piece to piece of writing—sometimes within the same piece.

In any given essay, a writer may try to say two things at the same time. Sometimes writers question their own assertions (as in the previous paragraph); sometimes they say one thing out loud and think another silently to themselves (again, see the previous paragraph); sometimes they say one thing that means two things; sometimes they express contradictions, paradoxes, or conundrums; and sometimes they establish that most of us have more than one voice with which to speak.

Double voices in a text may be indicated by parentheses—the equivalent of an actor speaking an "aside" on the stage (see what I mean?) or in a film. The internal monologue of a character may be revealed as voice-over or through printed subtitles while another action is happening on screen. Or by changes in the size or *type font*, a switch to *italics*, **boldface**, or CAPITAL LETTERS signaling a switch in the writer's voice. Or the double voice may occur without distinguishing markers at all. Or it may be indicated by simple paragraph breaks, as in the following selection from D. H. Lawrence in his critical essay on Herman Melville's *Moby Dick* from *Studies in Classic American Literature,* where he uses fragments, repetition, and double voice:

> Doom.
> Doom! Doom! Doom! Something seems to whisper it in the very dark trees of America.
> Doom of what?
> Doom of our white day. We are doomed, doomed. And the doom is in America. The doom of our white day.
> Ah, well, if my day is doomed, and I am doomed with my day, it is something greater than I which dooms me, so I accept my doom as a sign of the greatness which is more than I am.
> Melville knew. He knew his race was doomed. His white soul, doomed. His great white epoch, doomed. Himself doomed. The idealist, doomed. The spirit, doomed.

Here, Lawrence critiques Melville by carrying on a mock dialogue with himself, alternating his caricature of Melville's voice with his own whimsical acceptance of Melville's gloomy prophecy. Lawrence's essay seems written to provoke readers into reassessing their interpretations of literary classics, and so he provokes not only through the questions he raises but through his style as well. Note his poetic use of repetition and sentence fragments that contribute to his double-voice effect.

Double voice may also be offset spatially, in double columns or alternating paragraphs, as my student Paige Kaltsas did in trying to capture the experience of running a 26.2-mile marathon. She alternated voices in separate stanzas (one double stanza for each mile). The following example is from her third set of stanzas, where her first voice is in the race (present tense) while her second voice is remembering the training (past tense):

> Mile 3: Make sure you are going the right pace. Slow down a little, you're going too fast. Let people pass you. Don't worry, they will burn out and you'll glide by later. Don't make a mistake.

> My typical training week went like this:
> Monday: a two-mile warm-up at a slow pace; then three miles at a faster pace; finally, two miles slowing down, cooling off. After: a leg strength workout.
> Tuesday: a medium long run, ten miles at a medium pace.
> Wednesday: the dreaded speed workouts at the track, with mile runs alternating with quarter-mile sprints. Afterwards, nausea and arm and leg strength workouts.

ぐ COLLAGE

Collages are more often associated with visual than verbal art, but creative nonfiction writers borrow freely from other media, especially in this age of the Internet and its combined verbal-visual style. Even my own journal has elements of a collage when I use it as a scrapbook, taping in photos or clippings wherever I find white space, thus creating meaningful—if only to me—juxtapositions. However, collage writing has been used to more deliberate effect in the novels of John Dos Passos—a technique since borrowed by Tom Wolfe, Hunter S. Thompson, and others in what may be called "new journalism" or "literary nonfiction." Dos Passos, a writer of both fiction and nonfiction, began his essay "The Death of James Dean" (1959) with quotations taken from newspaper headlines, then continued using fragments, repetition, and double voices along the way.

> **TEEN-AGE DANCES SEEN THREATENED BY PARENTS FAILURE TO COOPERATE MOST OFFENDERS EMULATE ADULTS**
> James Dean is three years dead but the sinister adolescent still holds the headlines.
> James Dean is three years dead;
> but when they file out of the close darkness and breathed out air of the second-run motion picture theatres where they've been seeing James Dean's old films
> they still line up;
> the boys in the jackboots and the leather jackets, the boys in the skintight jeans, the boys in broad motorbike belts,
> before mirrors in the restroom
> to look at themselves
> and see
> James Dean;

Collage techniques may also be used to write collaboratively, with multiple authors contributing not only multiple voices but also multiple perspectives. In the following example, taken from a last-day-of-class writing exercise, my students wrote snapshots of their semester-long experience in this writing course, some of them looking all the way back to the first day of class, where, as an introductory exercise, we spent the entire time writing notes to each other in silence.

1. First day of class. I walk in early (I'm always early) only to find "Silence" written on the blackboard. Never had this Toby guy before. Instead of saying anything, he uses the overhead projector to tell us what to do: "On a sheet of paper, write a note to a classmate, introduce yourself, then pass it." Not a word spoken the whole class. Silent chaos. But also a way of showing us the importance of voice in writing. Weird, not one word the whole 75 minutes. (Paul)

2. Shhh! The instructor writes on the board "No talking." What the hell? I think to myself. A writing class and nobody talks, everybody writes. I get it. (Mary)

3. First day of class. I arrive ten minutes late, couldn't find the building. No one speaks. Everyone is writing. I get reprimanded for asking the girl next to me what's going on. I get reprimanded for asking the teacher what's going on. I figure it out and leave class partly annoyed, partly amused. (Brad)

✌ ELECTRONIC STYLE

Computer graphics programs included in conventional word-processing programs now allow writers to create varied page layouts with different fonts and styles. They make experimental writing easier, more possible, and more likely than at any other time in the history of print technology.

Clip art inserted from word-processing programs now enlivens many college papers and is something most instructors enjoy so long as it enhances or amplifies the content it accompanies and the content is solid, not fishy.

It is now common for instructors to receive college papers that resemble professionally published articles in sophisticated journals. In one example from a recent writing class about the phenomenon of "slam poetry," Gabe Krechmer designed his first page to resemble a page in popular journalism. (See page 284.)

What Is Slam Poetry, Anyway?
Gabriel Krechmer

There is something about poetry slam that can leave you utterly confused. I have done a good amount of public speaking in my life. I have been a teacher's assistant for several classes. I know Those eyes decide if you rise to the challenge or fail miserably; I understand the responsibility of having all eyes pointing at you, evaluating your every move. Those eyes can be especially taxing on your self-esteem. This is what flowed through my mind when I arrived at the Rhombus gallery for my introduction with Slam.

"Greetings!" the woman yells, her arms flinging up into the air, "and thank you all for returning, and thank you all for joining us. My name is Kathy, and I am fortunate enough to be your slam mistress tonight!"

Shouts of approval come from all around. Kathy has crazy, curly black hair and wears a green dress. She looks like she stepped out of a Grateful Dead show, complete with blood-shot eyes.

Finding the gallery itself was not an easy task. It's housed in one of gray stone buildings that you easily dismiss and pass on by, having no consequence to your daily life.

Much like the genre itself, the building remains incognito. It took me forty-five minutes, walking up and down College Street, to find the place. No big signs, no fancy lighted lettering, just a piece of eight-and-a-half-by-eleven piece of paper, colored hospital green, loosely taped next to the door, proclaiming the night's events:

> "We are at the testing zone of consciousness . . . we are at the moment of democratization of art. And it is the tongue of the slam that wants to tickle your inner-eardrum."
> —Bob Holman, New York City Slammaster

!! Poetry Slam !!
Rhombus Gallery
With Special Guests
Friday at 8 p.m.
Donations Appreciated.

On the first floor of the building resides a gift card shop, charmingly named *Initially Yours*. The Rhombus gallery lives and breathes on the upper floors. The gallery itself is located not on the second floor or first floor, but in between the two: on the mezzanine. This seemed to parallel the genre's place in the literary world. I was late and forced to wait a bit before I could get in. Intermission came and people left and I soon entered the threshold of the actual gallery, which was straight out of the 1960's. The walls were antiseptic white; pieces of local artwork hung here and there. The window trim was more of that hospital green. The ceiling was composed of one-by-one acoustic tiles with three sets of track lights, all focused on the stage: a five-foot-by-three-foot area painted black centered adjacent to the front wall. The philosophies of the gallery's era lent itself to the poets it hosted.

E-mail messages are also written in a loose style that I hesitate to call creative nonfiction but which includes similar elements of fun, such as emoticons ☺. And the convention of fast typing emerged with e-mail messages, which have come to be regarded as somewhere in between informal telephone messages and informal letters.

> the original word processing programs for e-mail correspondence were limited in their revision and editing capabilities, so that if you were typing on line 8 and you noticed an error in line 1, you had to erase all the text back to the error in order to fix it. so people just said, to heck with it, let the error go—which created a fast, loose, informal style in which typos, spelling mistakes, and lapses in punctuation simply came with the territory. the priority in e-mail correspondence has become speed and practicality, with most e-mail messages never being printed on paper at all. even with better and faster word-processing programs, the loose informal style of e-mail writing continues today. it is common, for example, for e-mail writers to use no capital letters or to invent unorthodox punctuation:)

ᴈ A WORD OF CAUTION

Wise writers will master both conventional and unconventional styles and formats, using each as occasion and audience demand. Proficiency in one format is a poor excuse for sloppiness or neglect of the other. Creative nonfiction or alternate-style techniques, used carefully and judiciously on selected writing tasks, are fun to write and enjoyable to read, and they may be able to convey the emotional and aesthetic dimension of ideas difficult to convey in conventional prose. At the same time, such stylistic devices are easy to overuse and exaggerate, resulting then in predictable, routine, or overly cute expressions that lose the very edge they are trying to achieve, and that made them effective in the first place. Check with your instructor before submitting an unconventional paper in response to a conventional assignment.

ᴈ SUGGESTIONS FOR WRITING AND RESEARCH

Individual

1. Select one or more of the following techniques to compose your next essay: lists, snapshots, fragment sentences, labyrinthine sentences, repetition, double voice, and collage.
2. Recast an essay previously written in one or more of the alternative style techniques above. Compare and contrast the effects created by each version.

Collaborative

As a class, compose a collaborative collage profile of your class, piecing together writing from each classmate in one thematically consistent text.

Chapter 25

Openings and Closings

I could start this essay just about anywhere at all, by telling you about the background, by stating the thesis, by telling a funny story, or even by rambling, which is what I usually do. Where do you think I should start?

WENDY

My advice, Wendy, is to lead with your best punch. Make your opening so strong your reader feels compelled to continue. Make your closing so memorable that your reader can't forget it.

Readers pay special attention to openings and closings, so make them work for you. Start with titles and lead paragraphs that grab readers' attention and alert them to what is to come; end with closings that sum up and reinforce where they've been. This chapter looks closely at how these special paragraphs function and how you can make these paragraphs stronger through skillful editing.

OPENINGS

Openings are first impressions. Your first paragraph—in fact, your whole first page—sets readers up for the rest of the paper. Here you provide the first clues about your subject, purpose, and voice and invite your audience to keep reading.

Good opening paragraphs are seldom written in first drafts. Often, it's not until you've finished a draft or two that you know exactly what your paper says and does. So when your paper is nearly finished, return to your first page, read it again, and edit carefully. The following examples of effective openings are taken from the student papers reproduced earlier in this textbook; all of them were rewritten at the editing stage.

Open with a Conflict

If your paper is about conflict, sometimes it's best to open by directly spelling out what that conflict is. Identifying a conflict captures attention by creating a kind of suspense: Will the conflict be resolved? How will it be resolved? Issa's exploration of the conflict between hikers and mountain bikers opens this way:

> The narrow, hard-packed dirt trail winding up the mountain under the spreading oaks and maples doesn't look like the source of a major environmental conflict, but it is. On the one side are hikers, environmentalists, and horseback riders who have traditionally used these wilderness trails. On the other side, looking back, are the mountain bike riders sitting atop their modern steeds who want to use them too. But the hikers don't want the bikers, so trouble is brewing.

This essay argues, ultimately, in favor of opening wilderness trails to mountain biking; however, Issa uses a delayed-thesis organizational pattern, exploring both sides of the conflict without giving away his own position on the conflict—the opening tells us that *trouble is brewing,* but we don't know how it will be resolved. (To read Issa's complete essay, see Chapter 12.)

Open with a Thesis

Argumentative and interpretive papers commonly open with a clear thesis statement that the rest of the paper will support. Opening with your thesis is the most direct way of telling your reader what the paper will be about. The reader may agree with you and want to see how you support your case, or the reader may disagree and want to find holes in your argument, or the reader may start off neutral and read to see whether or not your paper is believable and persuasive. Following is the first paragraph from Kelly's essay interpreting a poem by Gwendolyn Brooks:

> Gwendolyn Brooks writes "We Real Cool" (1963) from the point of view of the members of a street gang who have dropped out of school to live their lives hanging around pool halls—in this case "The Golden Shovel." These guys are semiliterate and speak in slangy street lingo that reveals their need for mutual support in their mutually rebellious attitude toward life. The speakers in the poem, "We," celebrate what adults would call adolescent hedonism—but they make a conscious choice for a short, intense life over a long, safe, and dull existence.

While you will need to read Brooks's short poem to fully understand Kelly's argument, this opening passage states Kelly's position about the subjects—who are also the narrators—of the poem: "they make conscious choices for a short intense life over a long, safe, and dull existence."

Thesis stating, like summary writing, is among the most difficult of all writing tasks and for that reason is most easily and clearly written only when you are thoroughly familiar with your subject, which is seldom the case with early drafts. Consequently, openings such as Kelly's above are commonly written after the rest of the paper is completed. (To read Kelly's complete interpretive essay, see Chapter 13.)

Open with a Story

Most readers enjoy stories. Professional writers often open articles with anecdotes to catch readers' attention. The following first paragraph is from Judith's reflective essay about finding a safe place to study in the library. Only in her last draft did she decide to start at the door, in present tense, locking up as she leaves to walk to the library:

> It is already afternoon. I fiddle with the key to lock the apartment door after me. I am not accustomed to locking doors. Except for the six months I spent in Boston, I have never lived in a place where I did not trust my neighbors. When I was little, we couldn't lock our farmhouse door; the wood had swollen, and the bolt no longer lined up properly with the hole, and nobody ever bothered to fix it. I still remember the time our babysitter, Rosie, hammered the bolt closed and we had to take the door off the hinges to get it open.

Stories not only need to catch attention; they need to set up or foreshadow the paper to follow. In Judith's case, the theme and content of her paper emerged through several drafts, so it was only in the final draft that she discovered a personal story about locked doors to anticipate the theme of safety in the rest of the paper. (To read Judith's complete essay, see Chapter 14.)

Open with a Specific Detail

Specific details appeal to readers' visual sense and help them see situations and settings. In early drafts, Beth opened her profile of Becky with Becky speaking; only in this latest version did she decide to set the physical scene first, letting Becky's manner and surroundings characterize her right from the start:

> Becky sits cross-legged at the foot of the bottom bunk on her pink and green homemade quilt. She leans up against the wall and runs her fingers through her brown shoulder-length hair. The sound of James Taylor's "Carolina on My Mind" softly fills the room. Posters of John Lennon, James Dean, and Cher look down on us from her walls. Becky stares at the floor and scrunches her face as if she were thinking hard.

Details need to be specific and interesting, but they also need to be purposeful and to advance the paper one way or another. It turns out that Becky is a nineties college student living in a room dominated by musical and visual icons from the fifties and sixties. The details anticipate an important theme in this profile—the subject's closeness to her mother and her mother's values, which are rooted in decades past rather than present. (To read Beth's complete profile, see Chapter 10.)

Open with a Quotation

Although Beth decided not to open with her subject talking, an opening quotation can be an effective hook. Readers enjoy hearing the voices of people on the subject of the piece. The following paragraph was Beth's original opening; she moved it to the second paragraph, adding the transitional first sentence to link the two paragraphs. (It would work equally well as her opening, since it introduces the reader to Becky's lively and interesting human voice, after which we expect more talk about her early years growing up "forever" with her mother and sister.)

> Finally, after minutes of silence [Becky] says, "I don't ever remember my father ever living in my house, really. He left when I was three and my sister was just a

baby, about a year old. My mom took care of us all. Forever, it was just Mom, Kate, and me. I loved it, you know. Just the three of us together."

Sometimes it's just a matter of personal preference in deciding where to start. That, of course, is what editing is all about: trying one thing, then another, looking at options, in the end selecting the one that you think is best.

Open with Statistics

Statistics that tell a clear story are another form of opening that suggests immediately that the writer has done his or her homework. Statistics catch attention when they assert something surprising or answer in numbers something the reader has been wondering about. The more dramatic your statistics are, the more useful they are to open your paper. In the following example, Elizabeth opens a research essay about unethical marketing practices with statistics more reassuring than dramatic.

> A recent Gallup poll reported that 75% of Americans consider themselves to be environmentalists (Smith & Quelch, 1993). In the same study, nearly half of the respondents said they would be more likely to purchase a product if they perceived it to be environmentally friendly or "green." According to Smith and Quelch (1993), since green sells, many companies have begun to promote themselves as marketing products that are either environmentally friendly or manufactured from recycled material. Unfortunately, many of these companies care more about appearance than reality.

Statistics are used here to establish a norm—that most Americans care about the environment—and then to examine the way in which clever, but not necessarily truthful, advertisers exploit that norm. (To read Elizabeth's complete research essay, see Chapter 21.)

Open with a Question

Questions alert readers to the writer's subject and imply that the answer will be forthcoming in the paper. Every paper is an answer to some implicit or explicit question. Look at the examples above: Issa's mountain bike essay asks, "Can mountain bikes peacefully coexist with hikers on wilderness trails?" Kelly's interpretative paper asks, "What is the meaning of the poem 'We Real Cool'?" Judith's reflection on the comfort of libraries asks, "What has scholarship to do with safety?" Beth's profile asks, "What is Becky like?" And Elizabeth's research essay asks, "How do advertisers exploit the environmental movement?" In other words, the question that drives the paper is implicit, not directly stated.

In contrast, sometimes it's most effective to pose your question explicitly in the title or first paragraph so that readers know exactly what the report is about or where the essay is going, as Gabe did by titling his paper "What Is Slam Poetry, Anyway?" (See the sample page in Chapter 24.)

Open at the Beginning

In narrative writing, sometimes the best opening is where you think things started. In the following autobiographical snapshot, Rebecca opens a portrait of herself with brief portraits of her parents:

> My mother grew up in Darien, Connecticut, a Presbyterian. When she was little, she gave the Children's Sermon at her church. My father grew up in Cleveland, Ohio, a Jew. When I went away to college, he gave me the Hebrew Bible he received at his Bar Mitzvah.

Since the theme of Rebecca's self-profile is her devotion to an eclectic blend of religion, it seems right that she begin by focusing on the mixed religious heritage derived from her parents. (To read Rebecca's complete essay, see Chapter 9.)

We have just looked at eight different openings from eight effective essays, each effective in its own way. All eight strategies are good ones, fairly commonly used. None is necessarily better or worse than another, as each reflects a different intention. If we examine eight more essays, we might find still another set of opening strategies. Keep in mind that the first opening you write is just that, an initial opening to get you started. If it's effective, you may decide to keep it, but don't be afraid to return to that opening once the essay is complete and see if a new approach might better suit the direction your essay has finally taken.

↜ WRITING 1

Recast the opening of a paper you are currently working on, using one or more of the suggestions in this section: a conflict, a thesis, a story, a specific detail, a quotation, an interesting statistic, or a provocative question. For your final draft, use the opening that pleases you most.

↜ CLOSINGS

Closings are final impressions. Your concluding sentences, paragraphs, or the entire last page are your final chance to make the point of your paper stick in readers' minds. The closing can summarize your main point, draw a logical conclusion, speculate about the issues you've raised, make a recommendation for some further action, or leave your reader with yet another question to ponder.

After writing and revising your paper, attend once more to the conclusion, and consider whether the final impression is the one you want. You may discover that an earlier paragraph makes a more suitable ending, or you may need to write a new one to conclude what you've started. The following examples of effective closings are from student papers reproduced earlier in this textbook.

Close with an Answer

If your paper began by asking a question, and the rest of the paper worked at answering that question, it makes good sense to close with an answer—or at least the summary of your answer, reminding readers where they started some pages ago. For example, consider Gabe's opening question, which is also the title of his paper: "What is slam poetry, anyway?" Following that question, Gabe explored answers by attending a slam poetry reading, interviewing people about what it meant, and searching the Internet for both its history and definitions. After digesting several different answers, Gabe concluded with this summary sentence:

> So, in the end, slam poetry proves to be simply another kind of poetry reading, with this difference: the audience and the performer are one and the same.

This conclusion closes the frame of Gabe's paper, satisfying the reader (and writer) by answering the paper's central question. Of course, the real exploration of the meaning of slam poetry occurs on the way to this one-sentence answer, the whole paper providing a more carefully detailed answer than any single sentence could provide.

Close by Completing a Frame

An effective way to end some papers is to return deliberately to the issue or situation with which you began—to frame the body of your paper with an opening and a closing that mirror each other. Keith's CD paper used as a frame a customer considering the cost of a CD. Judith uses a frame in her paper about personal safety, returning in the closing to the setting of the opening—her front door:

> Hours later—my paper started, my exam studied for, my eyes tired—I retrace the path to my apartment. It is dark now, and I listen closely when I hear footsteps behind, stepping to the sidewalk's edge to let a man walk briskly past. At my door, I again fumble for the now familiar key, insert it in the lock, open the door, turn on the hall light, and step inside. Here, too, I am safe, ready to eat, read a bit, and finish my reflective essay.

Close with a Resolution

Many argumentative papers present first one side of an issue and then the other side, then conclude by agreeing with one side or the other. This is the case in Issa's essay about the conflict between hikers and mountain bikers, where he concludes by suggesting that more education is the only sensible resolution to the problem:

> Education and compromise are the sensible solutions to the hiker/biker stand-off. Increased public awareness as well as increasingly responsible riding will open still more wilderness trails to bikers in the future. It's clear that mountain bikers don't want to destroy trails any more than hikers do. The surest way to preserve

America's wilderness areas is to establish strong cooperative bonds among the hikers and bikers, as well as those who fish, hunt, camp, canoe, and bird-watch, and to encourage all to maintain the trails and respect the environment.

The most successful resolution to many conflicts is not an either/or conclusion but an enlightened compromise where both sides gain. Issa manages to accomplish this by siding with mountain bikers but, at the same time, suggesting they need to change some of their actions to gain further acceptance by the opposition. In any case, it is especially important to edit carefully at the very end of your paper so that your last words make your best case.

Close with a Recommendation

Sometimes writers conclude by inviting their readers to do something—to support some cause, for example, or take some action. This strategy is especially common in papers that argue a position or make a case for or against something. Such papers seek not only to persuade readers to believe the writer but to act on that belief. In the following closing, Elizabeth asks consumers to be more careful:

> Consumers who are genuinely interested in buying environmentally safe products and supporting environmentally responsible companies need to look beyond the images projected by commercial advertising in magazines, on billboards, and on television. Organizations such as Earth First! attempt to educate consumers to the realities by writing about false advertising and exposing the hypocrisy of such ads ("Do people allow," 1994), while the Ecology Channel is committed to sharing "impartial, unbiased, multiperspective environmental information" with consumers on the Internet (Ecology, 1996). Meanwhile the Federal Trade Commission is in the process of continually upgrading truth-in-advertising regulations (Carlson et al., 1993). Americans who are truly environmentally conscious must remain skeptical of simplistic and misleading commercial advertisements while continuing to educate themselves about the genuine needs of the environment.

Along with asking consumers to be more careful in the future, this closing reveals that further action is being taken by watchdog groups, suggesting those who are careful will have powerful allies on their side. (To read Elizabeth's complete essay, see Chapter 21.)

Close with a Speculation

In papers relating personal experience, reflection, or speculation, the issues you raise have no clear-cut conclusions or demonstrated theses. In these papers, then, the most effective conclusion is often one in which you admit some uncertainty, as Zoe does at the end of her investigation of becoming a photographer's assistant:

> So I'll keep sitting around hoping for a break. I can't guarantee that this research method for landing an internship will work; it still remains to be tried and tested. To my knowledge, there is no foolproof formula for a successful start. Like everybody else before me, I'm creating my own method as I go along.

294 REVISING AND EDITING

The casualness in such a paragraph, as if it were written off the cuff, is often deceptive. While that is the effect Zoe wanted to create, this paragraph emerged only late in her drafting process, and when it did, she edited and reedited it to achieve just the effect she wanted.

Close by Restating Your Thesis

In thesis-driven papers, it's often a good idea to close by reminding readers where you began, what you were trying to demonstrate in the first place. In Andrew's literary research paper, he concludes by restating the power of Thoreau's concept of freedom:

> Practical or not, Thoreau's writings about freedom from government and society have inspired countless people to reassess how they live their lives. Though unable to live as he advocated, readers everywhere remain inspired by his ideal, that one must live as freely as possible.

Restating a thesis is especially important in longer research-based papers where a lot of information has to be digested along the way. (To read Andrew's complete paper, see Chapter 21.)

Close with a Question

Closing with a question, real or rhetorical, suggests that things are not finished, final, or complete. Concluding by admitting there are things you, the author, still do not know creates belief because most of us live with more questions than answers. In the following example, Rebecca ends with a question that is both tongue-in-cheek and serious at the same time:

> I wear a cross around my neck. It is nothing spectacular to look at, but I love it because I bought it at the Vatican. Even though I am not a Catholic, I am glad I bought it at the Pope's home town. Sometimes, when I sit in Hebrew class, I wonder if people wonder, "What religion is she, anyway?"

Throughout her essay, Rebecca has presented herself and her commitment to religion in cryptic terms, suggesting perhaps that she herself is still not sure what she is or where her mixed religious heritage will take her. Closing with the bemused question above seems appropriate to her intentions. (To read Rebecca's complete essay, see Chapter 9.)

As with openings, so with closings—there are no formulas to follow, just a variety of possibilities that you'll need to weigh with your audience and purpose in mind. Since closing is a last impression, be willing to revise and edit carefully, and be willing to try more than once to get it just right.

✐ WRITING 2

Recast the closing of a current draft using one or more of the suggestions in this section: a summary, a logical conclusion, a real or a rhetorical question, a speculation, a recommendation, or the completion of a frame. In the final draft, use the closing that pleases you most.

✧ FINDING THE TITLE

Finally, after revising and editing to your satisfaction, return to your title and ask, "Does it work?" You want to make sure it sets up the essay to follow in the best possible way, both catching readers' attention and providing a clue for the content to follow. One good strategy for deciding on a title is to create a list of five or ten possibilities and then select the most suitable one. Play with words, arranging and rearranging them until they strike you as just right. Many writers spend a great deal of concentrated time on this task because titles are so important.

In my own case, when I was casting about for a title for this book, I happened to be riding my motorcycle late one chilly October night on the interstate highway between Binghamton and Albany, New York. I was cold and knew I had a good hour before arriving at my destination, so to forget the cold, I set about brainstorming titles and came up with "Writing to Discover" or "The Discovering Writer." As I thought about them, I decided to keep the word *writing* but to move away from *discover* toward *work,* and so I played with "The Writer Working," "Working Writers," "The Writer at Work," "Work and the Writer," and finally *The Working Writer.* Work hard to find the words that seem about right, and play with them until they form a construction that pleases you.

The following suggestions may help you find titles:

- Use one good sentence from your paper that captures the essence of your subject.
- Ask a question that your paper answers.
- Use a strong sense word or image from your paper.
- Locate a famous line or saying that relates to your paper.
- Write a one-word title (a two-word, a three-word title, and so on).
- Make a title from a gerund (an *-ing* word, such as *working*).
- Make a title starting with the word *on* and the name of your topic.
- Remain open at all times to little voices in your head that make crazy suggestions and copy these down.

✧ WRITING 3

Using some of the strategies described in this section, write five titles for a paper you are currently working on. For the final draft, use the title that pleases you most.

Working Paragraphs

Paragraphs tell readers how writers want to be read.

WILLIAM BLAKE

W hile there are no hard-and-fast rules for editing, there are important points to keep in mind as your essay nears completion. Once you are satisfied with the general shape, scope, and content of your paper, it's time to stop making larger conceptual changes—to stop revising—and start attending to smaller changes in paragraphs, sentences, and words—to begin editing. When you edit, you shape these three elements (paragraphs, sentences, and words) so that they fulfill the purpose of the paper, address the audience, and speak in the voice you have determined is appropriate for the paper.

࿐ THE WORK OF PARAGRAPHS

Most texts of a page or more in length are subdivided by indentations or breaks—paragraphs—that serve as guideposts, or as Blake puts it, "that tell readers how writers want to be read." Readers expect paragraph breaks to signal new ideas or directions; they expect each paragraph to have a single focus as well and to be organized in a sensible way; and they expect clear transition markers to link one paragraph to the next.

In truth, however, there are no hard-and-fast rules for what makes a paragraph, how it needs to be organized, what it should contain, or how long it should be.

I could, for instance, start a new paragraph here (as I have just done), leaving the previous sentence to stand as a single-sentence paragraph and so call a little extra attention to it. Or I could connect both that sentence and these to the previous paragraph and have a single five-sentence paragraph to open this section.

Most experienced writers paragraph intuitively rather than analytically; that is, they indent almost unconsciously, without thinking deliberately about it, as they develop or emphasize ideas. Sometimes their paragraphs fulfill conventional expectations, presenting a single well-organized and -developed idea, and sometimes they serve other purposes—for example, creating pauses or breathing spaces or points of emphasis.

The following paragraphs, many of which are excerpts from student papers that have already appeared in previous chapters, do different kinds of work, and although each is a good example, none is perfect. As you study them, bear in mind that each is *illustrative* of various purposes and organization, not *definitive*.

☙ WRITING WELL-ORGANIZED PARAGRAPHS

Unity: Stick to a Single Idea

Paragraphs are easiest to write and easiest to read when each one presents a single idea, as most of the paragraphs in this textbook do. The following paragraph opens Kelly's interpretive essay:

> Gwendolyn Brooks writes "We Real Cool" (1963) from the point of view of the members of a street gang who have dropped out of school to spend their lives hanging around pool halls—in this case "The Golden Shovel." These guys are semiliterate and speak in slangy street lingo that reveals their need for mutual support in their mutually rebellious attitude toward life. The speakers in the poem, "We," celebrate what adults would call adolescent hedonism—but they make a conscious choice for a short, intense life over a long, safe, and dull existence.

The opening sentence focuses on "the point of view of the members of a street gang" and the sentences that follow continue that focus on "these guys" and "the speakers." (To read Kelly's complete essay, see Chapter 13.)

Focus: Write a Topic Sentence

One of the easiest ways to keep each paragraph focused on a single idea is to include a *topic sentence* in it, announcing or summarizing the topic of the paragraph, with the rest of the sentences supporting that main idea. Sometimes topic sentences conclude a paragraph, as in the previous example, where the topic sentence is also the thesis statement for the whole essay. More commonly, however, the topic sentence introduces the paragraph, as in the next example from Issa's essay on mountain biking:

> Mountain bikes have taken over the bicycle industry, and with more bikes come more people wanting to ride in the mountains. The first mass-produced mountain bikes date to 1981, when five hundred Japanese "Stumpjumpers" were sold; by 1983 annual sales reached 200,000; today the figure is 8.5 million. In fact, mountain biking is second only to in-line skating as the fastest growing sport in the nation: "For a sport to go from zero to warp speed so quickly is unprecedented," says Brian Stickel, director of competition for the National Off Road Bicycle Association (Schwartz 75).

The sentences after the first one support and amplify the topic sentence, explaining the rapid growth of mountain bikes, from *five hundred Japanese "Stumpjumpers"* to *8.5 million* today. (To read Issa's complete essay, see Chapter 12.)

Most of the following examples have topic sentences, and all focus on single subjects. Note, however, that not all paragraphs need topic sentences. For example, if a complicated idea is being explained, a new paragraph in the middle of the explanation will create a pause point. Sometimes paragraph breaks are inserted to emphasize an idea, like my own one-sentence paragraph earlier, which is a topic and a support sentence all in one. Additionally, paragraphs in a personal experience essay seldom have a deliberate topic sentence since these sorts of essays are seldom broken into neat topics (see, for example, Judith's paragraph below). Nevertheless, in academic writing, there is great reverence for topic sentences because they point to clear organization and your ability to perform as an organized and logical thinker within the discipline. Thus, when you write academic papers, attend to topic sentences.

> ✎ **WRITING 1**
>
> Examine the paragraphs in a recent draft, and pencil in brackets around those that stick well to a single idea. Put an *X* next to any sentences that deviate from the main idea in a paragraph, and note whether you want to delete that sentence or use it to start a new paragraph. Finally, underline each topic sentence. If a paragraph does not have one, should it? If so, write it.

Order: Follow a Recognizable Logic

On first drafts, most of us write sentences rapidly and paragraph intuitively. However, when we revise and edit, it pays to make certain that paragraphs work according to a recognizable logic. There are dozens of organizational patterns that make sense. Here we look at five of them: free association, rank order, spatial, chronological, and general to specific.

When ideas are organized according to *free association,* one idea triggers the next because it is a related one. Free association is especially common in advancing a narrative, as in a personal experience essay. It is quite fluid and suggestive and seldom includes topic sentences. In the following paragraph, Judith allows the first act of locking a door to trigger memories related to other locked doors.

> It is already afternoon. I fiddle with the key to lock the apartment door after me. I am not accustomed to locking doors. Except for the six months I spent in Boston, I have never lived in a place where I did not trust my neighbors. When I was little, we couldn't lock our farmhouse door; the wood had swollen, and the bolt no longer lined up properly with the hole, and nobody ever bothered to fix it. I still remember the time our babysitter, Rosie, hammered the bolt closed and we had to take the door off the hinges to get it open.

Notice that Judith uses a reverse chronological arrangement to order her associations; that is, she moves backward from the present—first to Boston, then to childhood—thereby using one pattern to strengthen another, helping us still further to follow her. (To read Judith's complete essay, see Chapter 14.)

When ideas are arranged by *rank order,* that is, order of importance, the most significant idea is reserved for the end of the paragraph. The writer leads with the idea to be emphasized least, then the next most important, and so on. This paragraph is commonly introduced by a topic sentence alerting readers that an orderly list is to follow. Here is a paragraph from Kelly's essay interpreting "We Real Cool" where he explains the values of the narrators of the poem:

> However, the most important element of their lives is being "cool." They live and love to be cool. Part of being cool is playing pool, singing, drinking, fighting, and messing around with women whenever they can. Being cool is the code of action that unites them, that they celebrate, for which they are willing to die.

Kelly explains three dimensions of being cool, making sure he concludes the paragraph with his most important point: first, being cool includes "[living] and [loving] to be cool," second, that it includes "playing pool, singing, drinking," and so on; and third, that it is their "code of action . . . for which they are willing to die." (To read Kelly's complete essay, see Chapter 13.)

When ideas are arranged *spatially,* each is linked to the next. Thus, the reader's eye is drawn through the paragraph as if through physical space. For example, a writer might describe a landscape by looking first at the field, then the forest, then the mountain, then the sky. In the following paragraph, Beth begins by showing Becky in the spatial context of her dormitory room; her description moves from bed to walls to floor:

> Becky sits cross-legged at the foot of the bottom bunk on her pink and green homemade quilt. She leans up against the wall and runs her fingers through her brown shoulder-length hair. The sound of James Taylor's "Carolina on My Mind" softly fills the room. Posters of John Lennon, James Dean, and Cher look down on us from her walls. Becky stares at the floor and scrunches her face as if she were thinking hard.

Becky's subtle, silent actions carry readers through the paragraph as Beth describes her sitting, leaning, listening, and staring—actions that set up the next paragraph in her paper in which Becky speaks. (To read Beth's complete essay, see Chapter 10.)

When ideas or facts are arranged *chronologically,* they are presented in the order in which they happened, with the earliest first. Sometimes it makes sense to use *reverse chronology,* listing the most recent first and working backward in time. The following paragraph from a collaborative research paper illustrates forward chronology; it begins with the first microbrewery and then moves to a full-fledged brewers' festival five years later.

> According to Shaw (1990) the home brewing revolution did not begin in Vermont until February 1987 when Stephan Mason and Alan Davis opened Catamount Brewery, which offered golden lager, an amber ale, and a dark porter as well as several seasonal brews. This was only the beginning. In September 1992, the first Vermont Brewers Festival was held at Sugarbush Resort. Sixteen breweries participated and the forty-plus beers present ranged from American light lagers to German-style bock and everything in between. The beers included such colorful names as Tall Tale Pale Ale, Black Fly Stout, Slopbucket Brown Ale, Summer Wheat Ale, Avid Barley Wee Heavy, and Hickory Switch Smoked Amber Ale.

Notice that starting in the middle of the paragraph, another supportive pattern is at work here: the pattern of general to specific. It is unlikely that the first draft of this paragraph contained these mutually supportive organizational patterns; careful editing made sure the final draft did.

A *general to specific pattern* begins with an overall description or general statement and moves toward a description of smaller, more specific details. In the preceding paragraph, the general idea is "all breweries"; the specific idea is "Catamount beer." Notice, too, the pattern in Elizabeth's essay on questionable environmental advertising:

> Some companies court the public by mentioning environmental problems and pointing out that they do not contribute to those problems. For example, the natural gas industry describes natural gas as an alternative to the use of ozone-depleting CFC's ("Don't you wish," 1994). However, according to Fogel (1985), the manufacture of natural gas creates a host of other environmental problems from land reclamation to the carbon-dioxide pollution, a major cause of global warming. By mentioning problems they don't cause, while ignoring ones they do, companies present a favorable environmental image that is at best a half truth, at worst an outright lie.

The opening sentence introduces the general category, "environmental problems," and the following sentences provide a particular example, "the natural gas industry." (To read Elizabeth's complete essay, see Chapter 21.)

Note, too, that *specific to general,* the reverse of the previous pattern, is also common and has a recognizable logic. For example, the environmental paragraph could have opened with the description of a specific abuse by a specific company and closed by mentioning the general problem illustrated by the specific example. The point here, as it is in all writing, is to edit carefully for a pattern that's recognizable and logical so that you lead your readers through the paper in ways that match their expectations.

✒ WRITING 2

Review a near-final draft of a paper you are working on, and identify the organizational pattern in each paragraph. Do you find a pattern to your paragraphing? Identify paragraphs that contain a single idea carefully developed and paragraphs that need to be broken into smaller paragraphs. What editing changes would you now make in light of this review?

✒ HELPING THE READER

So far, most of this discussion has focused on structures within paragraphs. When editing, it's important to know how to rewrite paragraphs to improve essay readability. However, you can improve readability in other ways as well. One of them is to break up lengthy paragraphs.

Paragraph breaks help readers pause and take a break while reading, allowing them, for example, to imagine or remember something sparked by the text and yet find their place again with ease. Breaking into a new paragraph can also recapture flagging attention, especially important in long essays, reports, or articles where detail sometimes overwhelms readers. And you can emphasize points with paragraph breaks, calling a little extra attention to what follows.

✂ TRANSITIONS BETWEEN PARAGRAPHS

Your editing is not finished until you have linked the paragraphs, so that readers know where they have been and where they are going. In early drafts, you undoubtedly focused on getting your ideas down and paid less attention to clarifying relationships between ideas. Now, as you edit your final draft, consider the elements that herald transitions: words and phrases, subheads, and white spaces.

Words and Phrases

Writers often use transitional expressions without consciously thinking about them. For example, in writing a narrative, you may naturally use sequential transition words to indicate a chronology: *first, second, third; this happened, next this happened, finally this happened; or last week, this week, yesterday, today.* Here are some other transitional words and phrases and their functions in paragraphs:

- Contrast or change in direction: *but, yet, however, nevertheless, still, at the same time, on the other hand*
- Comparison or similarity: *likewise, similarly*
- Addition: *and, also, then, next, in addition, furthermore, finally*
- Summary: *finally, in conclusion, in short, in other words, thus*
- Example: *for example, for instance, to illustrate, specifically, thus*
- Concession or agreement: *of course, certainly, to be sure, granted*
- Time sequence: *first, second, third; (1), (2), (3); next, then, finally, soon, meanwhile, later, currently, eventually, at last*
- Spatial relation: *here, there, in the background, in the foreground, to the left, to the right, nearby, in the distance, above, below*

There is no need to memorize these functions or words; you already know and have used all of them and usually employ them quite naturally. When reworking your final draft, though, be sure you have provided transitions. If you haven't, work these words in to alert your readers to what's coming next.

Alternative Transitional Devices

Other common devices that signal transitions include subheadings, lines, alternative typefaces, and white spaces. The first two are more common in textbooks and technical reports; the latter two may appear in any text, including literary-style essays. When you edit your final draft, consider whether using any of these techniques would make your ideas clearer.

Subheads

To call extra attention to material or to indicate logical divisions of ideas, some writers use subheads. They are more common in long research papers, technical reports, and laboratory analyses and less common in narrative essays. They are essential in textbooks, such as this one, for indicating divisions of complex material.

Lines

Blocks of text can be separated by either continuous or broken dashes (————) or asterisks (*****), which signify material clearly to be set off from other material. In technical writing, for example, material may even be boxed in by continuous lines to call special attention to itself. In a *New Yorker* research essay or short story, a broken line of asterisks may suggest a switch in time, place, or point of view.

Alternative Typefaces

Writers who use computers can change fonts with ease. When they use alternative typefaces, they are indicating a transition or a change. In a narrative or essay, *italics* may suggest someone talking or the narrator thinking. A switch to a smaller or larger typeface may signal information of less or more importance.

White Space

You can indicate a sharp break in thought between one paragraph and the next by leaving an extra line of space between them (although the space break does not tell the reader what to expect next). When I use a space break, I am almost always suggesting a change in direction more substantial than a mere paragraph indentation; I want readers to notice the break and be prepared to make a larger jump than a paragraph break signals. In a narrative essay, I may use the space to suggest a jump in time (the next day, week, or year); in argumentative writing, to begin a presentation of an opposing point of view; in an essay, to introduce another voice. White space, in other words, substitutes for clear transition words and subheadings but does not explicitly explain the shift.

When I work on early drafts, I may use some of these transition or separation devices to help me keep straight the different parts of what I'm writing. In final drafts, I decide which devices will help my readers as much as they have helped me, and I eliminate those that no longer work. In other words, paragraphs and transitions are as useful for me when I'm drafting as they will be later for my readers. (For more ideas on alternate transitions, see Chapter 24.)

ᴈ WRITING 3

When you edit a final draft, look carefully at your use of transitional devices. Identify those that are doing their work well; add new ones where appropriate.

Chapter

࿔ *27* ࿔

Working Sentences

Teachers are always nitpicking about little things, but I think writing is for communicating,
not nitpicking. I mean, if you can read it and it makes sense, what else do you want?

OMAR

Omar, editing is about nitpicking. It's about making your text read well, with the most possible sense. After the ideas are in order and well supported, your job is to polish the paragraphs, sentences, and individual words so that they shine. Then you correct to get rid of all the "nitpicky" errors in punctuation, spelling, and grammar. In other words, you attend to editing *after* your ideas are conceptually sound, carefully supported, skillfully organized, and fairly well aimed at your readers. (Even now, it doesn't hurt to review it once more to make sure it represents your voice and ideas in the best way possible.)

As in editing paragraphs, there is no one best way to go about editing sentences. You edit in such a way that you remain, as much as possible, in control of your text. (As you probably know by now, texts have a way of getting away from all of us at times. Editing is how we try to get control back!) At the same time you're wrestling for final control of a text for yourself, you're also anticipating reader needs. In this sense, sentence editing is your final balancing act, as you work to please yourself and your readers.

࿔ EDITING FOR CLARITY, STYLE, AND GRACE

To effect maximum communication, you should edit your sentences first for clarity, making sure each sentence clearly reflects your purpose. You also must edit to convey an appropriate style for the occasion, that is, the formality or informality of the language. And at perhaps the highest level, you must edit to convey grace—some sense that this text is not only clearly written, by you, but that it is also particularly well written—what we might call elegant or graceful.

While I can explain this loose hierarchy as if these several levels are easily distinguished, in fact, they are not, and they mix and overlap easily. For instance, in writing the chapters for this text I have tried to edit each chapter, paragraph, and sentence with all three goals in mind, demanding that all my language be clear, hoping that my style is friendly and that my sentences are also graceful—knowing that, in many cases, grace has proved beyond my reach. The remainder of this chapter will examine the fine tuning of words and phrases that make clear, stylistically appropriate, and sometimes graceful sentences.

♫ WRITING 1

Reread a near-final draft of one of your papers, and draw a straight vertical line next to places where your text seems especially clear. Draw a wavy line next to passages where the style sounds especially like you. And put an asterisk next to any passages that you think are especially graceful. Exchange drafts with a classmate and see if you agree with each other's assessment.

♫ THE WORK OF SENTENCES

Sentences are written in relation to other sentences, so most of our attention thus far has been on larger units of composition, from whole texts on down to individual paragraphs. This chapter focuses on strategies for strengthening sentences. In editing, you should first look at the effect of particular words within sentences, especially nouns, verbs, and modifiers. Second, you should consider the importance of rhythm and emphasis in whole sentences. And finally, you should learn to identify and avoid the common problems of wordiness, clichés, jargon, passive constructions, and biased language.

♫ WRITE WITH CONCRETE NOUNS

Nouns label or identify persons (*man, Gregory*), animals (*dog, golden retriever*), places (*city, Boston*), things (*book, The Working Writer*), or ideas (*conservation, Greater Yellowstone Coalition*). General nouns name general classes or categories of things (*man, dog, city*); concrete nouns refer to particular things (*Gregory, golden retriever, Boston*). Notice that concrete nouns (not just any dog, but a golden retriever) appeal more strongly to a reader's senses (I can see the dog!) than abstract nouns do and create a more vivid and lively reading experience.

Here is an example of a paragraph composed primarily of general nouns (underlined in the passage):

> Approaching the library, I see lots of people and dogs milling about, but no subjects to write about. I'm tired from my walk and go inside.

When Judith described a similar scene for her essay on personal safety, she used specific nouns (which are underlined) to let us see her story sharply:

> Approaching the library, I see skateboarders and bikers weaving through students who talk in clusters on the library steps. A friendly black dog is tied to a

bench watching for its master to return. Subjects to write about? Nothing strikes me as especially interesting and, besides, my heart is still pounding from the walk up the hill. I wipe my damp forehead and go inside.

Judith could have gone even further (writers always can) in using concrete nouns. She could have named the library, described some individual students, identified the dog, and described the bench. None of these modifications would have changed the essential meaning of the sentences, but each would have added a dimension of specific reality—one of the key ways writers convince readers that what they are writing about is true or really happened.

☞ WRITE WITH ACTION VERBS

Action verbs *do* something in your sentences; they make something happen. *Walk, stride, run, jump, fly, hum, sing, sail, swim, lean, fall, stop, look, listen, sit, state, decide, choose,* and *conclude*—all these words and hundreds more are action verbs. Static verbs, in contrast, simply *appear* to describe how something *is*—like the verb *is* in this sentence. Action verbs, like concrete nouns, appeal to the senses, letting readers see, hear, touch, taste, or smell what is happening. They create more vivid images for readers, drawing them more deeply into the essay.

In the following passage, the conclusion to Judith's reflective essay, notice how action verbs (underlined) help you see clearly what is going on:

> Hours later—my paper started, my exam studied for, my eyes tired—I retrace the path to my apartment. It is dark now, and I listen closely when I hear footsteps behind, stepping to the sidewalk's edge to let a man walk briskly past. At my door, I again fumble for the now familiar key, insert it in the lock, open the door, turn on the hall light, and step inside. Here, too, I am safe, ready to eat, read a bit, and finish my reflective essay.

Judith also uses several static verbs (*is, am*) in other places; these verbs describe necessary states of being, carrying a different kind of weight. When they are used among action verbs, they do good work. But the paragraph gets its life and strength from the verbs that show action.

Editing for action verbs is one of the chief ways to cut unneeded words, thus increasing readability and vitality. Whenever you find one of the following noun phrases (in the first column) consider substituting an action verb (in the second column):

reach a decision	decide
make a choice	choose
hold a meeting	meet
formulate a plan	plan
arrive at a conclusion	conclude
have a discussion	discuss
go for a run	run

ᴥ USE MODIFIERS CAREFULLY AND SELECTIVELY

Well-chosen modifiers can make both nouns and verbs more concrete and appealing to readers' senses. Words that modify—describe, identify, or limit—nouns are called *adjectives* (*damp* forehead); words that amplify verbs are called *adverbs* (listen *closely*). Modifiers convey useful clarifying information and make sentences vivid and realistic.

In the previous example paragraph, Judith could have added several more modifiers to nouns such as *man* (*tall, thin, sinister*) and *door* (*red, heavy wooden*). And she could have used modifiers with verbs such as *retrace* (*wearily, slowly*) and *fumble* (*nervously, expectantly*). Judith's writing would not necessarily benefit by these additions, but they are further possibilities for her to examine as she edits her near-final sentences. Sometimes adding modifiers to sentences distracts from rather than enhances a paragraph's purpose. And that's what editing is all about: looking carefully, trying out new possibilities, settling for the effect that pleases you most.

Not all modifiers are created equal. Specific modifiers that add descriptive information about size, shape, color, texture, speed, and so on appeal to the senses and usually make writing more realistic and vivid. General modifiers such as the adjectives *pretty, beautiful, good, bad, ugly, young,* or *old* can weaken sentences by adding extra words that do not convey specific or vital information. And the adverbs *very, really,* and *truly* can have the same weakening effect because they provide no specific clarifying information.

ᴥ WRITING 2

Review a near-final draft, and mark all concrete nouns (underline once), action verbs (underline twice), and modifiers (circle). Then place parentheses around the general nouns, static verbs, and general modifiers. Reconsider these words, and edit appropriately.

ᴥ FIND A PLEASING RHYTHM

Rhythm is the pattern of sound sentences make when you read them out loud. Some rhythms sound natural—like a person in a conversation. Such sentences are easy to follow and understand and are usually pleasing to the ear. Others sound awkward and forced, make comprehension difficult, and offend the ear. It pays to read your sentences out loud and see if they sound like a real human being talking. To make sentence clusters sound better, use varied sentence patterns and parallel construction.

Varied sentence patterns make sentence clusters clear and enjoyable for readers. Judith effectively varied her sentences—some long, some short, some simple, some complex. For example, note the dramatic effect of following a lengthy compound sentence with a short simple sentence (made up of short words) to end the paragraph above: "Nothing strikes me as especially interesting and, besides, my heart is still pounding from the walk up the hill. I wipe my damp forehead and go inside."

Parallelism, the repetition of a word or grammatical construction within a sentence, creates symmetry and balance, makes an idea easier to remember, and is pleasing to the ear. The following sentence from Brendan's essay demonstrates the pleasing rhythmic effect of parallel construction: "A battle is being waged between environmental conservationists, who support the reintroduction of wolves, and sheep and cattle farmers and Western hunters, who oppose it." The parallelism is established by repetition of the word *who* plus a verb; the verbs, opposite in meaning, provide additional dramatic effect.

In the following example, the repetition of the word *twice* establishes a rhythm and contributes as well to the writer's point about costs: "A CD may be twice as expensive as a cassette tape, but the sound is twice as clear and the disc will last forever."

✌ PLACE THE MOST IMPORTANT POINT LAST

As in paragraphs, the most emphatic place in sentences is last. You achieve the best effect by placing information that is contextual, introductory, or less essential earlier in the sentence and end with the idea you most want readers to remember. Sometimes you write first-draft sentences with emphatic endings, but often such emphasis needs to be edited in. Notice the difference in emphasis in the following version of the same idea:

> **Angel needs to start now if he wants to have an impact on his sister's life.**

> **If Angel wants to have an impact on his sister's life, he has to start now.**

The second sentence is much more dramatic, emphasizing the need for action on Angel's part.

The next two sentences also illustrate the power of placing what you consider important at the end of the sentence:

> **Becky stares at the floor and scrunches her face as if she were thinking hard.**

> **As if she is thinking hard, Becky stares at the floor and scrunches up her face.**

The first sentence emphasizes Becky's concentration. To end with Becky's scrunching up her face diminishes the emphasis on her thinking.

In the following sentence, Judith uses end-of-sentence emphasis for a transitional purpose: "I wipe my damp forehead and go inside." The ending forecasts the next paragraph—in which Judith goes inside the library. To reverse the actions would emphasize the damp forehead instead of Judith's entrance into the library.

One more example from Judith's essay suggests how emphasis at the end can increase and then resolve suspense: "It is dark now, and I listen closely when I hear footsteps behind, stepping to the sidewalk's edge to let a man walk briskly past." At first we are alarmed that footsteps are coming up behind the writer—as Judith wants us to be. Then we are relieved that a man passes harmlessly by—as Judith also wants us to be. The end of the sentence relieves the tension and resolves the suspense.

ↄ **WRITING 3**

Examine the sentences in a recent draft for rhythm and end-of-sentence emphasis by reading the draft out loud, listening for awkward or weak spots. Edit for sentence variety and emphasis as necessary.

ↄ EDIT WORDY SENTENCES

Cut out words that do not pull their weight or add meaning, rhythm, or emphasis. Sentences clogged with unnecessary words cause readers to lose interest, patience, and comprehension. Editing sentences for concrete nouns, action verbs, and well-chosen modifiers will help you weed out unnecessary words. Writing varied and emphatic sentences helps with this task too. Look at the following sentences, which all say essentially the same thing:

- In almost every situation that I can think of, with few exceptions, it will make good sense for you to look for as many places as possible to cut out needless, redundant, and repetitive words from the papers and reports, paragraphs and sentences you write for college assignments. (48 words)
- In most situations it makes good sense to cut out needless words from your college papers. (16 words)
- Whenever possible, omit needless words from your writing. (8 words)
- Omit needless words. (3 words)

The forty-eight-word-long first sentence is full of early-draft language; you can almost see the writer finding his or her way while writing. The sixteen-word sentence says much the same thing, with only one-third the number of words. Most of this editing simply cut out unnecessary words. Only at the end were several wordy phrases condensed: "from the papers and reports, paragraphs and sentences you write for college assignments" was reduced to "from your college papers."

That sixteen-word sentence was reduced by half by rephrasing and dropping the emphasis on college writing. And that sentence was whittled down by nearly two-thirds, to arrive at the core three-word sentence, "Omit needless words."

The first sentence was long-winded by any standard or in any context; each of the next three might serve well in different situations. Thus, when you edit to make language more concise, you need to think about the overall effect you intend to create. Sometimes the briefest construction is not the best one for your purpose. For example, the three-word sentence is more suited to a brief list than to a sentence of advice for this book. To fit the purposes of this book, in fact, I might write a fifth version on needless words, one including more of my own voice:

I prefer to read carefully edited papers, where every word works purposefully and pretty much pulls its own weight. (19 words)

In this sentence, I chose to include *I* to emphasize my own preference as a teacher and reader and to add the qualifying phrase *pretty much* to impart a conversational tone to the sentence.

In the following example, one of Judith's effective paragraphs has been deliberately padded with extra words, some of which might have existed in earlier drafts:

> It is now several hours later, almost midnight, in fact. I have finally managed to get my paper started and probably overstudied for my exam. My eyes are very tired. I get up and leave my comfortable chair and walk out of the library, through the glass doors again, and retrace the path to my apartment. Since it is midnight, it is dark, and I nervously listen to footsteps coming up behind me. When they get too close for comfort, I step to the sidewalk's edge, scared out of my wits, to let a man walk briskly past. When I am finally at my door, I again fumble for the now familiar key, insert it in the lock, open the door, turn on the hall light, and step inside. Here, too, I am safe, ready to eat leftover pizza, study some more for my exam, and finish my reflective essay.

Now compare this with Judith's final version for simplicity, brevity, smoothness, and power. (To see Judith's complete essay, see Chapter 14.)

> Hours later—my paper started, my exam studied for, my eyes tired—I retrace the path to my apartment. It is dark now, and I listen closely when I hear footsteps behind, stepping to the sidewalk's edge to let a man walk briskly past. At my door, I again fumble for the now familiar key, insert it in the lock, open the door, turn on the hall light, and step inside. Here, too, I am safe, ready to eat, read a bit, and finish my reflective essay.

The best test of whether words are pulling their own weight and providing rhythm, balance, and emphasis is to read the passage out loud and let your ear tell you what is sharp and clear and what could be sharper and clearer.

৵ EDIT CLICHÉS

Clichés are phrases we've heard so often before that they no longer convey an original or individual thought. In the wordy paragraph above, the phrase "scared out of my wits" is a cliché. As you edit, note whether you remember hearing the same exact words before, especially more than once. If so, look for fresher language that is your own. Common clichés to avoid including the following:

throwing the baby out with the bath water
a needle in a haystack
the last straw
better late than never
without further ado
the handwriting on the wall
tried and true

last but not least
lay the cards on the table
jump-start the economy

Each of these phrases was once new and original and attracted attention when it was used; now when we read or hear these phrases, we pay them no conscious mind and may even note that the writer or speaker using them is not very thoughtful or original.

✃ EDIT PASSIVE CONSTRUCTIONS

A construction is passive when something is done to the subject rather than the subject's doing something. *The ball was hit by John* is passive. *John hit the ball* is active. Not only is the first sentence needlessly longer by two words, it takes readers a second or two longer to understand since it is a roundabout way to make an assertion. Writing that is larded up with such passive construction lacks vitality and is tiresome to read.

Most of the example paragraphs in this book contain good examples of active constructions: *I retrace . . . I get up . . . Becky sits . . . Greg attributes. . . .*

✃ EDIT BIASED LANGUAGE

Your writing should not hurt people. As you edit, make sure your language doesn't discriminate against categories of people based on gender, race, ethnicity, or social class.

Eliminate Sexism

Language is sexist when it includes only one gender. The most common occurrence of sexist language is the use of the word *man* or *men* to stand for *human being* or *people*—which seems to omit *women* from participation in the species. Americans have been sensitized to the not-so-subtle bias against woman embedded in our use of language.

It is important to remember that many thoughtful and powerful English-language works from the past took masculine words for granted, using *man, men, he, him,* and *his* to stand for all members of the human race. Consider Thomas Jefferson's "All men are created equal" and Tom Paine's "These are the times that try men's souls." Today we would write "All people are created equal" or "These are the times that try our souls"—two of several possible fixes for this gender nearsightedness. When you read older texts, recognize that the composing rules were different then, and the writers are no more at fault than the culture in which they lived.

As you edit to avoid sexist language, you will notice that the English language does not have a gender-neutral third-person singular pronoun to match the gender-neutral third-person plural (*they, their, them*). We use *he* (*him, his*) for men and *she* (*her, hers*) for women. In the sentence "Everybody has his own opinion," the indefinite pronoun *everybody* needs a singular pronoun to refer to it. While it is grammatically correct to say

"Everybody has *his* own opinion," the sentence seems to exclude women. But it is grammatically incorrect to write "Everybody has *their* own opinion," although *their* is gender neutral. In editing, be alert to such constructions and consider several ways to fix them:

- Make the sentence plural so it reads *"People* have *their* own opinions."
- Include both pronouns: "Everybody has *his* or *her* own opinion."
- Eliminate the pronoun: "Everybody has *an* opinion."
- Alternate masculine and feminine pronouns throughout your sentences or paragraphs, using *she* in one paragraph and *he* in the next.

In my own writing, I have used all of these solutions at one time or another. The rule I most commonly follow is to use the strategy that makes for the clearest, most graceful writing.

Avoid Stereotypes

Stereotypes lump individuals into oversimplified and usually negative categories based on race, ethnicity, class, gender, sexual preference, religion, or age. You know many of these terms. The kindest are perhaps "Get out of the way, old man" and "Don't behave like a baby." I am willing to set these down in this book since we've all been babies and we're all growing older. The other terms offend me too much to write.

The mission of all institutions of higher learning is to teach students to read, write, speak, and think critically, which means treating each situation, case, problem, or person individually on its own merits and not prejudging it by rumor, innuendo, or hearsay unsupported by evidence or reason. To use stereotypes in academic writing will label you as someone who has yet to learn critical literacy. To write with stereotypes in any setting not only reveals your ignorance but hurts people.

✒ PROOFREAD

The last act of editing is *proofreading,* the process of reading your manuscript word for word to make sure it is correct in every way. Here are some tips to help you in this process:

- Proofread for typing and spelling errors first by using a spelling checker on your computer, if you have one. But be aware that computers will *not* catch certain errors, such as omitted words or mistyping (for example, *if* for *of*). So you must also proofread the old-fashioned way—by reading slowly, line by line, word by word.
- Proofread for punctuation by reading your essay out loud and looking for places where your voice pauses, comes to a full stop, questions, or exclaims. Let your verbal inflections suggest appropriate punctuation (commas, periods, question marks, and exclamation points, respectively). Also review Writer's Reference 5, the punctuation reference at the end of this book, paying special attention to the use of commas, the most common source of punctuation errors.

- Proofread the work of others, and ask others to proofread for you. It's easy when reading your own writing to fill in missing words and read through small errors; you're much more likely to catch such errors in someone else's writing. We are all our own worst proofreaders; ask somebody you trust to help you.
- Proofread as a whole class: Tape final drafts on the wall, and roam the class with pencils reading one another's final drafts, for both pleasure and correctness.

↬ WRITING 4

Examine a recent draft for wordiness, clichés, passive constructions, and biased language. Edit as necessary according to the suggestions in this section. Proofread before you hand in or publish the paper.

Chapter

ᔄ 28 ᔄ

Portfolios and Publishing

Revised and edited final drafts are written to be read. At the minimum, your audience is your instructor; at the maximum, it's the whole world—an audience for student writing now made possible by everyone's access to the Internet. In writing classes, the most common audience, in addition to the instructor, is the class itself. This final chapter explores three common avenues of presenting your work in final published form via writing portfolio, class book, and World Wide Web.

ᔄ DESIGNING DOCUMENTS

No matter where you "publish," prepare your paper thoughtfully. To design your document means deciding how to present your material visually for the strongest effect. Your design should simply, clearly, and unobtrusively display your writing in a manner appropriate to an academic audience and make its organization apparent. The paper you write on, the typeface you use, your spacing, titles, and headings—all contribute to your design.

Good design is transparent; that is, it should call attention to your work, not to itself. If the first thing someone notices about your paper is an unusual design element, the design is detracting from, not contributing to, readability.

Word processors place a vast array of design tools at your fingertips for desktop publishing—generating elaborately designed newsletters, brochures, advertising materials, or mass mailings. However, design for academic writing is much more conservative than that of, say, magazine articles. Even if your instructor requires a simple manuscript format achievable in longhand or with a typewriter, there are things you can do to make your paper visually appealing. Also note that for college papers, you should follow any format guidelines your instructor provides and consider the style of your discipline—MLA, APA, and so on.

Your design should be simple and clear, and it should help readers understand your paper's organization. The size of margins, the typefaces you select for text and titles, the spacing, and the placement of page numbers all contribute to the appearance of a page. Together they constitute a document's style, which should be consistent from page to page. A word processor can automatically carry these choices from page to page or document to document.

✑ CREATING A DESIGN STYLE

Typefaces

Pick a typeface that is easy to read and conventional. A *serif* typeface—one with little strokes at the end of each letterform, such as (New) Courier, (New) Times Roman, or (New) Century Schoolbook—is more traditional. *Sans serif* typefaces—those without serifs, like this—are more modern. Unless your instructor specifies, choose the typeface you find easiest to read. Save *italic* typefaces for special effects that suggest a particular emphasis or different voice in your text. Since modern word processors now offer dozens of typeface choices, be careful and select those that do not call unnecessary attention to themselves: the primary focus in academic writing will always be on content rather than style.

Spacing and Margins

White space, the area on your paper not filled by type, is an important design element. For example, the indentation at the beginning of a paragraph or an extra line of space before a heading gives readers a place to pause and helps them understand your organization.

Use a margin of at least one inch on the top, bottom, and sides of each page. Most word processors let you preview a page of text, reduced in size, to give you an idea of the proportion of type to white space.

Although most word processors can set type in two or more columns on a page, most academic papers use a single column. If you use a word processor, select a type size—10 point or 12 point—that is easy to read. Generally, the wider each line of type, the larger the point size you need for easy reading. Do not, however, select larger typefaces or wider margins in an attempt to conceal the brevity of a paper—such dodges are pretty obvious and reflect poorly on your work.

If you typewrite, your text will come out ragged right: The lines will be of unequal length. A word processor can be set to *justify* type so that all lines are of equal length, like the lines in this book. Justified type is pleasing, but make sure that the computer does not leave large spaces between words or within words, which can hinder reading. If you justify your text, also check to see that the computer has hyphenated correctly. Most word processors allow you to insert a hyphen where you want one.

Headings

Highlight the organization of your paper by using slightly different type styles for the title, headings, and subheads. On a word processor, you can vary the type size, use boldface, or a different font from the body of your paper.

Center and Capitalize Your Title

It is standard practice to center your title and capitalize all significant words. You can call further attention to your title by **boldfacing,** changing font style, changing font size, or any combination of these three; however, keep in mind that attracting too much attention is more distracting than helpful. My own most common practice is to type the entire manuscript in 11- or 12-point New Gothic font, then raise the font size to 14 point and boldface the title.

FIRST-LEVEL HEADING

First-level headings identify the major sections into which your paper is divided; if you used a formal outline, these would correspond to Roman numerals. On a word processor, first-level headings are usually typed on a line by themselves, and boldfaced.

Second-level heading

Second-level headings correspond to the *A*'s and *B*'s of a formal outline and indicate subsections of larger sections. On a word processor, second-level headings are commonly typed flush left, boldfaced, and followed by a period on the same line as the text to follow or on a separate line.

Organizing Text Visually

Word processors allow you to create papers with the same visual flair and interest as published articles in popular periodicals. When appropriate, you can add clip art from word processor files, scan in photographs, or create charts and graphics to show patterns or compare data. No matter what visual approach you use in academic papers, keep it clear and easy to read, and make sure to refer to each graphic illustration in your text so that it is more purposeful than decorative.

Another way to call special attention to points in your text is to present a list with bullets (centered dots, squares, or dashes), usually boldfaced, at the beginning of each item. If you use bullets, keep these points in mind:

- Use a simple bullet rather than a fancy dingbat that will attract attention to itself.
- Reserve bulleted lists for brief points you want to emphasize.
- Indent all lines of your bulleted list to set them off from the rest of your text.

✒ PREPARING WRITING PORTFOLIOS

In simplest terms, a *writing portfolio* is a collection of your writing contained within a single binder or folder. This writing may have been done over a number of weeks, months, or even years; it may be organized chronologically, thematically, or according to quality. A private writing portfolio may contain writing that you wish to keep for yourself; in this case you decide what's in it and what it looks like. However, a writing portfolio assigned for a class will contain writing to be shared with an audience to demonstrate your writing and reasoning abilities. One kind of writing portfolio, accumulated during a college course, presents a record of your work over a semester and will be used to assign a grade. Another type of portfolio presents a condensed, edited story of your semester's progress in a more narrative form. In addition, portfolios are often requested by prospective employers in journalism and other fields of professional writing; these samples of your best work over several years may determine whether or not you are offered a job as a writer or editor.

Course Portfolios

The most common type of portfolio assigned in a writing course contains the cumulative work collected over the semester plus a cover letter in which you explain the nature and value of these papers. Sometimes you will be asked to assign yourself a grade based on your own assessment. The following suggestions may help you in preparing a course portfolio:

1. **Make your portfolio speak for you.** If your course portfolio is clean, complete, and carefully organized, that's how you will be judged. If it's unique, colorful, creative, and imaginative, that, too, is how you'll be judged. So, too, will you be judged if your folder is messy, incomplete, and haphazardly put together. Before giving your portfolio to somebody else for evaluation, consider whether it reflects how you want to be presented.

2. **Include exactly what is asked for.** If an instructor wants three finished papers and a dozen sample journal entries, that's the minimum your course portfolio should contain. If an employer wants to see five samples of different types of writing, be sure to include five samples. Sometimes you can include more than asked for, but never include less.

3. **Add supplemental material judiciously.** Course portfolios are among the most flexible means of presenting yourself. If you believe that supplemental writing will present you in a better light, include that too, but only after the required material. If you include extra material, attach a note to explain why it is there. Supplemental writing might include journals, letters, sketches, or diagrams that suggest other useful dimensions of your thinking.

4. **Include perfect final drafts.** At least make them as close to perfect as you can. Show that your own standard for finished work is high. Final drafts should be printed double-spaced on one side only of high-quality paper, be carefully proofread, and follow the language conventions appropriate to the task—unless another format is requested.

5. **Demonstrate growth.** This is a tall order, of course, but course portfolios, unlike most other assessment instruments, can demonstrate positive change. The signal value of portfolios in writing classes is that they allow you to demonstrate how a finished paper came into being. Consequently, instructors commonly ask for early drafts to be attached to final drafts of each paper, the most recent on top, so they can see how you followed revision suggestions, how much effort you invested, how many drafts you wrote, and how often you took risks. To build such a record of your work, date every draft of each paper and keep it in a safe place.

6. **Demonstrate work in progress.** Course portfolios allow writers to present partially finished work that suggests future directions and intentions. Both instructors and potential employers may find such preliminary drafts or outlines as valuable as some of your

finished work. When you include such tentative drafts, be sure to attach a note explaining why you still believe it has merit and in which direction you want to take it.

7. **Attach a table of contents.** For portfolios containing more than three papers, attach a separate table of contents. For those containing only a few papers, embed your table of contents in the cover letter.

8. **Organize your work with clear logic.** Three methods of organization are particularly appealing:

- *Chronological order.* Writing is arranged in order, beginning the first week, ending the last week, with all drafts, papers, journal entries, letters, and such fitting in place according to the date written. Only the cover letter (see below) is out of chronological order, serving as an introduction to what follows. This method allows you to show the evolution of growth most clearly, with your latest writing—presumably the best—presented at the end.
- *Reverse chronological order.* The most recent writing is up front and the earliest writing at the back. In this instance, the most recent written document—the cover letter—is in place at the beginning of the portfolio. This method features your latest (and best) work up front and allows readers to trace back through the history of how it got there.
- *Best-first order.* You place your strongest writing up front and your weakest in back. Organizing a portfolio this way suggests that the work you consider strongest should count most heavily in evaluating the semester's work.

Unless otherwise specified, arrange your work in a three-ring binder as you collect it during the term, so you can add and delete as you choose, making the best possible case for yourself. Work collected this way can then be bound in a simple cover for presentation at term's end or left in the binder. In either case, put your name and address on the outside cover for easy identification.

9. **Write a careful and honest cover letter.** Include a cover letter. For many instructors, the cover letter will be the most important part of your course portfolio since it represents your own most recent assessment of the work you completed over the semester. A cover letter serves two primary purposes: (1) an introduction describing and explaining the portfolio's contents and organization, including accounts of any missing or unusual pieces to be found therein; and (2) a self-assessment of the work, from earliest to latest draft of each paper, and from earliest to latest work over the course of the semester. The following excerpt is from Kelly's letter describing the evolution of one paper:

> In writing the personal experience paper, I tried three different approaches, two different topics, and finally a combination of different approaches to my final topic. My first draft [about learning the value of money] was all summary and didn't show anything actually happening. My second draft wasn't focused because I was still trying to cover too much ground. At this point, I got frustrated and tried a new topic [the hospital] but that didn't work either. Finally, for my last draft, I returned to my original topic, and this time it worked. I described one scene in great detail and included dialogue, and I liked it better and so did you. I am pleased with the way this paper came out when I limited my focus and zeroed in close.

The following excerpt describes Chris's assessment of her work over the whole semester:

> As I look back through all the papers I've written this semester, I see how far my writing has come. At first I thought it was stupid to write so many different drafts of the same paper, like I would beat the topic to death. But now I realize that all these different papers on the same topic all went in different directions. This happened to some degree in the first paper, but I especially remember in my research project, when I interviewed the director of the Ronald McDonald House, I really got excited about the work they did there, and I really got involved in the other drafts of that paper.
>
> I have learned to shorten my papers by editing and cutting out needless words. I use more descriptive adjectives now when I'm describing a setting and try to find action verbs instead of "to be" verbs in all of my papers. I am writing more consciously now—I think that's the most important thing I learned this semester.

Guidelines for Creating Course Portfolios

1. Date, collect, and save in a folder all papers written for the course.
2. Arrange papers in chronological, reverse chronological, or qualitative order, depending on the assignment.
3. In an appendix, attach supplemental writing such as journal excerpts, letters, class exercises, quizzes, or other relevant writing.
4. Review your writing and compose a cover letter explaining the worth or relevance of the writing in the portfolio: Consider the strengths and weaknesses of each individual paper as well as those of the entire collection. Provide a summary statement of your current standing as a writer as your portfolio represents you.
5. Attend to the final presentation: Include all writing in a clean, attractive folder; organize contents logically; attach a table of contents; write explanatory memos to explain unusual materials; and make sure the portfolio meets the minimum specifications of the assignment.

Story Portfolios

A *story portfolio* is a shorter, more carefully edited and composed production than a cumulative course portfolio. Instead of including a cover letter and all papers and drafts written during the term as evidence for self-assessment, a story portfolio presents the evolution of your work and thought over the course of the semester in narrative form. In a story portfolio, you include excerpts of your papers insofar as they illustrate points in your development as a writer. In addition, you include excerpts of supplemental written records accumulated at different times during the semester, including the following:

- Early and dead-end drafts of papers
- Journal entries
- Lecture and discussion notes

- In-class writing and freewriting
- Comments on papers from your instructor
- Letters to or from your instructor
- Comments from classmates about your papers

In other words, to write a story portfolio, you conduct something like an archeological dig through the written remains of your work in a class. By assembling this evidence in chronological order and choosing the most telling snippets from these various documents, you write the story that explains, amplifies, or interprets the documents included or quoted. The best story portfolios commonly reveal a theme or set of issues that run from week to week or paper to paper throughout the semester. As you can see, a story portfolio is actually a small research paper, presenting a claim about your evolution as a writer with the evidence coming from your own written sources.

I encourage students to write their story portfolios using an informal voice as they might in a journal or letter. However, some students choose a more formal voice. Some prefer to write in the third person, analyzing the semester's work as if they did not know the writer (themselves). I also encourage them to experiment with the form and structure of their story portfolios, so that some present their work as a series of dated journal entries or snapshots while others write a more fluid essay with written excerpts embedded as they illustrate this or that point. Following are a few pages from Karen's story portfolio that illustrate one example of such a portfolio:

> When I entered English 1, I was not a confident writer and only felt comfortable writing factual reports for school assignments. Those were pretty straightforward, and personal opinion was not involved. But over the course of the semester I've learned that I enjoy including my own voice in my writing. The first day of class I wrote this in my journal:
>
>> *8/31 Writing has always been hard for me. I don't have a lot of experience writing papers except for straightforward things like science reports. I never did very well in English classes, usually getting B's and C's on my papers.*
>
> But I began to feel a little more comfortable when we read and discussed the first chapter of the book—a lot of other students besides me felt the same way, pretty scared to be taking English in college.
>
> Our first assignment was to write a paper about a personal experience that was important to us. At first, I couldn't think where to start, but when we brainstormed topics in class, I got some good ideas. Three of the topics listed on the board were ones I could write about:
>
> - excelling at a particular sport (basketball)
> - high school graduation
> - one day in the life of a waitress
>
> I decided to write about our basketball season last year, especially the last game that we lost. Here is a paragraph from my first draft:

> *We lost badly to Walpole in what turned out to be our final*
> *game. I sat on the bench most of the time.*

As I see now, that draft was all telling and summary—I didn't show anything happening that was interesting or alive. But in a later draft I used dialogue and wrote from the announcer's point of view, and the result was fun to write and my group said fun to read.

> *Well folks, it looks as if Belmont has given up, the coach is*
> *preparing to send in his subs. It has been a rough game for Bel-*
> *mont. They stayed in it during the first quarter, but Walpole has*
> *run away with it since then. Down by twenty with only six min-*
> *utes left, Belmont's first sub is now approaching the table.*

You were excited about this draft too, and your comment helped me know where to go next. You wrote:

> *Great draft, Karen! You really sound like a play-by-play an-*
> *nouncer—you've either been one or listened closely to lots of bas-*
> *ketball games. What would happen if in your next draft you*
> *alternated between your own voice and the announcer's voice?*
> *Want to try it?*

This next excerpt comes from a story portfolio that included twelve pages of discussion and writing samples and concluded with this paragraph:

> I liked writing this story portfolio at the end of the term because I can really see how my writing and my attitude have changed. I came into class not liking to write, but now I can say that I really do. The structure was free and we had plenty of time to experiment with different approaches to each assignment. I still have a long way to go, especially on my argumentative writing, since neither you nor I liked my final draft, but now I think I know how to get there: rewrite, rewrite, rewrite.

Guidelines for Creating Story Portfolios

1. Assemble your collected writing in chronological order, from beginning to end of each paper, from beginning to end of the semester.
2. Reread all your informal work (in journals, letters, instructor comments) and highlight passages that reflect the story of your growth as a writer.
3. Reread all your formal work (final papers, drafts) and highlight passages that illustrate your growth as a writer. Note especially if a particular passage had evolved over several drafts in the same paper—these would show you learning to revise.
4. Arrange all highlighted passages in order and write a story that shows (a) how one passage connects to another and (b) the significance of each passage.
5. Before writing your conclusion, reread your portfolio and identify common themes or ideas or concerns that have occurred over the semester; include these in your portfolio summary.

◇⁊ PUBLISHING CLASS BOOKS

Publishing a class book is a natural end to any writing class—or any class in which interesting writing has taken place. A class book is an edited, bound collection of student writing, usually featuring some work from each student in the class. The compilation and editing of such a book is commonly assigned to class volunteers, who are given significant authority in designing and producing the book. It is a good idea for editors to discuss book guidelines with the whole class so that consensus guidelines emerge. I suggest that class editors do the following:

Define editorial mission. Usually students ask for camera-ready copy from class-mates to simplify and speed the publishing process. However, editors may want to see near-final drafts and return with comments; or they may wish to set up class editorial boards to preview or screen near-final drafts; or they may wish to leave this role with the instructor.

Divide editorial responsibilities. Class books are best done by editorial teams consisting of two or more students who arrange among themselves the various duties described above.

Establish manuscript guidelines. Discuss with the class what each submitted paper should look like, arrive at a consensus, and ask for camera-ready copy to speed production and decrease the editors' work load. The editors should specify the following:

- double- or single-spaced typing
- type face
- font size
- margins: top and bottom, right and left
- justified or not
- title font and size
- position of author name

Set page limits. Each student may be allowed a certain number of single-or double-spaced pages. Since printing charges are usually made on a per-page basis, page length discussion is related to final publication cost.

Set deadlines. This is usually set so published copies can be delivered on the last class day or next-to-last class day (this latter allowing for reading and discussing the book on the last class day).

Organize the essays. Arrange the collected essays according to some ordering principle; this may depend upon how many and what kind of writing was done during the term. For example, students may have written (1) personal profiles, (2) feature stories, and (3) reflective essays; should there be a separate section for each? How should the essays be arranged within each section?

Write an introduction. The most significant editorial work is writing an introduction to contextualize the class book, explain its development, and describe the contents. Introductions vary in length from a paragraph to several pages.

Prepare a table of contents. Include essay title, author's name, and page number on which the essay is to be found. (And remember to paginate the manuscript, by hand if necessary.)

Ask the instructor to write an afterword or preface. Instructors may write about the assignment objectives, impressions of the essays, or reactions to the class; they may add any other perspective that seems relevant to the book.

Collect student writer biographies. Most class books conclude with short (fifty to one hundred word) biographies of the student writers, which may be serious, semiserious, or comical depending on class wishes.

Design a cover. The editor(s) may commission a classmate to do the artwork or take it on themselves. Covers can be graphic or printed, black and white or color. (Color and plastic covers usually cost more.)

Estimate publishing cost. The editors are responsible for exploring (with local print shops such as Kinko's or Staples or the college print shop) the cost of producing a certain number of copies of class books (for example, 60 pages at $0.05 each, plus color cover at $1 each, plus binding at $0.50 each equals $4.50 per book); often the local college print shop will sell the books to the students, therefore freeing the instructor from handling the money. Alternatively, the editors may ask the class to assemble the book—each student could be asked to bring in twenty copies of his or her essay to be collated and bound with the rest.

Lead a class discussion. The final responsibility of class editors is to stimulate a class discussion on the theme, format, and voice as represented by the class book or its various sections. I try to arrange for this to happen during the last class. Class responsibilities here include both a careful reading of their own book and bringing cookies and cider to facilitate the discussion.

Since editors do a significant amount of work in compiling their classmates' writing, I usually excuse them from making oral reports or leading class discussions—which the rest of the students do during the term.

Finally, when I assign editors to design and publish a class book, I make it clear that the paper published in the book is the final draft of that particular paper; when reading student portfolios at term's end, it is in the class book that I find the final drafts of one or more essays.

Publishing Books in a Computer-Equipped Classroom

If by any chance your class meets in a room equipped with computers, the whole business of collecting and editing manuscripts is greatly simplified, as students can provide editors with electronic copy that is easily and instantly modifiable—margins, font size and style, columns, and so forth can be changed at the press of a button. When I teach in such classrooms, we commonly produce a published anthology at the completion of each paper assignment—making it possible to publish two to four such "books" in a semester, giving many more students the opportunity to play the role of editor.

ᢩᡒ CONSTRUCTING WEB PAGES

Web sites, designed for personal use, as school assignments, or to promote a club or other organization, have become an almost essential way to communicate. And, although Web pages should be designed with the same regard to readable fonts, white space, and organization, their interactivity requires a slightly different approach than publishing on paper.

PREPARING GRAPHICS. Most spreadsheet programs include a simple graphic component that quickly converts data to charts or tables. The completed chart can be pasted into a document by a word-processing program.

Power of a Web Page

The interactivity of a Web page gives it a power that the printed page cannot have. It also places some restrictions on content that the printed page does not have.

A well-designed Web page gives instant access to related content, both within the same site and at other locations on the Internet. It allows the viewer to determine the sequence in which information is presented. It can offer supplemental information in the form of photos, illustrations, music, spoken words, or video. At its best, a Web page empowers the viewer.

The key to providing a high level of interactivity is to view the Web page as a three-dimensional object. The visible content provides the same two-dimensional view as a printed page. The links to other Web pages provide the third dimension.

Links can direct users to other pages within the site, to search engines, to e-mail windows, to photos, audio files, and video clips, and to other Web pages throughout the Internet. Links can take the form of graphic elements, words (hyperlinked text), or photographs.

Restrictions of Web Publishing

When designing a Web page, remember that the content of the page must travel through some kind of network. The bandwidth of the network and the speed of the receiving computer limit how quickly the content arrives on your computer.

Graphic elements, especially video and photographs, take longer to transmit than does text. Using large photographs, or many smaller ones, greatly slows transmission speed. Using animated art elements or complicated frame arrangements also makes a page slow to load.

⤳ DESIGN GUIDELINES

> **SIX STEPS OF WEB DESIGN**
>
> Decide content.
> Organize.
> Edit.
> Test.
> Publish.
> Update.

As with all document design, the content should determine how the page is designed. For Web pages, this is vital. Is your site aimed at family members, a club or organization, the general public or a specific audience? Do you want to present information, provide links, or maintain a team's records? Will your site contain lots of photographs, MP3s, or shareware?

Once you decide on the content, the next step it to *organize* it, with the goal of providing fast and easy access.

- If you are designing a Web page to display family photographs, do not place all the photographs on your main page. That would slow loading time for each viewer at each visit. Instead, use text links from a main page to direct viewers to pages of photographs, organized by content. This strategy allows viewers to decide whether they want to invest the transmission time to view the photographs.
- If you want to provide a list of links to related sites, organize them, by topic or alphabetically, so that viewers can quickly locate links they want to use.
- If you are creating a Web site for a club or team and have individual pages for each member, include a hypertext index of the members.

Each of these plans organizes the material and gives the viewer the choice of what material to access.

Web pages that require extensive scrolling discourage viewers. Edit content to fit on a single screen. Single-screen sites allow viewers to move quickly from page to page. They also encourage tighter organization and will generally allow faster uploading of the page.

Editing content does not mean eliminating it. If you have more links than fit on a screen, organize them into topics. Link from an index screen (see Example 1) to pages that contain lists for one topic. If you have photographs, again, organize them into topics and use thumbnail-size images that link to larger photos.

Example 1   A simple text index.

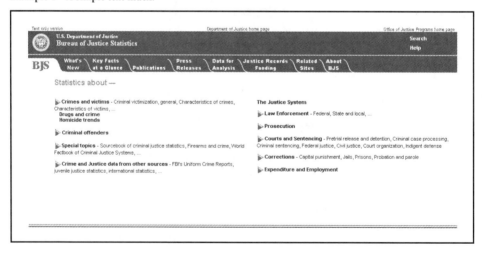

Once your pages are completed, test the links that allow viewers to move from page to page. Also test the links to other Web sites. Check that thumbnail photos do move to larger images. After your Web site is up and running, periodically check the links to make sure they remain valid.

After the design is completed and the testing is finished, publish the Web site. The method you use to publish the site depends on your Internet service provider, your Web page host, and the software you use to create your page.

Continue to update and maintain your site as long as it is on the Web. A static site, unchanged even for a few days, will soon cease to be interesting. When you update your site, leave obvious clues on the main entry page that changes have been made. In addition to providing fresh material, continue to check the validity of links.

The statistics page from the U.S. Department of Justice (Example 1) is tightly organized and easy to use. The designer used broad index links to help users quickly find the area of information they are seeking. The links are all contained on one display screen.

The designer also used color to help guide users. The bullets that separate and identify topics are tinted differently from the area at the top of the page, which serves as a name plate and a site index and helps users quickly return to the home page.

Example 2 Personal home page with icons.

The extremely functional page shown in Example 2, designed by a graduate student at North Carolina State University, includes a simple nine-point index. It follows the same design rules as the previous example. At the top, there is a strong graphic element, the

Example 3   Commercial Web page with icons.

photograph, and the name of the site. Instead of bullets, the designer used easily identifiable icons to illustrate links to other pages. The icons have small file sizes, so they quickly load onto the page. A photo gallery is included in this Web site, but at a location other than the home page. This organization gives users the power to decide whether they want to spend the time to download the photographs.

Although Example 3 (above) looks much different from the previous pages, it follows the same principles. The page is strongly identified with type and with an icon. In addition, several snapshots appear to have been tossed onto the page, a visual clue to the identity of the Web site (a film- and photograph-processing company). The index, on the right side of the page, demonstrates a simple progression from the bullets and icons used on the other pages.

WRITER'S REFERENCES

Writer's Reference 1: MLA Documentation Guidelines

Writer's Reference 2: APA Documentation Guidelines

Writer's Reference 3: Writing Letters and Résumés

Writer's Reference 4: Writing Essay Examinations

Writer's Reference 5: A Brief Guide to Punctuation

MLA Documentation Guidelines

The most common and economical form for documenting sources in research-based English papers is the MLA (Modern Language Association) system.

- All sources are briefly mentioned by author name in the text.
- A list of works cited at the end provides full publication data for each source named in the paper.
- Additional explanatory information written by the writer of the paper can be included in footnotes or endnotes.

The MLA system is explained in authoritative detail in the *MLA Handbook for Writers of Research Papers,* 5th ed. (New York: MLA, 1999) and on the MLA Web site <http://www.mla.org>.

✑ GUIDELINES FOR FORMATTING MANUSCRIPTS

The MLA guidelines for submitting college papers are fairly conservative and do not reflect the wealth of visually interesting options in fonts, type sizes, graphics, and other options available with most modern word processors. If your instructor requests MLA format, follow the guidelines below. If your instructor encourages more open journalistic formats, use good judgment in displaying the information in your text.

Paper and Printing

Print all academic assignments on 8½ × 11 white paper in a standard font such as Times New Roman or Courier. Use 10- or 12-point type and a good-quality printer.

Margins and Spacing

Allow margins of one inch all around. Justify the left margin only. Double-space everything, including headings, quoted material, and the Works Cited page. Indent five spaces for paragraphs. Indent ten spaces for prose quotations of five or more lines or poetry of more than three lines. (Do not use quotation marks around these indented quotations.)

Identification

On the first page, include your name, your instructor's name, the course title, and the date on separate lines, double-spaced, flush with the upper left margin.

Title

Center the title on the first page, capitalizing key words only. If your instructor asks for strict MLA style, avoid using italics, underlining, quotation marks, boldface type, unusual fonts, or large type for the title. (MLA does not require a title page or an outline.) Double-space to the first paragraph.

Page Numbers

Print page numbers in the upper right margin, one-half inch below the top. If following strict MLA format, include your last name before each page number to guarantee correct identification of stray pages (*Turner 1, Turner 2,* and so on).

Punctuation

One space is required after commas, semicolons, colons, periods, question marks, and exclamation points and between the periods in an ellipsis. (Double spacing is optional after end punctuation.) Dashes are formed by two hyphens, with no extra spacing on either side.

Visual Information

Label each table or chart as *Table 1, Table 2,* and so on. Label each drawing or photograph as *Figure 1* or *Fig. 2,* and so on. Include a clear caption for each figure, and place each visual aid as near as possible to the passage that refers to it.

✑ GUIDELINES FOR IN-TEXT CITATIONS

The following guidelines explain how to include research sources in the main body of your text using MLA style. Each source mentioned in your paper needs to be accompanied by a brief citation including the author's last name and the page number. These are placed either in the text itself or in parentheses. Each *in-text citation* refers readers to the alphabetical list of works cited at your paper's end, listing full publication information about each source. The following examples illustrate the most common types of in-text citations.

1.   Author Identified in a Signal Phrase

When you include the author's name in the sentence introducing the source, add only the specific page on which the material appeared, in parentheses following the information.

Carol Lea Clark explains the basic necessities for the creation of a page on the World Wide Web (77).

Do not include the word *page* or the abbreviation *p.* before the number. The parenthetical reference comes before the period.

For a work by two or three authors, include all authors' names:

Clark and Jones explain. . . .

For works with more than three authors, list all authors or use the first author's name and add *et al.* (Latin abbreviation for "and others") without a comma:

Britton et al. suggest. . . .

2.   Author Not Identified in a Signal Phrase

When you do not include the author's name in your text, add it in parentheses along with the source page number. Do not punctuate between the author's name and the page number(s).

Provided one has certain "basic ingredients," the Web offers potential worldwide publication (Clark 77).

For a work by two or three authors, include all authors' last names:

(Clark and Jones 15)

(Smith, Web, and Beck 210)

For works with more than three authors, list all authors' last names or list the first author only, adding *et al.*:

(White et al. 95)

3.   Two or More Works by the Same Author

If your paper refers to two or more works by the same author, each citation needs to identify the specific work. Either mention the title of the work in the text or include a shortened version of the title (usually the first one or two important words) in the parenthetical citation. Here are three correct ways to do this:

> According to Lewis Thomas in <u>Lives of a Cell</u>, many bacteria become dangerous only if they manufacture exotoxins (76).

> According to Lewis Thomas, many bacteria become dangerous only if they manufacture exotoxins (<u>Lives</u> 76).

> Many bacteria become dangerous only if they manufacture exotoxins (Thomas, <u>Lives</u> 76).

Identify the shortened title by underlining or quotation marks, as appropriate. Put a comma between the author's last name and the title.

4. Unknown Author

When the author of a work is unknown, give either the complete title in the text or a shortened version in the parenthetical citation, along with the page number.

> According to <u>Statistical Abstracts</u>, in 1990 the literacy rate for Mexico stood at 75 percent (374).

> In 1990 the literacy rate for Mexico stood at 75 percent (<u>Statistical</u> 374).

5. Corporate or Organizational Author

When no author is listed for a work published by a corporation, a foundation, an organization, or an association, indicate the group's full name either in the text or in parentheses:

> (Florida League of Women Voters 3)

If the name is long, it is best to cite it in the sentence and put only the page number in parentheses.

6. Authors with the Same Last Name

When you cite works by two or more different authors with the same last name, include the first initial of each author's name in the parenthetical citation:

> (C. Miller 63; S. Miller 101-04)

7. Works in More Than One Volume

Indicate the pertinent volume number for each citation before the page number, and follow it with a colon and one space:

> (Hill 2: 70)

If your source is one volume of a multivolume work, do not specify the volume number in your text, but specify it in the Works Cited list.

8.   One-Page Works

When you refer to a work one page long, do not include the page number since that will appear in the Works Cited list.

9.   Quotation from a Secondary Source

When a quotation or any information in your source is originally from another source, use the abbreviation *qtd. in:*

> Lester Brown of Worldwatch feels that international agricultural production has reached its limit (qtd. in Mann 51).

10.   Poem or Play

In citing poems, name part (if divided into parts) and line numbers; include the word *line* or *lines* in the first such reference. This information will help your audience find the passages in any source where those works are reprinted, which page references alone cannot provide.

> In "The Mother," Gwendolyn Brooks remembers "the children you got that you did not get" (line 1).

When you cite up to three lines from a poem in your text, separate the lines with slash marks:

> Emily Dickinson describes being alive in a New England summer: "Inebriate of air am I / And debauchee of dew / Reeling through endless summer days" (lines 6-8).

When you cite more than three lines, indent ten spaces for a block quotation.

Cite verse plays using act, scene, and line numbers, separated by periods. For major works such as *Hamlet,* use identifiable abbreviations:

> (Ham. 4.4.31-39)

11.   More Than One Work in a Citation

To cite two or more works, separate them with semicolons:

> (Aronson, Golden Shore 177; Didion 49-50)

In this case, more than one work by Aronson is cited in the paper, so a shortened form of the title is given as well.

12. Long Quotation Set Off from Text

To set off quoted passages of four or more lines, indent one inch or ten spaces from the left-hand margin of the text; double-space, and omit quotation marks. The parenthetical citation follows end punctuation (unlike citations for shorter, integrated quotations) and is not followed by a period.

> Fellow author W. Somerset Maugham had this to say about Austen's dialogue:
>> No one has ever looked upon Jane Austen as a great stylist. Her spelling was peculiar and her grammar often shaky, but she had a good ear. Her dialogue is probably as natural as dialogue can ever be. To set down on paper speech as it is spoken would be very tedious, and some arrangement of it is necessary. (434)

13. Electronic Texts

The MLA guidelines on documenting online sources are explained in detail online at <http://www.mla.org/set_stl.htm>.

Electronic sources are cited in the body of the text the same way as print sources: by author, title of text or Web site, and page numbers. If no page numbers appear in the source, include section (*sec.*) number or title and/or paragraph (*par.*) numbers:

> The Wizard of Oz "was nominated for six Academy Awards, including Best Picture" (Wizard par. 3).

However, Web pages commonly omit page and section numbers and are not organized by paragraphs. In such cases, omit numbers from your parenthetical references. (For a document downloaded from the Web, the page numbers of a printout should normally not be cited since pagination may vary in different printouts.)

> In the United States, the birthrate per 1,000 people has fallen steadily from 16.7 in 1995 to 14.6 in 1998 (Statistical).

✌ GUIDELINES FOR ENDNOTES AND FOOTNOTES

MLA style uses notes primarily to offer comments, explanations, or additional information (especially source-related information) that cannot be smoothly or easily accommodated in the text of the paper. In general, however, you should omit additional information, outside the main body of your paper, unless it is necessary for clarification or justification. If a note is necessary, insert a raised (superscript) numeral at the reference point in the text. Introduce the note itself with a corresponding raised numeral, and indent it.

Text with Superscript

The standard ingredients for guacamole include avocados, lemon juice, onion, tomatoes, coriander, salt, and pepper.[1] Hurtado's poem, however, gives this traditional dish a whole new twist (lines 10-17).

Note

[1] For variations see Beard 314, Egerton 197, and Eckhardt 92. Beard's version, which includes olives and green peppers, is the most unusual.

Any published reference listed in the notes also appears in the Works Cited list.

Notes may be included as *footnotes* at the bottom of the page on which the citation appears or as *endnotes,* double spaced on a separate page at the end of your paper. Endnote pages should be placed between the body of the paper and the Works Cited list, with the title *Note* or *Notes.*

✌ GUIDELINES FOR THE LIST OF WORKS CITED

Every source mentioned in the body of your paper should be identified in a list of works cited attached to the end of the paper using the following format:

- Center the title *Works Cited,* with no quotation marks, underlining, or bold-face, one inch from the top of a separate page following the final page of the paper. (If asked to include works read but not cited, attach an additional page titled *Works Consulted.*)
- Number this page, following in sequence from the last text page of your paper. If the list runs more than a page, continue the page numbering in sequence, but do not repeat the title *Works Cited.*
- Double-space between the title and first entry and within and between entries.
- Begin each entry at the left-hand margin, and indent subsequent lines the equivalent of a paragraph indention (five spaces or one-half inch).

Order of Entries

Alphabetize entries according to authors' last names. If an author is unknown, alphabetize according to the first word of the title (ignoring an initial *A, An,* or *The*).

Entry Formats

Each item in the entry begins with a capital letter and is followed by a period. Each period is followed by one space. Capitalize all major words in the book and article titles. Underline published titles (books, periodicals); put quotation marks around titles of chapters, articles, stories, and poems within published works. Do not underline volume and issue numbers or end punctuation. The four most common variations on general formats are the following:

BOOKS

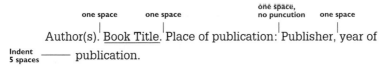

JOURNAL ARTICLES

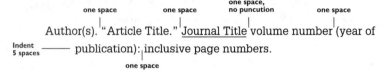

MAGAZINE AND NEWSPAPER ARTICLES

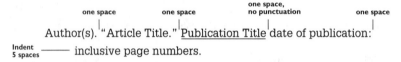

ELECTRONIC SOURCES

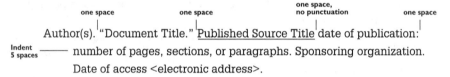

AUTHORS

- List the author's last name first, followed by a comma and then the rest of the name as it appears on the publication, followed by a period. Never alter an author's name by replacing full spellings with initials or by dropping middle initials.
- For more than one author, use a comma rather than a period after the first author; list the other authors' full names, first names first, separated by commas. Spell out the word *and* before the final author; do not use an ampersand (&). Put a period at the end.
- For more than one work by the same author, use three hyphens for the name after the first entry.

TITLES

- List full titles and subtitles as they appear on the title page of a book or in the credits for a film, videotape, or recording. Separate title and subtitles with colons (followed by one space).
- Underline titles of entire books and periodicals.
- Use quotation marks around the titles of essays, poems, songs, short stories, and chapters or other parts of a larger work.
- Put a period after a book or article title. Use no punctuation after a journal, magazine, or newspaper title.

PLACES OF PUBLICATION

- Places of publication are given for books and pamphlets, not for journals or magazines.
- Give the city of publication from the title page or copyright page. If several cities are given, use only the first.
- Use a colon to separate the place of publication from the publisher.
- For electronic sources, include the Internet address at the end of the entry in angle brackets (< >).

PUBLISHERS

- The name of the publisher is given for books and pamphlets.
- Shorten the publisher's name as described below under Abbreviations. If a title page indicates both an imprint and a publisher (for example, Arbor House, an imprint of William Morrow), list both shortened names, separated by a hyphen (*Arbor-Morrow*).
- Use a comma to separate the publisher's name from the publication date.

DATE

- For books, give the year of publication followed by a period.
- For other publications, give the year of publication within parentheses followed by a colon.
- For newspapers, put the day before the month and year (*25 May 1954*) with no commas separating the elements.
- For magazines and newspapers, put a colon after the date.
- For electronic sources, include the date the site was accessed.

PAGE NUMBERS

- Page numbers are included for all publications other than books.
- Use a hyphen, not a dash, between inclusive page numbers, with no extra space on either side.
- Use all digits for ending page numbers up to 99 and the last two digits only for numbers above 99 (*130-38*) unless the full number is needed for clarity (*198-210*).
- If subsequent pages do not follow consecutively, use a plus sign after the last consecutive page number (*39+; 52-55+*).
- If no page numbers are available for electronic sources, include paragraph or section numbers.

ABBREVIATIONS

- To shorten a publisher's name, drop the words *Press, Company,* and so forth in the publisher's name (use *Blair* for *Blair Press*). Use the abbreviation *UP* for University Press (*Columbia UP; U of Chicago P*).
- Use only the first name if the publisher's name is a series of names (*Farrar* for *Farrar, Straus & Giroux*). Use only the last name if the publisher's name is a person's name (*Abrams* for *Harry N. Abrams*).

- If no publisher or date of publication is given for a source, use the abbreviations *n.p.* ("no publisher") or *n.d.* ("no date").
- For periodicals, abbreviate months using the first three letters followed by a period (*Apr., Dec.*) except for *May, June,* and *July.* If an issue covers two months, use a hyphen to connect the months: *Apr.-May, June-Aug.* (See model 20.)

Documenting Books

1. Book by One Author

Thomas, Lewis. <u>Lives of a Cell: Notes of a Biology Watcher</u>. New York: Viking, 1974.

2. Book by Two or Three Authors

Fulwiler, Toby, and Alan R. Hayakawa. <u>The Blair Handbook</u>. 3rd ed. Upper Saddle River: Blair-Prentice, 2000.

Second and third authors are listed first name first. Do not alphabetize the authors' names within an individual Works Cited entry. The final author's name is preceded by *and;* do not use an ampersand (*&*). A comma always follows the inverted ordering of the author's first name.

3. Book by More Than Three Authors

Britton, James, et al. <u>The Development of Writing Abilities 11-78</u>. London: Macmillan, 1975.

With more than three authors, you have the option of using the abbreviation *et al.* (Latin for "and others") or listing all the authors' names in full as they appear on the title page of the book. Do not alphabetize the names within the Works Cited entry.

4. Book by a Corporation, an Organization, or an Association

U.S. Coast Guard Auxiliary. <u>Boating Skills and Seamanship</u>. Washington: Coast Guard Auxiliary National Board, 1997.

Alphabetize by the name of the organization.

5. Revised Edition of a Book

Hayakawa, S. I. <u>Language in Thought and Action</u>. 4th ed. New York: Harcourt, 1978.

6. Edited Book

Hoy, Pat C., II, Esther H. Shor, and Robert DiYanni, eds. <u>Women's Voices: Visions and Perspectives</u>. New York: McGraw, 1990.

7.   Book with an Editor and Author

Britton, James. Prospect and Retrospect. Ed. Gordon Pradl. Upper Montclair: Boyn-
    ton, 1982.

The abbreviation *Ed.* when followed by a name replaces the phrase "Edited by" and
cannot be made plural.

8.   Book in More Than One Volume

Waldrep, Tom, ed. Writers on Writing. 2 vols. New York: Random, 1985-88.

When separate volumes were published in different years, use inclusive dates.

9.   One Volume of a Multivolume Book

Waldrep, Tom, ed. Writers on Writing. Vol. 2. New York: Random, 1988.

When each volume has its own title, list the full publication information for the volume you used first, followed by information on the series (number of volumes, dates).

Churchill, Winston S. Triumph and Tragedy. Boston: Houghton, 1953. Vol. 6 of The
    Second World War. 6 vols. 1948-53.

10.   Translated Book

Camus, Albert. The Stranger. Trans. Stuart Gilbert. New York: Random, 1946.

11.   Book in a Series

Magistrate, Anthony. Stephen King, The Second Decade: Danse Macabre to The
    Dark Half. Twayne American Authors Series 599. New York: Twayne, 1992.

A book title appearing within another book's title is not underlined. Add series information just before the city of publication.

12.   Reprinted Book

Hurston, Zora Neale. Their Eyes Were Watching God. 1937. New York: Perennial-
    Harper, 1990.

Add the original publication date after the title; then cite information for the current
edition.

13.   Introduction, Preface, Foreword, or Afterword in a Book

Selfe, Cynthia. Foreword. Electronic Communication Across the Curriculum. Ed.
    Donna Rice et al. Urbana: NCTE, 1998. ix-xiv.

Atwell, Nancie. Introduction. <u>Coming to Know: Writing to Learn in the Intermedi-ate Grades</u>. Ed. Nancie Atwell. Portsmouth: Heinemann, 1990. xi-xxiii.

14. Work in an Anthology or Chapter in an Edited Collection

Donne, John. "The Canonization." <u>The Metaphysical Poets</u>. Ed. Helen Gardner. Baltimore: Penguin, 1957. 61-62.

Gay, John. <u>The Beggar's Opera</u>. 1728. <u>British Dramatists from Dryden to Sheridan</u>. Ed. George H. Nettleton and Arthur E. Case. Carbondale: Southern Illinois UP, 1975. 530-65.

Lispector, Clarice. "The Departure of the Train." Trans. Alexis Levitin. <u>Latin American Writers: Thirty Stories</u>. Ed. Gabriella Ibieta. New York: St. Martin's, 1993. 245-58.

Use quotation marks around the title of a poem, a short story, an essay, or a chapter. For a work originally published as a book, underline the title. Add inclusive page numbers for the selection at the end of the entry.

When citing two or more selections from one anthology, you may list the anthology separately under the editor's name.

Gardner, Helen, ed. <u>The Metaphysical Poets</u>. Baltimore: Penguin, 1957.

All entries within that anthology will then include only a cross-reference to the anthology entry.

Donne, John. "The Canonization." Gardner 61-62.

15. Essay or Periodical Article Reprinted in a Collection

Emig, Janet. "Writing as Mode of Learning." <u>College Composition and Communi-cation</u> 28 (1977): 122-28. Rpt. in <u>The Web of Meaning</u>. Ed. Janet Emig. Upper Montclair: Boynton, 1983. 123-31.

Gannet, Lewis. Rpt. of "John Steinbeck's Way of Writing." <u>Steinbeck and His Crit-ics: A Record of Twenty-five Years</u>. Ed. E. W. Tedlock, Jr., and C. V. Wicker. Al-buquerque: U of New Mexico P, 1957. 23-37. Introduction. <u>The Portable Steinbeck</u>. New York: Viking, 1946. 1-12.

Include the full citation for the original publication, followed by *Rpt. in* (for "Reprinted in") and the publication information for the book. Add inclusive page numbers for the article or essay found in the collection; add inclusive page numbers for the original source when available. If the original piece was published under a different title, provide information for the new publication first, as in the second example above.

16. Article in a Reference Book

"Behn, Aphra." <u>The Concise Columbia Encyclopedia</u>. 1998 ed.

Miller, Peter L. "The Power of Flight." <u>The Encyclopedia of Insects</u>. Ed. Christopher O'Toole. New York: Facts on File, 1986. 18-19.

For a signed article, begin with the author's name. For commonly known reference works, full publication information and editors' names are not necessary. For entries arranged alphabetically, page and volume numbers are not necessary.

17.   Anonymous Book

The World Almanac and Book of Facts. New York: World Almanac-Funk, 2000.

Alphabetize by title, excluding an initial *A, An,* or *The.*

18.   Government Document

United States. Central Intelligence Agency. National Basic Intelligence Fact Book.
     Washington: GPO, 1999.

If the author is identified, begin with that name. If not, begin with the government (country or state), followed by the agency or organization. The U.S. Government Printing Office is abbreviated *GPO.*

19.   Dissertation

Kitzhaber, Albert R. "Rhetoric in American Colleges." Diss. U of Washington, 1953.

Use quotation marks for the title of an unpublished dissertation. Include the university name and the year. For a published dissertation, underline the title and give publication information as you would for a book, including the order number if the publisher is University Microfilms International (UMI).

Documenting Periodicals

20.   Article, Story, or Poem in a Monthly or Bimonthly Magazine

Linn, Robert A., and Stephen B. Dunbar. "The Nation's Report Card Goes Home."
     Phi Delta Kappan Jan. 2000: 127-43.

"From Beans to Brew." Consumer Reports Nov. 1999: 43-46.

Abbreviate all months except *May, June,* and *July.* Hyphenate months for bimonthlies (*July-Aug. 1993*). Do not list volume or issue numbers. If the article is unsigned, alphabetize by title.

21.   Article, Story, or Poem in a Weekly Magazine

Ross, Alex. "The Wanderer." New Yorker 10 May 1999: 56-63.

Note that when the day of the month is specified, the publication date is inverted.

22. Article in a Daily Newspaper

Brody, Jane E. "Doctors Get Poor Marks for Nutrition Knowledge." New York Times
    10 Feb. 1992: B7.

"Redistricting Reconsidered." Washington Post 12 May 1999: B2.

For an unsigned article, alphabetize by the title. Give the full name of the newspaper as it appears on the masthead, but drop any introductory *A, An,* or *The.* If the city is not in the name, it should follow in brackets: *El Diario [Los Angeles].*

With the page number, include the letter that designates any separately numbered sections. If sections are numbered consecutively, list the section number (*sec. 2*) before the colon, preceded by a comma.

23. Article in a Journal Paginated by Volume

Harris, Joseph. "The Other Reader." Journal of Advanced Composition 12 (1992):
    34-36.

If the page numbers are continuous from one issue to the next throughout the year, include only the volume number (always in Arabic numerals) and year. Do not give the issue number or the month or season. Note that there is no space between the end parenthesis and the colon.

24. Article in a Journal Paginated by Issue

Tiffin, Helen. "Post-Colonialism, Post-Modernism, and the Rehabilitation of Post-
    Colonial History." Journal of Commonwealth Literature 23.1 (1998): 169-81.

If each issue begins with page 1, include the volume number followed by a period and then the issue number (both in Arabic numerals, even if the journal uses Roman). Do not give the month of publication.

25. Editorial

"Gay Partnership Legislation a Mixed Bag." Editorial. Burlington Free Press 5 April
    2000: A10.

If the editorial is signed, list the author's name first.

26. Letter to the Editor and Reply

Kempthorne, Charles. Letter. Kansas City Star 26 July 1999: A16.

Massing, Michael. Reply to letter of Peter Dale Scott. New York Review of Books 4
    Mar. 1993: 57.

27. Review

Kramer, Mimi. "Victims" Rev. of 'Tis Pity She's a Whore. New York Shakespeare
Festival. New Yorker 20 Apr. 1992: 78-79.

Lane, Anthony. Rev. of The Mummy by Stephan Sommers. New Yorker 10 May
1999: 104.

Documenting Electronic Sources

Electronic sources include both databases, available in portable forms such as CD-ROM, diskette, or magnetic tape, and online sources accessed with a computer connected to the Internet.

Databases

The Works Cited entries for electronic databases (newsletters, journals, and conferences) are similar to entries for articles in printed periodicals: cite the author's name; the article or document title in quotation marks; the newsletter, journal, or conference title; the number of the volume or issue; the year or date of publication (in parentheses); and the number of pages, if available.

Portable databases are much like books and periodicals. Their entries in Works Cited lists are similar to those for printed material except that you must also include the following items:

- The medium of publication (*CD-ROM, diskette, magnetic tape*).
- The name of the vendor, if known. (This may be different from the name of the organization that compiled the information, which must also be included.)
- The date of electronic publication, in addition to the date the material originally may have been published (as for a reprinted book or article).

28. Periodically Updated CD-ROM Database

James, Caryn. "An Army as Strong as Its Weakest Link." New York Times 16 Sep.
1994: C8. New York Times Ondisc. CD-ROM. UMI-Proquest. Oct. 1994.

If a database comes from a printed source such as a book, periodical, or collection of bibliographies or abstracts, cite this information first, followed by the title of the database (underlined), the medium of publication, the vendor name (if applicable), and the date of electronic publication. If no printed source is available, include the title of the material accessed (in quotation marks), the date of the material if given, the underlined title of the database, the medium of publication, the vendor name, and the date of electronic publication.

29. Nonperiodical CD-ROM Publication

"Rhetoric." The Oxford English Dictionary. 2nd ed. CD-ROM. Oxford: Oxford UP,
1992.

List a nonperiodical CD-ROM as you would a book, adding the medium of publication and information about the source, if applicable. If citing only part of a work, underline the title of the selected portion or place it within quotation marks, as appropriate (as you would the title of a printed short story, poem, article, essay, or similar source).

30. Diskette or Magnetic Tape Publication

Lanham, Richard D. <u>The Electronic Word: Democracy, Technology, and the Arts</u>. Diskette. Chicago: U of Chicago P, 1993.

Doyle, Roddy. <u>The Woman Who Walked into Doors</u>. Magnetic tape. New York: Penguin Audiobooks, 1996.

List these in the Works Cited section as you would a book, adding the medium of publication (for example, *Diskette* or *Magnetic tape*).

Online Sources

Documenting a World Wide Web (WWW) or other Internet source follows the same basic guidelines as documenting other texts: you need to cite who said what, where, and when. However, important differences need to be noted. In citing online sources from the World Wide Web or electronic mail (e-mail), two dates are important: the date the text was created (published) and the date you found the information (accessed the site). When both publication and access dates are available, provide both.

Many WWW sources are often updated or changed, leaving no trace of the original version, so always provide the access date, which documents that this information was available on that particular date. Thus, most electronic source entries will end with an access date immediately followed by the electronic address: *23 Dec. 1999* <*http://www.cas.usf.edu/english*>. The angle brackets < > identify the source as the Internet.

The following guidelines are derived from the MLA Web site <http://www.mla.org>. To identify a WWW or Internet source, include, if available, the following items in the following order, each punctuated by a period except the date of access:

- **Author** (or editor, compiler, or translator). Give the full name of the author, last name first. Include an alias, if available, for an unknown author.
- **Title.** Include titles of poems, short stories, and articles in quotation marks. Include titles of postings to a discussion list or forum in quotation marks, followed by *Online posting*. Underline the titles of published sources (books, magazines, films, recordings).
- **Editor, compiler, or translator.** Include the name, if not cited earlier, followed by the appropriate abbreviation (*Ed., Com., Trans.*).
- **Print source.** Include the same information as for a printed citation.
- **Title** of the scholarly project, database, or personal or professional Web site (underlined); if there is no title, include a description such as *Home page*. Include the name of the editor, if available.
- **Identifying number.** For a journal, include volume and issue numbers.
- **Date of electronic publication.**

- **Discussion list** information. Include the full name or title of the list or forum.
- **Page, paragraph, or section numbers.**
- **Sponsorship or affiliation.** Include the name of any organization or institution sponsoring the site.
- **Date of access.** Include the date you visited the site.
- **Electronic address.** Include the electronic address (URL) within angle brackets < >. To interrupt an electronic address at the end of a line, hit the return key. Break a URL only after a slash; do not hyphenate.

31. Published Web Site

Beller, Jonathon L. "What's Inside The Insider?" Pop Matters Film. 1999. 21 May
    2000 <http://popmatters.com/film/insider.html>.

32. Personal Web Site

Fulwiler, Toby. Home page. 2 Apr. 2000 <http://www.uvm.edu/~tfulwile>.

The ~ mark indicates a personal Web page.

33. Professional Web Site

Yellow Wall-Paper Site. 1995. U of Texas. 12 Dec. 1999
      <http://www.cwrl.utexas.edu/,daniel/amlit/wallpaper/>.

34. Book

Twain, Mark. The Adventures of Tom Sawyer. Internet Wiretap Online Library. 4
    Jan. 1998. Carnegie-Mellon U. 4 Oct. 1998 <http://www.cs.cmu.edu/Web/
    People/rgs/sawyr-table.html>.

35. Poem

Poe, Edgar Allan. "The Raven." American Review. 1845. Poetry Archives 8 Sep.
    1998 <http://www.tqd.advanced.org/3247/cgi-bin/
    dispoem.cgi?poet=poe.edgar&poem>.

36. Article in a Journal

Erkkila, Betsy. "The Emily Dickinson Wars." Emily Dickinson Journal 5.2 (1996): 14
    pars, 8 Nov. 1998 <http://www.colorado.edu/EDIS/journal/index.html>.

37. Article in a Reference Database

"Victorian." Britannica Online. Vers. 97.1. 1 Mar. 1997. Encyclopaedia Britannica. 2
    Dec. 1998 <http://www.eb.com:180>.

38. Posting to a Discussion List

Beja, Morris. "New Virginia Woolf Discussion List." Online posting. 22 Feb. 1996.
    The Virginia Woolf Society. Ohio State U. 24 Mar. 1996
    <gopher://dept. English.upenn.edu:70/0r0-1858-?Lists/20th/vwoolf>.

39. E-mail or Listserv Message

Fulwiler, Toby. "A Question About Electronic Sources." E-mail to the author.
    23 May 2001.

Harley, Robert. "Writing Committee Meeting." E-mail to a UCLA distribution list.
    24 Jan. 1999.

40. Newsgroup (Usenet) Message

Answerman (Mathes, Robert). "Revising the Atom." 2 Mar. 1997. 4 July 1997
    <alt.books.digest>.

If you quote a personal message sent by somebody else, be sure to get permission before including his or her address on the Works Cited page.

Documenting Other Sources

41. Cartoon, Titled or Untitled

Davis, Jim. "Garfield." Cartoon. Courier [Findlay, OH] 17 Feb. 1996: E4.

Roberts, Victoria. Cartoon. New Yorker 13 July 2000: 34.

42. Film or Videocassette

Casablanca. Dir. Michael Curtiz. Perf. Humphrey Bogart and Ingrid Bergman.
    Warner Bros., 1942.

Fast Food: What's in It for You. Prod. Center for Science. Videocassette. Los Angeles: Churchill, 1988.

Begin with the title, followed by the director, the studio, and the year released. You may also include the names of lead actors, the producer, and the like between the title and the distribution information. If your essay is concerned with a particular person's work on a film, lead with that person's name, arranging all other information accordingly.

Lewis, Joseph H., dir. Gun Crazy. Screenplay by Dalton Trumbo. King Bros., 1950.

43. Personal Interview

Holden, James. Personal interview. 12 Jan. 2001.

Begin with the interviewee's name and specify the kind of interview and the date. You may identify the interviewee's position if relevant to the purpose of the interview.

Morser, John. Professor of Political Science, U of Wisconsin–Stevens Point. Telephone interview. 15 Dec. 2000.

44. Published or Broadcast Interview

Sowell, Thomas. "Affirmative Action Programs." All Things Considered. Natl. Public Radio. WGTE, Toledo. 5 June 1990.

Steingass, David. Interview. Counterpoint 7 May 1970: 3-4.

For published or broadcast interviews, begin with the interviewee's name. Include appropriate publication information for a periodical or book and appropriate broadcast information for a radio or television program.

45. Print Advertisement

Cadillac DeVille. Advertisement. New York Times 21 Feb. 2000: A20.

Begin with the name of the product, followed by the description *Advertisement* and publication information for the source.

46. Unpublished Lecture, Public Address, or Speech

Graves, Donald. "When Bad Things Happen to Good Ideas." National Council of Teachers of English Convention. St. Louis. 21 Nov. 1989.

Begin with the speaker, followed by the title (if any), the meeting (and sponsoring organization, if needed), the location, and the date. If there is no title, use a descriptive label (such as *Speech*) with no quotation marks.

47. Personal or Unpublished Letter

Friedman, Paul. Letter to the author. 18 Mar. 1998.

Personal letters and e-mail messages are handled nearly identically in Works Cited entries. Begin with the name of the writer, identify the type of communication (for example, *Letter*), and specify the audience. Include the date written, if known, or the date received.

To cite an unpublished letter from an archive or private collection, include information that locates the holding (for example, *Quinn-Adams Papers. Lexington Historical Society. Lexington, KY*).

48. Published Letter

King, Jr., Martin Luther. "Letter from Birmingham Jail." 28 Aug. 1963. <u>Civil Disobedience in Focus</u>. Ed. Hugo Adam Bedau. New York: Routledge, 1991. 68-84.

Cite published letters as you would a selection from an anthology. Specify the audience in the letter title (if known). Include the date of the letter immediately after its title. Place the page number(s) after the publisher information. If you cite more than one letter from a collection, cite the entire collection in the Works Cited list, and indicate individual dates and page numbers in your text.

49. Map

<u>Ohio River: Foster, KY, to New Martinsville, WV</u>. Map. Huntington: U.S. Army Corps of Engineers, 1985.

Cite a map as you would a book by an unknown author. Underline the title, and identify the source as a map or chart.

50. Performance

Bissex, Rachel, perf. <u>Folk Songs</u>. Flynn Theater, Burlington, VT. 14 May 1990.

<u>Rumors</u>. By Neil Simon. Dir. Gene Saks. Broadhurst Theater, New York. 17 Nov. 1988.

Identify the pertinent details such as title, place, and date of performance. If you focus on a particular person, such as the director or conductor, lead with that person's name. For a recital or individual concert, lead with the performer's name.

51. Audio Recording

Young, Neil, comp., perf. <u>Mirror Ball</u>. In part accompanied by members of Pearl Jam. Burbank: Reprise, 1995.

Marley, Bob, and the Wailers. "Buffalo Soldier." <u>Legend</u>. Audiocassette. Island Records, 1984.

Depending on your focus, begin with the artist, composer, or conductor. Enclose song titles in quotation marks, followed by the recording title, underlined. Do not underline musical compositions identified only by form, number, and key. If you are not citing a compact disc, specify the recording format. End with the company label, the catalog number (if known), and the date of issue.

52. Television or Radio Broadcast

"Emissary." <u>Star Trek: Deep Space Nine</u>. Teleplay by Michael Pillar. Story by Rick Berman and Michael Pillar. Dir. David Carson. Fox. WFLX, West Palm Beach. 9 Jan. 1993.

If the broadcast is not an episode of a series or the episode is untitled, begin with the program title, underlined. Include the network, the station and city, and the date of the broadcast. The inclusion of other information such as narrator, writer, director, or performers depends on the purpose of your citation.

53.   Work of Art

Holbein, Hans. Portrait of Erasmus. The Louvre, Paris. The Louvre Museum. By
   Germain Bazin. New York: Abrams, n.d. 148.

Begin with the artist's name. Follow with the title, and conclude with the location. If your source is a book, also give pertinent publication information.

For two sample papers in MLA style, see Chapter 21.

APA Documentation Guidelines

Most disciplines in the social sciences and related fields use the name-and-date system of documentation put forth by the American Psychological Association (APA). The disciplines of education and business also use this system. This citation style highlights dates of publication because the currency of published material is of primary importance in these disciplines. Because collaborative authoring is common in the social sciences, listing the first six authors is the most up-to-date standard. For more about the foundations and purposes of the APA system, see the *Publication Manual of the American Psychological Association,* 5th ed. (Washington, DC: APA, 2001) or the APA Web site <http://www.apastyle.org/

⋗ GUIDELINES FOR FORMATTING MANUSCRIPTS

The APA guidelines for submitting college papers have been fairly fairly conservative and have not encouraged the use of the wealth of visually interesting options in fonts, type sizes, graphics, and other options available on most modern word processors. APA now recommends the use of the options available on most word processors: italic typeface (rather than underlining), bold face, the long (or em) dash, and so on. If your instructor requests APA format, follow the guidelines below. Check, however, whether your instructor allows or prefers underlining and double hyphen. For the dash when your typing equipment makes italics and other characters or features difficult or unavailable. If your instructor encourages more creative formats, use good judgment in displaying the information in your text. The following guidelines describe the preparation of the main body of your paper.

Paper and Printing

Print all academic assignments on 8½ × 11 inch white paper in a standard serif font (such as Times New Roman or Courier which have small strokes that project from the main lines) and type size (11 or 12 points) using a good-quality printer.

Margins and Spacing

Allow margins of one inch all around. Justify the left margin only. Double-space everything, including headings, quoted material, and the References page. Indent five spaces or one-half inch for paragraphs.

For prose quotations of more than forty words, indent five spaces or one-half inch from the left margin. Do not use quotation marks to mark the beginning and ending of quoted passages; include page numbers in parentheses at the end of the passage after any ending punctuation. (*pp. 34–41*)

Page Numbers

Print page numbers in the upper right margin of all pages (including the title page and abstract) one-half inch below the top. APA format requires a shortened title (2–3 words) above or five full spaces to the left of each page number to guarantee correct iden- tification of stray pages (*Green Is 1, Green Is 2*).

Title Page

Attach a numbered title page to your paper. Center the paper title between the left and right margins and about fifteen lines from the top; immediately below, type your name, your instructor's name, the course name, and the date.

On the first page of your text, center the title, capitalizing key words only. If your in- structor asks for strict APA style, avoid using italics, underlining, quotation marks, boldface, unusual fonts, or large type for the title. Double-space to the first paragraph.

If you are not using a title page, include on page 1 your name, your instructor's name, the course title, and the date double-spaced on separate lines, flush with the upper left margin.

Abstract

Write a paragraph of seventy-five to one hundred words (never exceed 120 words) that states your thesis and the main supporting points in clear, concise, descriptive lan- guage. Avoid statements of personal opinion and inflammatory judgment. Attach the ab- stract immediately following the title page; center the word *Abstract* one inch from the top of the page; double-space. (Outlines are not required in APA format, but if your instructor requires one, it should follow the Abstract.)

Punctuation

One space is required after periods, commas, semicolons, colons, question marks, and exclamation points and between the periods in an ellipsis. Dashes are formed by the em-dash keystrokes on your computer or by two hyphens, with no extra spacing on either side.

Visual Information

APA style requires the labeling of all tables (charts, graphs) and figures (drawings, photographs) included in the text: *Table 1, Figure 2,* and so on. Include a clear caption for each, and place each graphic element in the text as close to the passage to which it refers as is possible. In your text, be sure to discuss the most important information or feature in each table or figure that has been included.

✄ GUIDELINES FOR IN-TEXT CITATIONS

The following guidelines illustrate how to cite source information in the main body of your paper using APA style.

1.   Single Work by One or More Authors

Whenever you quote, paraphrase, or summarize material in your text, give both the author's last name and the date of the source. For direct quotations, provide specific page numbers. (Specific page reference are recommended for paraphrases but are not required.) Page references in the APA system are always preceded, in text or in the reference list for newspapers and parts of books (not for articles in journals or magazines), by the abbreviation *p.* or *pp.* to designate single or multiple pages.

Supply authors' names, publication dates, and page numbers (when listed) in parentheses following the cited material. Do not repeat any of these elements if you identify them in the text preceding the parenthetical citation.

> Exotoxins make some bacteria dangerous to humans (Thomas, 1974).

> According to Thomas (1974), "Some bacteria are only harmful to us if they make exotoxins" (p. 76).

> We need fear some bacteria only "if they make exotoxins" (Thomas, 1974, p. 76).

For a work by two authors, cite both names:

> Smith and Rogers (1990) agree that all bacteria that produce exotoxins are harmful to humans.

> All known exotoxin-producing bacteria are harmful to humans (Smith & Rogers, 1990).

The authors' names are joined by *and* within your text, but APA convention requires an ampersand (&) to join authors' names in parentheses.

For a work by three to five authors, identify all the authors by last name the first time you cite a source. In subsequent references, list only the first author followed by *et al.* ("and others").

> The most recent study supports the belief that alcohol abuse is on the rise (Dinkins, Dominic, & Smith, 1989).

> When homeless people were excluded from the study, the results were the same (Dinkins et al., 1989).

If you are citing a source by six or more authors, identify only the first author in all in-text references, followed by *et al.* In the References list, the first six authors are identified by name; *et al.* is used to shorten the listing beginning with the seventh author.

2. Two or More Works by an Author Published in the Same Year

To distinguish between two or more works published in the same year by the same author or team of authors, place a lowercase letter (*a, b, c,* and so on) immediately after the date. This letter should correspond to that one used on the References page, where the entries will be alphabetized by title. If two entries appear in one citation, repeat the year:

(Smith, 1992a, 1992b)

3. Unknown Author

To cite the work of an unknown author, identify the first two or three words of the entry as listed on the References page. If the words are from the title, enclose them in quotation marks or italicize them, as appropriate:

Many researchers now believe that treatment should not begin until other factors have been dealt with ("New Evidence Suggests," 1987).

Statistical Abstracts (1991) reports the literacy rate for Mexico at 75% for 1990, up 4% from census figures 10 years earlier.

4. Corporate or Organizational Author

Spell out the name of the authoring agency for a work by a corporation, an association, an organization, or a foundation. If the name is long and can be abbreviated and remain identifiable, spell out the name the first time and put the abbreviation immediately after it, in brackets. For subsequent references you may use only the abbreviation:

(American Psychological Association [APA], 1993)

(APA, 1994)

5. Authors with the Same Last Name

To avoid confusion in citing two or more authors with the same last name, include each author's initials in every citation:

(C. L. Clark, 1995)

(J. M. Clark, 1994)

6. Quotation from an Indirect Source

Use the words *as cited in* to indicate that you are using quotations or information which you found in another source, not in the original source. (It is the indirect source that is cited on the References page.)

Lester Brown of Worldwatch believes that international agricultural production has reached its limit and that "we're going to be in trouble on the food front before this decade is out" (as cited in Mann, 1993, p. 51).

7. More Than One Work in a Citation

List two or more sources within a single parenthetical reference in the same order in which they appear in your reference list. If you refer to two or more works by the same author, list them in chronological order, with the author's name mentioned once and the dates separated by commas:

(Thomas, 1974, 1979)

List works by different authors in alphabetical order by the author's last name, separated by semicolons:

(Miller, 1990; Webster & Rose, 1988)

8. Web Site

When citing an entire Web site, not specific text or information, identify the site and give the electronic address (URL) in your text:

To locate information about faculty at the University of Vermont, visit the school's Web site (http://www.uvm.edu).

When the site's name and address are included in the text, no reference entry is needed.

9. Specific Information from a Web Site

Cite specific information (author, figure, table, paraphrased or quoted passage) from a Web site as you would a print source, by including the brief author/date information in the text or in parentheses, followed by complete information on the References page. With a direct quotation, cite the page number, if provided, or substitute the paragraph number (using the paragraph symbol (¶) or the abbreviation *para.*): (*May, 1990, para 3*). You can also identify the section of the document you are quoting from: (*Conley, Method Section, para. 2*).

10. Long Quotation Set Off from Text

Start quotations of forty or more words on a new line, and indent the block five spaces or one-half inch from the left-hand margin. Indent the first line of the second or any subsequent paragraphs (but not the first paragraph) five additional spaces. Double-space all such quotations, omit quotation marks, and place the parenthetical citation after any end punctuation, with no period following the citation.

11. Footnotes

Footnotes provide additional information of interest to some readers but are also likely to slow down the pace of your text or obscure your point for other readers. Make footnotes as brief as possible. When the information you want to add is extensive, present it in an appendix.

Number footnotes consecutively in a list on a page headed *Footnotes* that follows the References; double-space, and indent the first line of each footnote as you would a paragraph. Number the footnotes to correspond to their numbers in your text.

◊ GUIDELINES FOR THE APA REFERENCES PAGE

All works mentioned in a paper should be identified in a reference list according to the following general rules of the APA documentation system.

Format

After the final page of the paper, title a separate page *References* with no underlining, italics, or quotation marks. Center the title one inch from the top of the page. Number the page in sequence with the last page of the paper.

Double-space between the title and the first entry. Also double-space both between and within entries. Set the first line flush with the left-hand margin.

Indent the second and all subsequent lines of an entry five spaces from the left margin in a hanging indent. (If the hanging indent is not possible, you may use the paragraph indent you have st and used in your text.) APA journals use hanging indents for all published manuscripts.

If your reference list exceeds one page, continue on an additional page or pages, but do not repeat the title *References*.

Alphabetize the list of references according to authors' last names, using the first author's last name for works with multiple authors. For entries by an unknown author, alphabetize by the first word of the title, excepting insignificant words (*A, An, The*).

(*Note:* The 2001 APA *Publication Manual* permits the scholarly use of the special typeface and character options available on most word processors. This means that italic typeface rather than underlining is recommended. You will also note other standard publishing features in the models such as the long (or em) dash instead of two hyphens.)

Entry Formats

Each item in the entry begins with a capital letter and is followed by a period. Each period is followed by one space. Only the first word is capitalized in book and article titles; capitalize all major words in a journal title. The titles of books, journals, newspapers, Web sites, and stand-alone publications such as reports are italicized. The most common general forms are the following:

BOOKS

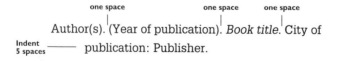

JOURNAL ARTICLES

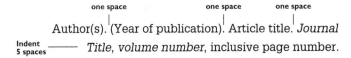

one space   one space   one space

Author(s). (Year of publication). Article title. *Journal*
Indent
5 spaces ——— *Title, volume number,* inclusive page number.

MAGAZINE AND NEWSPAPER ARTICLES

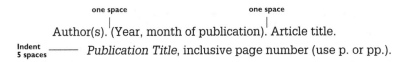

one space   one space

Author(s). (Year, month of publication). Article title.
Indent
5 spaces ——— *Publication Title,* inclusive page number (use p. or pp.).

AUTHORS

- List the author's last name first, followed by a comma and the author's initials (not first name).
- When a work has more than one author, list all authors up to and including the sixth in this way, separating the names with commas. Use *et al.* to shorten the authors when there are seven or more.
- For two to six authors of a single work, place an ampersand (&) before the last author's name. (Where there are seven or more, no ampersand precedes et al.)
- Place a period after the last author's name, unless this part of the entry ends with one already in *et al.*

TITLES

- List the complete titles and subtitles of books and articles, but capitalize only the first word of the title and any subtitle, as well as all proper nouns.
- Use italic type (for underlining when not using a word processor) for the titles of books, journals, Web sites, magazines, and newspapers, but do not italicize article or Web page titles or place quotation marks around them.
- Place a period after the title. (Make sure punctuation is underscored, if you must use underlining in place of italics.)

EDITION AND VOLUME NUMBERS

- For books, include the edition number, in parentheses, immediately following the title. (This is not italicized.)
- For periodicals, include volume number, also in italic type, immediately following the title. (If you must use underscoring, use a continuous underline under the title to the volume number and including and following punctuation marks.) Issue numbers are not italicized.

PUBLISHERS

- List publishers' names in a shortened form, omitting words such as *Company*. Retain *Book* and *Press*.
- Spell out the names of university presses and organizations in full.
- For books, use a colon to separate the place of publication from the publisher's name.

DATES AND PAGE NUMBERS

- For magazines and newspapers, use a comma to separate the year from the month and day, and enclose the publication date in parentheses: (*1954, May 25*).
- Give full sequences for pages and dates (*361–375,* not *361–75*), separating inclusive page numbers by an en-dash (long hyphen) if your computer offers this character key option or a hyphen with no spaces.
- If pages do not follow consecutively (as in newspapers), include subsequent page numbers after a comma: *pp. 1, 16.* Note that *pp.* precedes the page numbers for newspaper articles but not for journal articles.

ABBREVIATIONS

- Abbreviate state names, but not months, or countries.
- Use U.S. postal abbreviations (*NY, VT, WI*) for state abbreviations, but omit them for cities well known for publishing: Baltimore, Boston, Chicago, Los Angeles, New York, Philadelphia, and San Francisco.

Documenting Books

Following are examples of the reference list format for a variety of source types using standard APA hanging indent format.

1. Book by One Author

Benjamin, J. (1988). *The bonds of love: Psychoanalysis, feminism, and the problem of domination.* New York: Prometheus.

2. Book by Two or More Authors

Zweigenhaft, R. L., & Domhoff, G. W. (1991). *Blacks in the white establishment? A study of race and class in America.* New Haven, CT: Yale University Press.

Include all authors' names in the reference list, up to and including six authors. The newest APA guidelines (2001) recommend shortening the author list on the References page beginning with the seventh author.

3. More Than One Book by the Same Author

List two or more works by the same author (or the same author team listed in the same order) chronologically by year, earliest work first. Arrange any such works published in the same year alphabetically by title, placing lowercase letters after the dates. In either case, give full identification of author(s) for each reference listing.

> Bandura, A. (1969). *Principles of behavior modification.* New York: Holt, Rinehart and Winston.

> Bandura, A. (1977a). Self-efficacy: Toward a unifying theory of behavioral change. *Psychological Review, 84,* 191–215.

> Bandura, A. (1977b). *Social learning theory.* Englewood Cliffs, NJ: Prentice Hall.

If the same author is named first but listed with different coauthors, alphabetize by the last name of the second author. Works by the first author alone are listed before works with coauthors.

4. Book by a Corporation, an Association, an Organization, or a Foundation

> American Psychological Association. (2001). *Publication manual of the American Psychological Association* (5th ed.). Washington: Author.

Alphabetize corporate authors by the corporate name, excluding the articles *A, An,* and *The.* When the corporate author is also the publisher, designate the publisher as *Author.*

5. Revised Edition of a Book

> Peek, S. (1993). *The game inventor's handbook* (Rev. ed.). Cincinnati: Betterway.

6. Edited Book

> Schaefer, Charles E., & Reid, S. E. (Eds.). (1986). *Game play: Therapeutic use of childhood games.* New York: Wiley.

Place *Ed.* or *Eds.,* capitalized, after the editor(s) of an edited book.

7. Book in More Than One Volume

> Waldrep, T. (Ed.). (1985–1988). *Writers on writing* (Vols. 1–2). New York: Random House.

For a work with volumes published in different years, indicate the range of dates of publication. In citing only one volume of a multivolume work, indicate only the volume cited.

Waldrep, T. (Ed.). (1988). *Writers on writing* (Vol. 2). New York: Random House.

8.   Translated or Reprinted Book

Freud, S. (1950). *The interpretation of dreams* (A. A. Brill, Trans.). New York: Modern Library–Random House. (Original work published 1900)

The date of the translation or reprint is in parentheses after the author's name. Place the original publication date in parentheses at the end of the citation, with no period. In the parenthetical citation in your text, include both dates: (*Freud 1900/1950*).

9.   Chapter or Article in an Edited Book

Telander, R. (1996). Senseless crimes. In C. I. Schuster & W. V. Van Pelt (Eds.), *Speculations: Readings in culture, identity, and values* (2nd ed., pp. 264–272). Upper Saddle River, NJ: Prentice Hall.

The chapter or article title is not italicized or placed in quotation marks. Editors' names are listed in standard reading order (surname last). Inclusive page numbers, in parentheses, follow the book title.

10.   Anonymous Book

Stereotypes, distortions and omissions in U.S. history textbooks. (1977). New York: Council on Interracial Books for Children.

11.   Government Document

U.S. House of Representatives, Committee on Energy and Commerce. (1986). *Ensuring access to programming for the backyard satellite dish owner* (Serial No. 99-127). Washington, DC: U.S. Government Printing Office.

For government documents, provide the higher department or governing agency only when the office or agency that created the document is not readily recognizable. If a document number is available, list it in parentheses after the document title. Write out the name of the printing agency in full rather than using the abbreviation *GPO*.

This model can be adapted for reports or technical bulletins from universities or private organizations.

Documenting Periodicals

In citing periodical articles, use the same format for listing author names as for books.

12.   Article in a Journal Paginated by Volume

Hartley, J. (1991). Psychology, writing, and computers: A review of research. *Visible Language, 25,* 339–375.

If page numbers are continuous throughout volumes in a year, use only the volume number, underlined, following the title of the periodical.

13. Article in a Journal Paginated by Issue

Lowther, M. A. (1977). Career change in mid-life: Its impact on education. *Innovator, 8*(7), 1, 9–11.

Include the issue number in parentheses if each issue of a journal is paginated separately; do not use the abbreviation *p.* or *pp.*

14. Magazine Article

Garreau, J. (1995, December). Edgier cities. *Wired*, 158–163, 232–234.

For nonprofessional periodicals, include the year and month (not abbreviated) after the author's name; do not use the abbreviation *p.* or *pp.*

15. Newspaper Article

Finn, P. (1995, September 27). Death of a U-Va. student raises scrutiny of off-campus drinking. *The Washington Post*, pp. D1, D4.

If an author is listed for the article, begin with the author's name, then list the date (spell out the month); follow the article title with the title of the newspaper. If there is a section number or letter, combine it with the page or pages, including continued page numbers, using the abbreviation *p.* or *pp.*

Documenting Electronic Media

Formats for documenting electronic media have been changing and will continue to evolve. The 2001 APA *Publication Manual* (pp. 268–281) especially) and the APA Web page for electronic document citations (http://www.apastyle.org/elecref.html) are the best sources for up-to-date advice on standard models. The newest APA style recommendations divide electronic sources between two large categories: (1) sources on the Internet (divided into two types: periodicals and nonperiodicals [books, stand-alone documents, reports, etc.]) and (2) other electronic sources (CD-ROMs, databases, etc.).

Online Sources

APA provides the following general format for listing online sources, which closely follow the conventions for their printed equivalent—*who, when,* said *what, where.*

FORMAT FOR ONLINE PUBLICATIONS

Author(s) (date of publication). Article or chapter title. *Title of periodical, book, or site.* Inclusive page or paragraph numbers. Retrieved month, day, year, from [name of database source:] electronic address

There are three important vibrations to this model:

- If it is helpful to the reader, name the Web site (e.g., a complex university or government site) after the word *from* and followed by a colon.
- If you obtained your source from an aggregated database (not matter its format—CD-ROM, library server, or Internet supplier), provide the correct full name of the database after the word *from* and end the entry with a period.
- When you are citing the online version of a printed document, and the document has not been changed in any way (lost its pagination or had links and database added, e.g.), then follow the model for a printed document of its kind but add the words *Electronic version* in square brackets after the title and before the end period.

16.   Periodicals

Green, C.D. (1992) Of immortal mythological beasts: Operationism in psychology [Electronic version]. *Theory and Psychology, 2,* 292–320.

Many online articles are provided from the print version (e.g., original page numbers are indicated, no data or links have been added, etc.), then cite the same as you would the print text. However, add [*Electronic version*] after the title to complete the citation.

When the source has been altered from its print version in any way (making the online version to the extent unique), then add a retrieval statement and the URL to complete the citation. If your source is an Internet-only periodical or if you retrieved the article via file transfer protocol (ftp), you would follow the same format, as in model 18 below.

17.   Internet–only Journal Articles

Kapadia, S. (1995, November). *A tribute to Mahatma Gandhi: His views* on women and social exchange. *Journal of South Asia Women's Studies, 1*(1) [19 paragraphs]. Retrieved December 2, 1995, from b: http://www.shore.net/ ~india/jsaws

Indicate the number of paragraphs in brackets after the journal title and volume and issue numbers. Do not add a period at the end of the electronic address.

18.   Multipage Document Authored by a Private Organization

American Psychiatric Association. (2001) *Coping with a national tragedy—Resource, Tools, and Other Links.* September 20, 2001. Retrieved January 6, 2002, from http://psych.org/public__info/copydisaster92001.cfm

If the document has no revision or publication date, insert the abbreviation (*n.d.*) after the title and before the retrieval statement.

19.   Document Available on a University Project Web Site

Pierce, C. S. (1869, November 25). The English doctrine of ideas. Nation 9, 461–462. Retrieved January 16, 2002, from IUPUI (Indiana University and Purdue University, Indianapolis) Pierce Edition Project Web site: http://www.iupui.edu/%7Epierce/web/writings/v2/w2__30/v2__30.htm

The example was retrieved from a Web site for the Pierce Edition Project (dedicated to the works of philosopher Charles Sander Pierce), a large university project. The URL allows the reader to reach the cited article directly, without navigating the complex site. An online book, report, or data survey can be cited using this model or model 22.

20.   E-mail Messages

Personal electronic conversations are not listed on the References page. Cite e-mail messages in the text as you would personal letters or interviews. The new APA *Manual* recognizes that electronic mailing lists (often called listservs), online forums and discussion groups, and newsgroups are increasing in use and accessibility. APA guidelines permit the use of such sources only if they are archived and the citation serves a scholarly purpose in your essay.

R. W. Williams (personal communication, January 4, 1998) agrees with this statement.

21.   Newsgroups, Online Forums, and Electronic Mailing Lists

Trehub, A. (2002, January 28). The conscious access hypothesis [Msg 18]. Message posted to University ofouston Psyche Discussion Forum: http//listserv.uh.edu/cgi-bin/wa?A2=ind0201&L=psyche-b&F=&S=2334.

Other Electronic Sources

22.   Aggregated Database

A searchable, "aggregated database' is a selected group of resources stored in an electronic form for simplified, focused electronic access. You are not required to document how you accessed the database—via portable CD-ROM, on a library server, via a supplier World Wide Web site—but a retrieval statement that correctly names the source (in this case, the database) and gives the date of retrieval is always necessary. (If you include an item or accession number, place it in parenthesew after the title of your document.) Give the URL only when the information will help your reader locate the specific material. Do not italicize the names of databases.

Freud, S. (1913) *Interpretation of dreams* (A. A. Brill, Trans.). New York: Macmillian. Retrieved March 1, 2002, fro, Classics in the History of Psychology database.

This online book can be obtained on the Internet. A title or author keyword search will obtain the document in the database. If the database were obtained on CD-ROM or a library server, the citation would be the same.

23.   Electronic Copy of an Article, Abstract, or Data File

Harnad, S. (1992). Other bodies, other minds: A machine incarnation of an old philosophical problem. *Minds and Machines 1*, 43–54. Retrieved February 25, 2002, from Cognitive Science at Southhampton E-Print Archive database: http//cogsci.soton.ac.uk/harnad/genpub.html

The URL given to direct the reader to the document cited, without having to navigate the database.

24. Computer Software, Program Language

Commonly used commercial software and programming languages (Microsoft Word 5.1, e.g.) do not require a References page entry. Identify the correct name of the software, program, or language along with the version number in your text. If the software is not widely distributed or unfamiliar, cite according to this model.

HyperCard (Version 2.2) [Computer software]. (1993). Cupertino, CA: Apple Computer.

Documenting Other Sources

25. Motion picture, Audio recording, and Other Nonprint Media

Curtiz, M. (Director). (1942). *Casablanca* [Motion picture]. United States: Warner Bros.

Alphabetize a motion picture listing by the name of the person or persons with primary responsibility for the product. Other identifying information about this person should appear in parentheses. Identify the medium in brackets following the title, and indicate both the country location and name of the distributor (as publisher).

26. Interviews and Other Field Sources

These are identified in the text in parentheses (name, place, date) but are not listed on the References pages. (See model 20.)

For a sample paper in APA style, see Chapter 21.

Inside Address

Type the recipient's address two or more lines below the heading (depending on how much space is needed to center the letter on the page) flush with the left margin. Include the person's full name (and a courtesy title, if appropriate), followed by his or her position (if needed), the name of the division within the company, the company name, and the full street, city, and state address.

When writing to an unknown person, always try to find out the name, perhaps by calling the company switchboard. If doing this is impossible, use an appropriate title (*Personnel Director* or *Claims Manager,* for example) in place of a name.

Greeting

Type the opening salutation two lines below the inside address (*Dear Dr. Jones, Dear James Wong*) followed by a colon. If you and the recipient are on a first-name basis, it is appropriate to use only the first name. If you do not know the recipient's name, use *To Whom It May Concern* or some variation of *Dear Claims Manager* or *Attention: Director of Marketing* (the latter without a second colon). Avoid the old-fashioned *Dear Sir* or *Dear Sir or Madam.*

Body

Begin the main section of the letter two lines below the greeting. Single-space within paragraphs; double-space between paragraphs. If your reason for writing is clear and simple, state it directly in the first paragraph. If it is absolutely necessary to detail a situation, to provide background, or to supply context, do so in the first paragraph or two, and then move on to describe your purpose in writing.

If your letter is more than one page long, type the addressee's last name, the date, and the page number flush with the right margin of each subsequent page.

Closing

Type the complimentary closing two lines below the last line of the body of the letter. The most common closings are *Sincerely, Cordially, Yours truly, Respectfully yours* (formal), and *Best regards* (informal). Capitalize only the first word of the closing; follow it with a comma.

Signature

Type your full name, including any title, four lines below the closing. Sign the letter in blue or black ink in the space above your typed name; use your full name if you do not know the person well; you may use only your first name if you have addressed the recipient by first name.

Additional Information

You may provide additional brief information below your signature, flush with the left margin. Such information may include recipients of copies of the letter (*cc: Jennifer Rodriguez*); the word *Enclosures* (or the abbreviation *enc.*) to indicate you are also enclosing additional material mentioned in the letter; and, if the letter was typed by someone other than the writer, both the writer's and the typist's initials (*TF/jlw*).

Use your common sense when you write business letters; write on good-quality paper; proofread with your computer and with your own eye; correct all errors on screen before printing out; save a copy in a file that's easy to find. A sample business letter appears on page 366.

❧ RÉSUMÉS

A résumé is a brief summary of an applicant's qualifications for employment. It outlines education, work experience, and other activities and interests so a prospective employer can decide quickly whether or not an applicant is a good prospect for a particular job. Try to tailor your résumé for the position you are seeking by emphasizing experience that is most relevant to the position. Preparing a résumé on a computer lets you revise it easily and quickly.

Résumés should be brief and to the point, preferably no more than a page long (if relevant experience is extensive, more than one page is acceptable). Résumé formats vary in minor ways, but most include the following information.

Cover Letter

While not part of the résumé itself, a cover letter is a standard accompaniment to a résumé. Generally a cover letter introduces the applicant, indicates the position applied for, and offers additional information that cannot be accommodated on the résumé itself. While there is no formula for writing cover letters, it's a good idea to keep them brief, especially when you suspect your audience may be reading many such letters (see the figure on page 369). At the same time, a cover letter may convey something of your voice, including tone, style, values, and personality. Be clear and direct; avoid being too cute or pretentious. Again, use good judgment here; the cover letter introduces you to your audience, and first impressions are important.

Personal Information

Résumés begin with the applicant's name, address, and phone number, as well as e-mail and fax address if available. Center these at the top of the page, using one-inch margins (see the figure on page 370).

```
405 Martin Street
Lexington, Kentucky 40508      HEADING
February 10, 200X

Barbara McGarry, Director
Kentucky Council on the Arts   INSIDE ADDRESS
953 Versailles Road
Box 335
Frankfort, Kentucky 40602

Dear Ms. McGarry:      GREETING            BODY, UNINDENTED
John Huff, one of my professors at the University
of Kentucky, recommended that I write to you regarding
openings in the Council's internship program this
summer. I would like to apply for one of these
positions and have enclosed my résumé for your
consideration.

As you will note, my academic background combines a
primary concentration in business administration
with a minor in the fine arts. My interest in the
arts goes back to childhood when I first heard a
performance by the Lexington Symphony, and I have
continued to pursue that interest ever since. My
goal after graduation is a career in arts administration,
focusing on fund-raising and outreach for a major
public institution.

I hope you'll agree that my experience, particularly
my work with the local Community Concerts association,
is strong preparation for an internship with the
Council. I would appreciate the opportunity to
discuss my qualifications with you in greater
detail.

I will call your office within the next few weeks
to see about setting up an appointment to meet with
you. In the meantime, you can reach me at the above
address or by phone at 555-4033.

Thank you for your attention.

Sincerely,                     CLOSING

Chris Aleandro                 SIGNATURE

Chris Aleandro

enc.                           ADDITIONAL INFORMATION
```

Figure 2   Sample application letter

Chris Aleandro
405 Martin Street
Lexington, Kentucky 40508
(606) 555-4033

Objective: Internship in arts administration.

Education

University of Kentucky: 1999 to present.
Currently a sophomore majoring in business
administration with a minor in art history.
Degree expected May 2004.
Henry Clay High School (Lexington, KY): 1995 to 1999.
College preparatory curriculum, with emphasis
in art and music.

Related Work Experience

Community Concerts, Inc.: 1999 to present.
Part-time promotion assistant, reporting to local
director. Responsibilities include assisting with
scheduling, publicity, subscription/ticketing
procedures, and fund-raising. Position involves
general office duties as well as heavy contact
with subscribers and artists.
Habitat for Humanity: September to November 1999.
Co-chaired campus fund-raising drive that
included a benefit concert, raising $55,000.
Art in the Schools Program: 1997-1999.
Volunteer, through the Education Division of the
Lexington Center for the Arts. Trained to conduct
hands-on art appreciation presentations in
grade school classrooms, visiting one school a
month.

Other Work Experience

Record City: 1996 to 1998 (part time and summers).
Salesclerk and assistant manager in a music
store.

Special skills: Word processing (Windows-Mac), Adobe
Photo Delux, desktop design of brochures, programs,
and other materials.

References

Professor John Huff
School of Business
  Administration
The University of Kentucky
Lexington, KY 40506
(606) 555-3110

Ms. Joan Thomas
Community Concerts,
  Inc.
1200 Fayette Street
Lexington, KY 40513
(606) 555-2900

Figure 3   Sample résumé

Objective

Some college résumés include a line summarizing the applicant's objective, either naming the specific job sought or describing a larger career goal; others provide this information in the cover letter, keeping the résumé itself more general.

Education

Most first-time job applicants list their educational background first, since their employment history is likely to be fairly limited. Name the last two or three schools attended (including dates of attendance and degrees), always listing the most recent school attended first. Indicate major areas of study, and highlight any relevant courses. Also consider including grade point average, awards, and anything else that shows you in a good light. When employment history is more detailed, educational background is often placed at the end of the résumé. If you've written a senior or master's thesis, include its title and the name of your thesis director.

Work Experience

Starting with the most recent, list all relevant jobs, including company name, dates of employment, and a brief job description. Use your judgment about listing jobs where you had difficulties with your employer.

Special Skills or Interests

It is often useful to mention special skills, interests, or activities that provide additional clues about your abilities and personality.

References

Provide the names, addresses, and phone numbers of two or three people—teachers, supervisors, employers—whom you trust to give a good reference for you. (Make sure you get their permission first.) Alternatively, you may want to conclude with the line "References available on request."

᠅ MEMOS

Memo is short for *memorandum,* which is defined as a short note or reminder to someone to do something. Memos are used in business and college offices to suggest that actions be taken or to alert one or more people about a change in policy or an upcoming meeting. Part of the value of the memo is that the memo writer retains a record (a memory) that something was communicated to certain people on such and such a date.

Memo format is simple and direct. Names of both receiver and sender, along with the date and subject, are included at the top of the page so that the receiver knows instantly what this message is about:

Date (at right margin)

To: name of person or persons to whom it will be sent (flush left)
From: name of memo sender
Re: regarding what subject

Following is a typical memo:

January 25, 200X

To: Paul, Sue, Tony, and Jean
From: Toby
Re: Curriculum Committee meeting

Our last meeting will be Tuesday (1/30) at 3:30 p.m. in Old Mill, room 117. The agenda is to plan discussion items for the spring semester. Sue will provide refreshments.

Writing Essay Examinations

Essay examinations are common writing assignments in the humanities, but they are important in the social and physical sciences as well. Such exams require students to sit and compose responses to instructors' questions about information, issues, and ideas covered in the course. Instructors assign essay exams instead of "objective" tests (multiple choice, matching, true/false) because they want students to go beyond identifying facts and to demonstrate mastery of concepts covered in the course.

The best preparation for taking an essay exam is to acquire a thorough knowledge of the subject matter. If you have attended all the classes, done all the assignments, and read all the texts, you should be in a good position to write such essays. If you have also kept journals, annotated your text, discussed course material with other students, and posed possible essay exam questions, you should be in even better shape for such writing. Equally important is your strategic thinking about the course and its syllabus. If the course was divided into different topics or themes, think of a general question on each one; if it has been arranged chronologically, create questions focusing on comparisons or cause-and-effect relations within a particular period or across periods. Consider, too, the amount of class time spent on each topic, and pay proportionately greater attention to emphasized areas.

While there is no substitute for careful preparation, using certain writing strategies will enhance your presentation of information in virtually any exam. This chapter outlines suggestions for writing under examination pressure.

ॐ UNDERSTANDING QUESTIONS

The following suggestions will help you operate effectively in the examination room.

Read the Whole Examination

Before answering a single question, quickly read over the whole exam to assess its scope and focus. Answering three of four questions in fifty minutes requires a different approach than answering, say, five of eight questions in seventy-five minutes. If you are given a choice among several questions, select questions that together will demonstrate your knowledge of the whole course rather than answering two that might result in repetitious writing. Finally, decide which questions you are best prepared to answer, and respond to those first. Budget your time, however, so you can deal fully with the others later.

Starting with the questions you know you can answer relaxes you, warms you up intellectually, and often triggers knowledge about the others in the process.

Attend to Direction Words

Once you decide which questions you will answer, take a moment to analyze each one before you begin to write. Then focus closely on one particular question; read it several times. Underline the direction word that identifies the task you are to carry out, and understand what it is telling you to do.

Define or *identify* asks for the distinguishing traits of a term, subject, or event but does not require an interpretation or judgment. Use appropriate terminology learned in the course. For example, the question "Define John Locke's concept of *tabula rasa*" is best answered by using some of Locke's terminology along with your own.

Describe may ask for a physical description ("Describe a typical performance in ancient Greek theater"), or it may be used more loosely to request an explanation of a process, phenomenon, or event ("Describe the culture and practices of the mound builders"). Such questions generally do not ask for interpretation or judgment but require abundant details and examples.

Summarize asks for an overview or a synthesis of the main points. Keep in mind that "Summarize the impact of the Battle of Gettysburg on the future conduct of the war" asks only that you hit the highlights; avoid getting bogged down in too much detail.

Compare and contrast suggests that you point out both similarities and differences, generally between two subjects but sometimes among three or more. Note that questions using other direction words may also ask for comparison or contrast: "Describe the differences between the works of Monet and Manet."

Analyze asks that you write about a subject in terms of its component parts. The subject may be concrete ("Analyze the typical seating plan of a symphony orchestra") or abstract ("Analyze the ethical ramifications of Kant's categorical imperative"). In general, your response should examine one part at a time.

Interpret asks for a discussion or analysis of a subject based on internal evidence and your own particular viewpoint: "Interpret Flannery O'Connor's short story 'Revelation' in terms of your understanding of her central religious and moral themes."

Explain asks what causes something or how something operates. Such questions may ask for an interpretation and an evaluation. "Explain the function of color in the work of Picasso," for example, clearly asks for interpretation of the artist's use of color; although it does not explicitly ask for a judgment, some judgment might be appropriate.

Evaluate or *critique* asks for a judgment based on clearly articulated analysis and reasoning. "Evaluate Plato's concept of the ideal state" and "Critique the methodology of this experiment," for example, ask for your opinions on these topics. Be analytical and lead up to a final statement, but don't feel that your conclusion must be completely one-sided. In many cases, you will also want to cite more experienced judgments to back up your own.

Discuss or *comment on* is a general request, which allows you considerable latitude. Your answers to questions such as "Discuss the effects of monetarist economic theories on current Third World development" often let you demonstrate what you know especially well. Use terms and ideas as they have been discussed during the semester, and add your own insights with care and thoughtfulness.

ᛉ WRITING GOOD ANSWERS

Instructors give essay exams to find out not only how much students know about course content, but how thoroughly they understand and can discuss it. If instructors were interested in testing only for specific facts and information, they could give true/false or multiple choice tests. Therefore, the best essay answers will be accurate but also highly focused, carefully composed, and easy to follow. The following are two examples of answers to an essay question from a music history examination. Which do you think is the better answer?

Explain the origin and concept of neoclassicism, and identify a significant composer and works associated with the development of this music.

ESSAY 1

Neoclassicism in music is a return to the ideas of the classical period of earlier centuries. It is dry and emphasizes awkward and screeching sounds and does not appeal to the listener's emotions. It does not tell a story, but presents only a form. It is hard to listen to or understand compared to more romantic music such as Beethoven composed. Neoclassical music developed in the early part of the twentieth century. Stravinsky is the most famous composer who developed this difficult music.

ESSAY 2

Neoclassical music developed as a reaction against the romantic music of the nineteenth century. Stravinsky, the most famous neoclassical composer, took his style and themes from the eighteenth-century classical music of Bach, Handel, and Vivaldi rather than Beethoven or Brahms. Stravinsky emphasizes technique and form instead of story or image, with his atonal compositions appealing more to the intellect than the emotions. "Rite of Spring" (1913) and "Symphony of Psalms" (1940) are good examples.

Both answers are approximately the same length, and both are approximately correct. However, the second answer is stronger for the following reasons: It is more carefully organized (from general to specific); it includes more information (names, works, dates); it uses more careful disciplinary terms (*technique, form, image*); and it answers all parts of the question (the first answer omits the titles of works). It also does not digress into the writer's personal value judgments (that neoclassical music "is hard to listen to"), which the question did not ask for.

The following strategies will help you write more carefully composed answers.

Plan and Outline

Take one or two minutes per question to make a potential outline of your answer. For example, if asked to compare and contrast three impressionist painters, decide in advance which three you will write about and in which order. While ideas will come as you start writing, having a plan of organization at the beginning allows you to write more effectively. If you create a quick outline in the margins of your paper or even just hold it in your head, your writing will include more focused information, presented in a more logical order (as in essay 2 above) rather than scattered randomly throughout (as in essay 1 above).

Lead with a Thesis

The surest way to receive full credit on an essay question is to answer the question briefly and directly in your first sentence. In other words, state your answer in a thesis statement, which the rest of your essay explains, supports, and defends. Essay 2 opens with a thesis statement: "Neoclassical music developed as a reaction against the romantic music of the nineteenth century." The rest of the essay explains and supports this statement.

Write with Specific Detail, Examples, and Illustrations

Remember that most good writing contains specific information that lets readers see for themselves the evidence for your position. Use as many supportive specifics as you can; memorize names, works, dates, and ideas as you prepare for the exam so you can recall them accurately if they are needed. Individual statistics alone are not worth much, but when used as evidence along with strong reasoning, these specifics make the difference between good and mediocre answers.

Provide Context

In answering a question posed by an instructor who is an expert in the field, it is tempting to assume your instructor does not want a full explanation and thus to answer too briefly. However, you are being asked to demonstrate how much you understand, so view each question as an opportunity to show how much you know about the subject. Briefly explain any concepts or terms that are central to your answer. Take the time to fit any details into the larger scheme of the subject. In essay 2, for instance, it is clear that the second writer understood the relation of each musical movement to the century that produced it.

Use the Technical Terminology of the Discipline

Be careful not to drop in names or terms gratuitously, unless these names and terms have been an integral part of the course. Make sure you define any other terms, use them appropriately, and spell them correctly. Essay exams also test your facility with the language and concepts peculiar to a particular discipline. In the music example, it pays to know the correct terms for historical periods (*classical, romantic, neoclassical*) as well as technical terms used in discussing music (*image, tone*).

Stay Focused

Answer what the question asks. Attend to all parts of an answer, cover those parts, and once you have done that, do not digress or add extraneous information. While it may seem interesting to hear your other ideas on the subject at hand, some instructors may consider this digressing as reflecting unfocused attention.

↜ STRATEGIES FOR WRITING ESSAY EXAMINATIONS

1. Skim the whole examination and block your time. Read quickly through the whole exam so that you know what you're being asked to do, and allot blocks of time for tackling each section.
2. Choose your essay questions carefully. The essay questions you answer should allow you to write on what you know best. Choose a mix of answers to show your range of knowledge.
3. Focus on direction words. For each question you have chosen, it is important to recognize what your instructor is really asking, for this understanding enables you to answer the question successfully.
4. Plan and outline each essay. Prepare a rough outline of your answer by identifying the key points you need to make and organizing them well.
5. Write thesis-first essays. Doing this illustrates your confidence in knowing the answer and setting out to prove it.
6. Include specifics—details, examples, illustrations. Backing up statements with evidence shows your mastery of the subject matter. Include short, accurate, powerful quotations where relevant, citing each by author, title, and date. To help you remember, focus on key words and jot them in the margins near your answer.
7. Use the discipline's terms and methods. Enter into the conversation of a particular discipline by using its accepted terms and methods.
8. Provide context but stay focused. Explain all your points as if your audience did not have the understanding your teacher does, but keep all your information focused on simply answering that one question. If you know more than time allows you to tell, end your answer with an outline of key points that you would discuss if you had more time.
9. Proofread your answers in the last five minutes before handing in the finished exam. Even this short step back from composing will allow you to spot errors and omissions.

A Brief Guide to Punctuation

The following guide explains the most common uses of the most common punctuation marks. If you have further questions or need more detail, consult a grammar handbook or a dictionary.

Period **stops sentences and abbreviates words.**

1. Use a period to end a sentence that is a statement, a mild command, or an indirect question.

    The administration has canceled classes.

    Do not attempt to drive to school this morning.

    We wondered who had canceled classes.

2. Use a period for certain abbreviations.

    Dr. Joan Sharp

    Ms. Amy Bowen

    6:30 A.M.

3. Do not use periods to abbreviate most words in formal writing or in acronyms.

    He made $200 per week [*not* wk.].

    He worked for the FBI [*not* F.B.I.].

Question mark **ends a direct question and indicates uncertainty in dates.**

1. Use a question mark to end a direct question.

    Where is Times Square?

    She asked, "What time is it?"

2. Use a question mark to indicate uncertainty in a date.

> The plays of Francis Beaumont (1584?–1616) were as popular as Shakespeare's plays.

Exclamation point ends an emphatic or emotional sentence.

Use an exclamation point to end a sentence that is emphatic or conveys strong emotion.

> What a mess!

> "Ouch! That hurts!" he screamed.

Comma alerts readers to brief pauses within sentences.

1. Use a comma before a coordinating conjunction joining independent clauses.

> We must act quickly, or the problem will continue.

2. Use a comma after an introductory element.

> After we attend class, we'll eat lunch.

> Whistling, he waited for his train.

3. Use a comma around nonrestrictive modifiers (modifiers that are not essential to the meaning of the sentence).

> Cats, which are nocturnal animals, hunt small rodents.

4. Use a comma between items in a series.

> He studied all of the notes, memos, letters, and reports.

5. Use a comma to set off parenthetical elements or transitional expressions.

> Surprisingly enough, none of the bicycles was stolen.

6. Use a comma to set off attributory words with direct quotations.

> "Time will prove us right," he said.

7. Use a comma with numbers, dates, titles with names, and addresses.

> The sign gave the city's population as 79,087.

> She was born on June 19, 1976.

> Joyce B. Wong, M.D., supervised the CPR training.

> His new address is 169 Elm Street, Boston, Massachusetts 02116.

Semicolon joins independent clauses and connects items in a complex series.

1. Use a semicolon between independent clauses not joined with a coordinating conjunction.

    The storm raged all night; most of us slept fitfully, if at all.

2. Use semicolons between items in a series that contain internal commas.

    The candidates for the award are Maria, who won the essay competition; Elaine, the top debater; and Shelby, who directed several student productions.

Colon introduces lists, summaries, and quotations and separates titles and subtitles.

1. Use a colon to introduce a list.

    Writers need three conditions to write well: time, ownership, and response.

2. Use a colon before a summary or explanation.

    He had only one goal left: to win the race.

3. Use a colon to formally introduce a quotation in text or to introduce a long indented quotation.

    He quoted Puck's final lines from *A Midsummer Night's Dream:* "Give me your hands, if we be friends, / And Robin shall restore amends."

4. Use a colon to separate a title and subtitle.

    Blue Highways: A Journey into America

Apostrophe indicates possession, forms certain plurals, and forms contractions.

1. Use an apostrophe to indicate possession in a noun or indefinite pronoun.

    Jack's brother

    anyone's guess

2. Use an apostrophe to form the plural of a word used as a word and the plural of letters.

    She wouldn't accept any *if*'s, *and*'s, or *but*'s.

    The word *occurrence* is spelled with two *r*'s.

3. Use an apostrophe to replace missing letters in contractions.

    I can't means I won't.

Quotation marks indicate direct quotations and certain titles.

1. Use quotation marks around words quoted directly from a written or spoken source.

    She said, "It really doesn't matter anymore."

    Who wrote, "Fourscore and seven years ago"?

2. Use quotation marks for titles of stories, short poems, book chapters, magazine articles, and songs.

    "Barn Burning" (short story)

    "To an Athlete Dying Young" (poem)

    "Finding Your Voice" (book chapter)

    "Symbolism in Shakespeare's Tragedies" (magazine article)

    "A Day in the Life" (song)

3. Place periods and commas inside closing quotation marks. Place semicolons and colons outside closing quotation marks. Place question marks and exclamation points inside or outside quotation marks depending on the meaning of the sentence.

    After Gina finished singing "Memories," Joe began to hum "The Way We Were."

    The sign read "Closed": there would be no cold soda for us today.

    "Would you like some fruit?" Phil asked.

    I can't believe you've never read "The Lottery"!

Parentheses enclose nonessential or digressive but useful information.

Use parentheses to enclose nonessential information: explanations, examples, asides.

In 1929 (the year the stock market crashed) he proposed to his first wife.

He graduated with high honors (or so he said) and found a job immediately.

Dashes enclose nonessential information and indicate abrupt changes of direction.

Use dashes to enclose nonessential information and to indicate contrast or a pause or change of direction.

At first we did not notice the rain—it began so softly—but soon we were soaked.

Nothing is as exciting as seeing an eagle—except maybe seeing two eagles.

Ellipsis points indicate the omission of words in a direct quotation.

Use ellipsis points (three dots) to indicate where you have omitted words in a direct quotation from a written or spoken source.

"We the People of the United States, in Order to form a more perfect Union . . . do ordain and establish this Constitution for the United States of America."

Brackets indicate changes to or comments on a direct quotation.

Use brackets when you make changes to or comments on a direct quotation.

E. B. White describes a sparrow on a spring day: "Any noon, in Madison Square [in New York City], you may see one pick up a straw in his beak."

✌ Index ✌